Remote Sensing of Biophysical Parameters

Remote Sensing of Biophysical Parameters

Editors

Francisco Javier García-Haro
Manuel Campos Taberner
Hongliang Fang

MDPI • Basel • Beijing • Wuhan • Barcelona • Belgrade • Manchester • Tokyo • Cluj • Tianjin

Editors

Francisco Javier García-Haro
Universitat de València
València
Spain

Manuel Campos Taberner
Universitat de València
València
Spain

Hongliang Fang
Institute of Geographic
Sciences and Natural
Resources Research
Chinese Academy of Sciences
Beijing
China

Editorial Office
MDPI
St. Alban-Anlage 66
4052 Basel, Switzerland

This is a reprint of articles from the Special Issue published online in the open access journal *Remote Sensing* (ISSN 2072-4292) (available at: https://www.mdpi.com/journal/remotesensing/special_issues/Biophysical_Parameters).

For citation purposes, cite each article independently as indicated on the article page online and as indicated below:

LastName, A.A.; LastName, B.B.; LastName, C.C. Article Title. *Journal Name* **Year**, *Volume Number*, Page Range.

ISBN 978-3-0365-4901-9 (Hbk)
ISBN 978-3-0365-4902-6 (PDF)

© 2022 by the authors. Articles in this book are Open Access and distributed under the Creative Commons Attribution (CC BY) license, which allows users to download, copy and build upon published articles, as long as the author and publisher are properly credited, which ensures maximum dissemination and a wider impact of our publications.

The book as a whole is distributed by MDPI under the terms and conditions of the Creative Commons license CC BY-NC-ND.

Contents

About the Editors vii

Preface to "Remote Sensing of Biophysical Parameters" ix

Lauren E. H. Mathews and Alicia M. Kinoshita
Urban Fire Severity and Vegetation Dynamics in Southern California
Reprinted from: *Remotesensing* 2021, *13*, 19, doi:10.3390/rs13010019 1

Jiabin Pu, Kai Yan, Guohuan Zhou, Yongqiao Lei, Yingxin Zhu, Donghou Guo, Hanliang Li,
Linlin Xu, Yuri Knyazikhin and Ranga B. Myneni
Evaluation of the MODIS LAI/FPAR Algorithm Based on 3D-RTM Simulations: A Case Study of Grassland
Reprinted from: *Remotesensing* 2020, *12*, 3391, doi:10.3390/rs12203391 19

Benjamin Brede, Jochem Verrelst, Jean-Philippe Gastellu-Etchegorry, Jan G.P.W. Clevers, Leo
Goudzwaard, Jan den Ouden, Jan Verbesselt and Martin Herold
Assessment of Workflow Feature Selection on Forest LAI Prediction with Sentinel-2A MSI, Landsat 7 ETM+ and Landsat 8 OLI
Reprinted from: *Remotesensing* 2020, *12*, 915, doi:10.3390/rs12060915 37

Sadeed Hussain, Kaixiu Gao, Mairaj Din, Yongkang Gao, Zhihua Shi and Shanqin Wang
Assessment of UAV-Onboard Multispectral Sensor for Non-Destructive Site-Specific Rapeseed Crop Phenotype Variable at Different Phenological Stages and Resolutions
Reprinted from: *Remotesensing* 2020, *12*, 397, doi:10.3390/rs12030397 65

Siyuan Chen, Liangyun Liu, Xiao Zhang, Xinjie Liu, Xidong Chen, Xiaojin Qian, Yue Xu and
Donghui Xie
Retrieval of the Fraction of Radiation Absorbed by Photosynthetic Components ($FAPAR_{green}$) for Forest Using a Triple-Source Leaf-Wood-Soil Layer Approach
Reprinted from: *Remotesensing* 2019, *11*, 2471, doi:10.3390/rs11212471 85

Francisco Javier García-Haro, Fernando Camacho, Beatriz Martínez, Manuel
Campos-Taberner, Beatriz Fuster, Jorge Sánchez-Zapero and María Amparo Gilabert
Climate Data Records of Vegetation Variables from Geostationary SEVIRI/MSG Data: Products, Algorithms and Applications
Reprinted from: *Remotesensing* 2019, *11*, 2103, doi:10.3390/rs11182103 105

Marlena Kycko, Elżbieta Romanowska and Bogdan Zagajewski
Lead-Induced Changes in Fluorescence and Spectral Characteristics of Pea Leaves
Reprinted from: *Remotesensing* 2019, *11*, 1885, doi:10.3390/rs11161885 131

Luke A. Brown, Booker O. Ogutu and Jadunandan Dash
Estimating Forest Leaf Area Index and Canopy Chlorophyll Content with Sentinel-2: An Evaluation of Two Hybrid Retrieval Algorithms
Reprinted from: *Remotesensing* 2019, *11*, 1752, doi:10.3390/rs11151752 153

Muhammad Bilal, Majid Nazeer, Janet E. Nichol, Max P. Bleiweiss, Zhongfeng Qiu, Evelyn
Jäkel, James R. Campbell, Luqman Atique, Xiaolan Huang and Simone Lolli
A Simplified and Robust Surface Reflectance Estimation Method (SREM) for Use over Diverse Land Surfaces Using Multi-Sensor Data
Reprinted from: *Remotesensing* 2019, *11*, 1344, doi:10.3390/rs11111344 171

Deepak Upreti, Wenjiang Huang, Weiping Kong, Simone Pascucci, Stefano Pignatti, Xianfeng Zhou, Huichun Ye and Raffaele Casa
A Comparison of Hybrid Machine Learning Algorithms for the Retrieval of Wheat Biophysical Variables from Sentinel-2
Reprinted from: *Remotesensing* **2019**, *11*, 481, doi:10.3390/rs11050481 195

Matthias Wocher, Katja Berger, Martin Danner, Wolfram Mauser and Tobias Hank
Physically-Based Retrieval of Canopy Equivalent Water Thickness Using Hyperspectral Data
Reprinted from: *Remotesensing* **2018**, *10*, 1924, doi:10.3390/rs10121924 217

Yao Wang and Hongliang Fang
Estimation of LAI with the LiDAR Technology: A Review
Reprinted from: *Remotesensing* **2020**, *12*, 3457, doi:10.3390/rs12203457 235

About the Editors

Francisco Javier García-Haro

F. Javier García obtained his Ph.D ('97) in quantitative remote sensing from the University of Valencia. He is currently a Full Professor in the Department of Earth Physics and Thermodynamics, University of Valencia. His main research interest are canopy radiative transfer modeling and retrieval vegetation properties using satellite, including applications such as agro-meteorology, land and soil resources, agriculture, and forestry. He is responsible for the design and scientific validation of LSA SAF vegetation products from EUMETSAT satellites (https://landsaf.ipma.pt). His scientific production includes 60 papers, over 200 conference proceedings, and numerous technical reports. He is involved in several validation networks and exploitation programs of satellite missions and received several research awards.

Manuel Campos Taberner

Manuel Campos Taberner obtained his Ph.D ('17) in quantitative remote sensing from the University of Valencia. At present, he is a researcher in the Department of Earth Physics and Thermodynamics, University of Valencia. His main research interests include the retrieval of vegetation properties using satellite data, canopy radiative transfer modeling, and machine learning and deep learning for both regression and classification remote-sensing applications He is involved in several calibration/validation activities, and exploitation programs of satellite missions.

Hongliang Fang

Hongliang Fang (Professor) obtained his Ph.D ('03) in quantitative remote sensing from the Department of Geographical Sciences, University of Maryland, College Park. He is now a professor in the Institute of Geographic Sciences and Natural Resources Research, Chinese Academy of Sciences (CAS). His main research interest lies in estimation of land surface biophysical parameters, radiative transfer modellings, calibration and validation studies, and in situ measurements. He is serving as an editorial board member of the Remote Sensing of Environment, an associate editor of the IEEE Geoscience and Remote Sensing Letters and Acta Geographica Sinica. He is also serving as the biophysical focus area lead of the CEOS/WGCV/LPV. He is a senior member of IEEE and a member of AGU. (http://sourcedb.igsnrr.cas.cn/yw/zjrck/200910/t20091028_2638835.html)

Preface to "Remote Sensing of Biophysical Parameters"

This book reviews the state of the art in the retrieval of biophysical vegetation parameters using field, satellite and airborne data, as well as the assimilation of remote sensing data with vegetation models and its usage in a wide variety of applications in remote sensing. The following is a brief summary of the topics and applications that comprise the book.

In the first contribution of the book, *Mathews and Kinoshita* highlighted the different burn severity and green canopy loss patterns in urban Mediterranean riverine systems, which are altered by invasive riparian vegetation. A combination of satellite and field-based observations was used to investigate the impact of fuel conditions, fire behavior, and vegetation regrowth patterns.

The study of *Pu et al.* analyzed the uncertainty of the MODIS LAI/FPAR estimates caused by different sources, such as inherent model uncertainty, input uncertainty (BRF and biome classification), clumping effect, and scale dependency.

Regarding the feature selection on forest LAI prediction, *Brede et al.* developed a workflow for Sentinel-2 and Landsat harmonised biophysical products' retrieval, assessing the impact of multiple properties: the machine learning regression algorithm, a prior knowledge of leaf chemistry, the radiative transfer model, the addition of a noise in training data and the use of Sun Zenith Angle (SZA) as an additional feature.

The study of *Hussain et al.* demonstrated the capabilities of multispectral sensors onboard unmanned aerial vehicles (UAV) for retrieving the biophysical characteristics of rapeseed crops (leaf area index (LAI), leaf mass per area (LMA) and specific leaf area (SLA)) at different growth stages, using empirical methods based on optimal spectral vegetation indices.

Chen et al. developed a generic algorithm for the retrieval of FAPAR that performed well in separating the green and woody components, which is of great importance for obtaining a better understanding of the energy exchange within the canopy.

Dealing with the satellites of the EUMETSAT constellation, *García-Haro et al.* presented a methodology that was developed in LSA SAF to generate biophysical variables from the SEVIRI sensor on board MSG 1-4 (Meteosat 8-11) satellites. This study provides expert knowledge and evaluates the potential of the SEVIRI/MSG vegetation products, including both climate data records (CDRs) and near-real-time observations of FVC, LAI and FAPAR.

Kycko et al. investigated the influence of lead ions on the growth of pea plants, which caused noticeable changes in the physical properties of plants. This study demonstrates the potential of hyperspectral techniques and chlorophyll fluorescence measurements to detect the effect of heavy metals and monitoring of contaminated areas.

In the work by *Brown et al.*, the performances of two hybrid retrieval algorithms in the estimation of LAI and CCC from MSI data were evaluated. The work also highlights the importance of selecting the radiative transfer models that can most accurately describe the structure of the canopy in forest environments.

Bilal et al. evaluated different hybrid methods to estimate biophysical variables (LAI, FAPAR, FVC, chlorophyll content) in wheat crops from Sentinel-2 data, including a variety of machine learning (kernel-based and non-kernel-based) algorithms.

In the work of *Wocher et al.*, a simple yet effective, physically based model was developed to retrieve water content at the leaf and canopy scales using hyperspectral data, allowing for insights into the physical, and proving the transferability of the model to different sites and crop types.

Finally, *Wang and Fang* complete the book with an excellent review of LiDAR technology for LAI retrieval, different validation methods and impact factors.

Francisco Javier García-Haro, Manuel Campos Taberner, and Hongliang Fang
Editors

Article

Urban Fire Severity and Vegetation Dynamics in Southern California

Lauren E. H. Mathews and Alicia M. Kinoshita *

Department of Civil, Construction & Environmental Engineering, San Diego State University, San Diego, CA 92182-1326, USA; lmathews@sdsu.edu
* Correspondence: akinoshita@sdsu.edu

Abstract: A combination of satellite image indices and in-field observations was used to investigate the impact of fuel conditions, fire behavior, and vegetation regrowth patterns, altered by invasive riparian vegetation. Satellite image metrics, differenced normalized burn severity (dNBR) and differenced normalized difference vegetation index (dNDVI), were approximated for non-native, riparian, or upland vegetation for traditional timeframes (0-, 1-, and 3-years) after eleven urban fires across a spectrum of invasive vegetation cover. Larger burn severity and loss of green canopy (NDVI) was detected for riparian areas compared to the uplands. The presence of invasive vegetation affected the distribution of burn severity and canopy loss detected within each fire. Fires with native vegetation cover had a higher severity and resulted in larger immediate loss of canopy than fires with substantial amounts of non-native vegetation. The lower burn severity observed 1–3 years after the fires with non-native vegetation suggests a rapid regrowth of non-native grasses, resulting in a smaller measured canopy loss relative to native vegetation immediately after fire. This observed fire pattern favors the life cycle and perpetuation of many opportunistic grasses within urban riparian areas. This research builds upon our current knowledge of wildfire recovery processes and highlights the unique challenges of remotely assessing vegetation biophysical status within urban Mediterranean riverine systems.

Keywords: riparian; invasive vegetation; burn severity; canopy loss; wildfire

1. Introduction

Across the world, wildfires are increasing in frequency and magnitude under a changing climate and increased human interaction, which in turn impacts natural resources, infrastructure, and millions of people [1]. Continuous landscape conversion due to the expansion of the human population in southern California has fragmented chaparral ecosystems and encouraged urbanization to spread into the wildlands. This has increased the potential for ignition and damages to human communities and surrounding ecosystems [2,3].

Riparian environments serve as corridors that provide habitat connectivity for flora and fauna throughout wildland and urban riverine systems. These corridors starkly differ from the upland vegetation in species composition, functional type, canopy cover, and moisture content [4,5]. The immediate riparian zone surrounding Mediterranean riverine systems (Med-sys) is associated with different vegetation types with high relative humidity and cool temperatures, which can act as a barrier to fire spread. These riparian areas often exhibit the rapid recovery of pre-fire vegetation biomass in comparison to upland chaparral [4,6]. Hydrologic disturbances caused by urbanization, especially the increase of dry weather base-flow due to impermeable land cover ("urban drool"), significantly alter native riparian vegetation density and community structure [7]. Compounded by higher nutrient loads and heightened disturbance from more frequent flash floods, urban riparian environments encourage invasions and rapid settlements of opportunistic and invasive vegetation species [8–11].

Since 2002, the number of fires under 5 km² in the urban riparian environment has increased in southern California [12]. This can be attributed to human ignition sources from transportation corridors, recreation, powerlines, and people experiencing homelessness [2,13,14]. Further, invasive plant infestations increase the density of vegetation biomass, which profoundly alters riverine hydrology and geomorphology and also impacts fire behavior and frequency within these systems [13–15]. The universal infestation of non-native vegetation throughout the stream and river systems of coastal California in conjunction with human ignition sources has arguably instituted a new regime of invasive grass–fire feedback cycles within the urban environment [14–16].

The impact of invasive species infestation on fire behavior, specifically in the urban environment, has not been previously documented. Studies indicate that as climate and other anthropogenic alterations, such as drought, human ignition sources, vegetation type conversion, and fuel accumulation, intensify, riparian environments in southern California are changing from flood-defined to fire-defined ecosystems [10,15,17]. However, further research is needed to describe the impact of invasive species on fire patterns and ecosystem recovery throughout the urban landscape, and to provide information for identifying and prioritizing management techniques.

There is an innate and irreplaceable value in in-field surveys; however, modern monitoring approaches will improve our ability to capture universal patterns of post-fire vegetation dynamics across urban riparian environments, as well as providing rapid assessments after fires. One metric that provides context for the biophysical disturbance of vegetation, as well as the socio-economic impact of a fire, is burn severity [16]. Two widely used indices to measure the effect of fire on biomass are the differenced normalized burn ratio (dNBR) and the differenced normalized difference vegetation index (dNDVI). Generally, both remote sensing indices are used to quantify the loss of biomass or organic matter with respect to the pre-fire conditions of an ecosystem. Almost all previous studies found that dNBR had the strongest relationship with in-field observations of burn severity [18]. However, the development of these metrics focused on the measurement and definition of burn severity within boreal forests and upland chaparral environments [18], while the effect of fire on riparian ecosystems is generally underrepresented [4,19].

There is a need to understand the measurement and sensitivity of burn severity metrics in Mediterranean riparian areas, most notably in urban and wildland–urban interface (WUI) environments, which are capable of rapid biophysical changes under climate and anthropogenic influences. Developing quantitative assessments for to evaluate the effect of fire on diverse riverine habitats within urban areas will improve the management and mitigation of fire's impacts on human safety and ecosystems. We hypothesized that the presence of invasive vegetation species would alter the relationship between the upland, riparian, and invasive zones' burn severity and canopy loss. The main objectives of this study are to 1) measure and compare dNBR (burn severity) and dNDVI (canopy loss) between upland and riparian zones in urban environments, 2) determine if the presence of invasive vegetation species alters the relationship between the upland and riparian zones, burn severity, and canopy loss over time, and 3) determine if the current burn severity and canopy loss indices are valid within the urban Med-sys.

2. Methods

2.1. Study Area

Southern California has a semi-arid, Mediterranean climate, where a prolonged dry season from late spring to late fall is interrupted by a relatively short wet period from December to March [20]. The southern California Mediterranean type ecosystem (MTE) is characterized by chaparral shrub- and scrub-dominated hillslopes that feed into downslope temperate woodland riparian corridors. This study focuses on riparian areas of southern California that exist in or adjacent to the urban environment that are exhibiting ecological shifts towards vegetation cover dominated by non-native species. We also focus on one of the most prolific species, *Arundo donax*, in urban and WUI southern California drainages, and its role in the grass–fire cycle.

To represent a range of urban and Mediterranean vegetation and fire conditions, data from eleven fires were collected from the southern California region. Ten fires were selected from the California Department of Forestry and Fire Protection (CalFire) Fire and Resource Assessment Program (FRAP) 2018 database and one additional fire, estimated from satellite imagery, used as a case study (Figure 1). All fires selected have the following characteristics: occurred in southern California from 2007 to 2018, the burned area was less than 40 km^2, included a river or creek [21] and therefore a measurable riparian corridor, and the fire was near or encompassed urban land-use [22]. Seven fires included invasive vegetation cover [14], and three control fires (Colina and West Fires) had no invasive cover (Table 1). The 2018 Del Cerro fire was a small urban fire that was incorporated as the eighth invasive fire, and the eleventh fire in total. In June 2018, the Del Cerro fire burned a substantial portion of the riparian zone along Alvarado Creek, a perennial and channelized tributary of the San Diego River in California. This human-ignited brush fire was fueled primarily by the presence of non-native and highly invasive *Arundo donax* and *Washingtonia* spp.

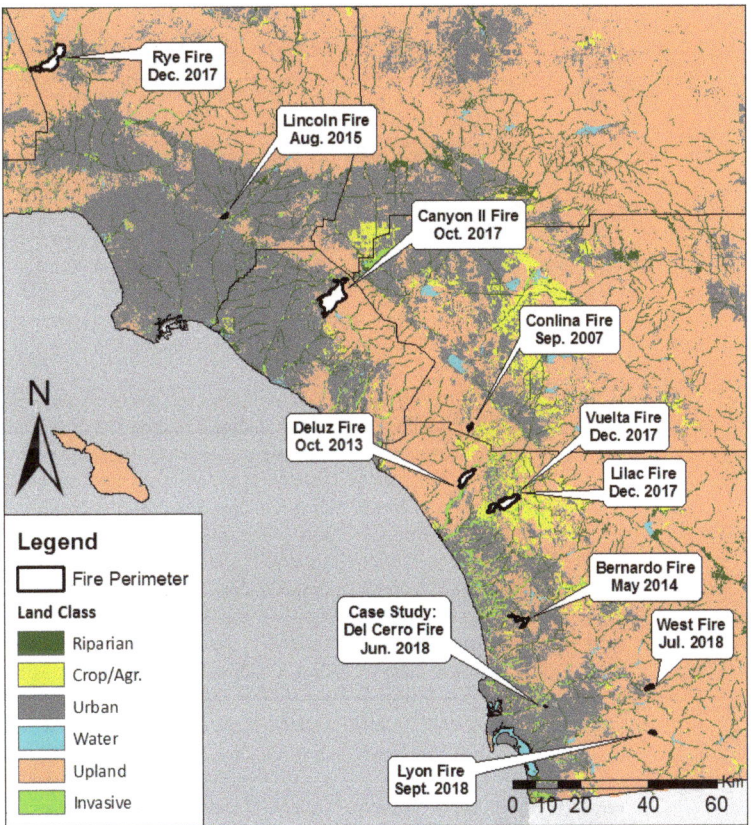

Figure 1. Eleven urban fires in southern California. The land class is shown at 30-m resolution within each county.

Table 1. Characteristics for the eleven selected fires in order of ascending total acres burned. Note that only the percentages of invasive, riparian, and upland cover, which are relevant to this study, are reported.

Name	Ignition Date (m/d/y)	Containment Date (m/d/y)	County	Impacted River	Ignition type	Total Area Burned (km^2)	Percent Invasive Cover	Percent Riparian Cover	Percent Upland Cover
Del Cerro	6/3/2018	6/4/2018	San Diego	Alvarado Creek	Juveniles	0.15	8.02%	37.04%	53.09%
Colina Fire	9/10/2007	9/10/2007	Riverside	Deluz Creek	Vehicle Accident	0.87	0.00%	17.93%	81.29%
Lincoln Fire	8/16/2015	Unknown	Los Angeles	Rio Hondo	Transient Camp	0.95	14.23%	41.17%	37.25%
Lyon Fire	9/9/2013	9/12/2013	San Diego	Wilson Creek	Accidental	1.06	0.00%	9.87%	90.13%
West	7/6/2018	7/10/2018	San Diego	Viejas Creek	Unknown	2.04	0.00%	17.65%	33.98%
Bernardo	5/13/2014	5/18/2014	San Diego	Lusardi Creek	Sparks from power equipment	6.26	1.25%	29.43%	64.99%
Vuelta Fire	6/16/2007	6/18/2007	San Diego	San Luis Rey River	Unknown- transient camp suspected	9.02	37.64%	58.01%	4.03%
Deluz Fire	10/5/2013	10/14/2013	San Diego	Santa Margarita River	Unknown	9.05	4.85%	15.73%	76.91%
Lilac Fire	12/7/2017	12/16/2017	San Diego	San Luis Rey River	Unknown	16.5	6.23%	12.19%	30.25%
Rye Fire	12/5/2017	12/13/2017	Los Angeles	Santa Clara	Unknown	24.5	1.89%	6.51%	78.44%
Canyon 2	10/9/2017	10/17/2017	Orange	Santa Anna River and Santiago Creek	Embers from Canyon 1 Fire	37.3	0.14%	11.86%	78.53%

2.2. Vegetation Classification

The CalFire FRAP, in cooperation with the California Department of Fish and Wildlife VegCamp program and USDA Forest Service Region 5 Remote Sensing Laboratory (RSL), compiled land cover data available for California into a single statewide data set. These data span from approximately 1990 to 2014. During this period, the most current, detailed, and consistent data were collected and compiled into the common classification scheme, the California Wildlife Habitat Relationships (CWHR) system. This vegetation dataset, FVEG [23], in coordination with the National Hydrology Dataset (NHD) [21] was used to classify land types relevant to this study, including: (1) riparian, (2) upland, (3) cropland, (4) urban, and (5) water.

Non-vegetated land covers "water" and "urban" were classified by fveg, but vegetation-based land cover, such as cropland, riparian and upland area, were not pre-defined. Thus, we used vegetated areas categorized into the fveg database by the specific Wildlife Habitat Relationship Name (WHRname), which represents major vegetative complexes at a scale sufficient to predict wildlife–habitat relationships. The following WHRnames were combined as "riparian" land cover from the fveg dataset: *Valley foothill riparian, Fresh emergent wetland, Saline emergent wetland, Wet meadow, Desert wash, Desert riparian, Marsh, Estuarine,* and *Riverine*. The riparian cover created from the fveg dataset was often not contiguous through urban corridors, so the NHD dataset was used to identify additional riparian environments.

From 2008 to 2010, Cal-IPC mapped *Arundo donax* and other invasive plant species at high resolution in all coastal watersheds in California from Monterey to San Diego [14]. These high-resolution data were generalized to include all invasive species, resampled at 30-m resolution to match the Landsat raster grid, and appended to the fveg classifications. In all eleven fire perimeters, invasive vegetation was located solely in the riparian region and classified independently from either riparian or upland classes within the burn perimeter (Table 1). The final six land classifications in this study included (1) riparian, (2) upland, (3) cropland (4) urban, (5) water, and (6) invasive plants (Figure 1). Cropland, urban, and water classifications were omitted from all calculations to reduce noise from non-vegetation features and land classes that were not present.

2.3. Burn Severity and Vegetation Metrics

Differenced normalized burn ratio (dNBR) identifies areas that have changed in the amount of charred plant material and soil, while differenced normalized difference vegetation index (dNDVI) identifies the change in the presence or absence of green vegetation. Landsat data images were acquired and processed prior to calculating dNBR and dNDVI metrics at a 30-m resolution (Table 2). Collection 1 level 1 data from the Landsat 5 Thematic Mapper™ or the Landsat 8 Operational Land Imager (OLI) and Thermal Infrared Sensor (TIRS) image data were collected for each fire based on the date of ignition. The Landsat digital number (DN) was converted into a top of atmosphere (TOA) reflectance or the amount of light reflected to the satellite. The calculation of TOA from DN corrects for atmospheric conditions and the position of the sun, which mitigates the effects of light scattering in the atmosphere and results in a reduced haze and less wavelength distortion [24]. Further atmospheric correction was not required as the images selected for this study contained minimal cloud cover (less than 5% cover) or other obstructions. Acquisition dates for each image are noted in Table 2.

Table 2. Landsat imagery dates for each time-point condition for each fire.

Fire Name	Immediately Pre-Fire (1)	Immediately Post-Fire (2)	One-Year Post-Fire (3)	Three-Years Post-Fire (4)
Del Cerro	16 May 2018	21 June 2018	19 June 2019	N/A
Colina	25 July 2007	11 September 2007	27 July 2008	17 July 2010
Lincoln	7 August 2015	10 October 2015	9 August 2016	15 August 2018
Lyon	10 August 2013	27 September 2013	29 August 2014	3 September 2016
West	21 June 2018	13 July 2018	8 June 2019	N/A
Bernardo	19 May 2014	29 May 2014	12 May 2015	1 May 2017
Vuelta	6 May 2007	25 June 2007	25 June 2008	20 May 2010
Deluz	11 September 2013	14 November 2013	30 September 2014	2 September 2016
Lilac	25 November 2017	27 December 2017	12 November 2018	N/A
Rye	2 December 2017	4 February 2018	26 November 2018	N/A
Canyon 2	8 October 2017	24 October 2017	27 October 2018	N/A

2.4. Differenced Normalized Burn Ratio (dNBR)

Normalized burn ratio (NBR) is an effective measure of burn severity in a variety of landscapes ranging from forest to chaparral [25]. This index can be related to the severity of a wildfire in the ecosystem by quantifying the transition from vegetated terrain to dry, ashy soil that is interspersed with blackened vegetation [26]. The NBR is calculated by using the relative difference in reflectance between the near infrared (NIR) and short-wave infrared (SWIR) (Equation (1)) from Landsat 5 TM or Landsat 8 OLI/TIRS. Equation (1) is based on the physical properties of vegetation, where green plant growth reflects NIR well, while dry, burned soil reflects highly in the SWIR [24,26]. NBR is the ratio of the difference in percent reflectance between the two spectra and ranges between −1 and 1.

NBR was calculated for the pre-fire, immediately post-fire, and one-year post-fire conditions for each fire (Table 2). Immediate images measure the fire impact on the landscape, while NBR after one year serves as a metric for longer-term and indirect ecosystem effects of fire [25]. Differencing the pre-fire and either immediate or one-year post-fire NBR (Equation (2)) creates a measure of burn severity. High dNBR values indicate high severity burn damage, and negative to low values indicate low burn severity to increasing vegetation productivity. dNBR for each fire was calculated using the three Landsat dates to capture the two time-point conditions (Table 2). See Equations (1) and (2) below:

$$NBR = (NIR - SWIR) * (NIR + SWIR)^{-1}, \qquad (1)$$

$$dNBR = NBR_{Pre\text{-}fire} - NBR_{Post\text{-}fire}. \qquad (2)$$

Burn severity approximated by dNBR, relates the change in reflectance from pre-fire to post-fire conditions to the surveyed ecological and socio-economic impact of the fire [18,26,27]. We utilized the burn severity levels established by Lutes et al. [18] and Key and Benson [26]: enhanced regrowth (−500 to −101), unburned (−100 to 99), low severity (100 to 269), moderate severity (270 to 659), and high severity (660 to 1300).

2.5. Differenced Normalized Difference Vegetation Index (dNDVI)

Plant health, NDVI, was estimated through the relationship of chlorophyll light absorption for photosynthesis in the red (Red) wavelength with high reflectance in the NIR [28] (Equation (3)). A lower NDVI represents bare soil to sparse vegetation (0.025–0.09) and a higher NDVI represents green vegetation ($\geq$0.25) [24]. The vegetation index was used to estimate the vegetation or biomass change before and after fire. This approach illuminates the spatial and ecological shifts in vegetation distribution within the sample fires and estimates vegetation health through "greenness" [29,30]. In forested environments, dNDVI has a weaker relationship with field-based measurements of burn severity than dNBR [26]. However, in Tran et al. [31], it was an effective measure of fire severity in riparian environments (Equation (4)). See Equations (3) and (4) below:

$$NDVI = (NIR - Red) * (NIR + Red)^{-1}, \qquad (3)$$

$$dNDVI = NDVI_{Pre\text{-}fire} - NDVI_{Post\text{-}fire}. \qquad (4)$$

Keeley and Keeley [32] showed that the total canopy cover for chaparral systems stabilizes three years after a fire. This study used the three years after a fire to represent a "stabilized" state of vegetation. To monitor both the immediate and longer-term post-fire vegetation trends, the dNDVI for each fire was calculated using pixel to pixel analysis between four Landsat dates (pre-fire, post-fire, one-year, three-years) to describe three time-point conditions (Table 2). Sparks et al. [33] related fire radiative energy density (FRED: $MJ \cdot m^{-2}$) and in-field measurements of percent canopy loss to remotely sensed dNDVI observations. Based on Sparks et al. [33], we categorized levels of canopy loss (approximated by dNDVI) as unburned (<0.005), low (0.005 to 0.049), moderate (0.05 to 0.199) and high (>0.2).

2.6. Statistical Analysis

To compare between vegetation classifications and between fires with invasive vegetation and without, control fires, fires that did not contain invasive cover, were separated into "upland control" and "riparian control" vegetation areas or "classes." Non-control fires, referred to as "invasive fires" through the rest of this study, were separated into "invasive," "riparian," and "upland" vegetation classes. Satellite image indices, dNBR and dNDVI (burn severity and green canopy loss), were compared between time-point conditions (Table 2) by each land class for each control or invasive fire using average values derived from 30-m resolution Landsat data. The average vegetation class values of burn severity and green canopy loss by each fire were also averaged across all control or invasive fires.

To compare the burn severity and green canopy loss between upland and riparian zones in urban environments (objective 1), we tested the hypothesis that riparian and upland vegetation would be different for both burn severity and green canopy loss across all fires (hypothesis 1). To determine if the presence of invasive vegetation species alters the relationship between the upland and riparian burn severity and the canopy loss immediately and over time (objective 2), we hypothesized that riparian and upland burn severity and green canopy loss would remain different between each time-point, showing a disparity between recovery efficacies (hypothesis 2; see Table 2). It was also hypothesized that the invasive vegetation class would experience higher immediate burn severity and canopy loss than all other vegetation classes in both invasive and control fires (hypothesis 3). Lastly, to determine if the current burn severity and canopy loss indices are valid within the urban Med-sys (objective 3), it was hypothesized that the relationship between dNBR versus dNDVI would be linear for each vegetation classification of both invasive and control fires (hypothesis 4). It was also hypothesized that, for Med-sys, green canopy loss and burn severity may have a stronger linear relationship in the riparian class than the upland class for both invasive and control fires (hypothesis 5).

All hypotheses regarding burn severity and canopy loss within all vegetation classes (invasive, riparian, riparian control, upland, and upland control) and between each fire were compared using unpaired two-tailed *t*-tests in Matlab (Table 2). The null hypothesis for each test was that the two-population means were equal and rejected if the resulting *p*-values were less than 0.05. Significant results are defined by *p*-values less than 0.05.

3. Results

To quantify the unique impacts of fire and the vegetation recovery dynamics of differing vegetation classifications in the Mediterranean riverine landscape, we discretized the invasive, riparian, and upland regions of eleven southern California urban fires and tracked dNBR and dNDVI for the first three years after fire.

3.1. Immediate Burn Severity and Canopy Loss by Fire

The Del Cerro and Colina fires had the smallest fire areas (0.15 km^2 and 0.87 km^2), and the Canyon 2 and Rye fires had the largest areas (37.3 km^2; Table 1). The percent cover of both riparian and invasive classes was calculated for each fire (Table 1). The three control fires, Colina, Lyon, and West, had no *Arundo* cover, while invasive vegetation in the riparian ranged from 0.14% (Canyon 2 Fire) to 38% (Vuelta). The Vuelta, Lincoln, and Del Cerro fires had the highest invasive to riparian cover ratios (0.65, 0.51, and 0.35, respectively), while the Canyon 2 and Bernardo fires had the smallest ratios (0.01 and 0.04).

To test if the immediate (0-year) burn severity was different between all vegetation classes and between control and invasive fires, the dNDVI was averaged for the upland, riparian, and invasive classes immediately following each fire (Figure 2A). The three control fires only tested upland and riparian vegetation and showed that the burn severity for each vegetation class was statistically different. The average 0-year burn severities for both the Lyon (0.48 ± 0.13) and West fires (0.39 ± 0.21) were moderate and were over two times higher than the low burn severity in Colina (0.18 ± 0.10). The riparian burn severities in the Lyon and West fires were 18% and 30% higher than in the upland vegetation, while the riparian burn severity was only 8.9% higher than the upland in the Colina fire.

In general, the riparian areas tended to have a higher burn severity than the upland regions across both invasive and control fires (9 of 11 fires; Figure 2A). The average 0-year burn severity across all eight fires that had invasive cover was generally classified as moderate (dNBR = 0.289; Figure 2A). The highest 0-year burn severity averaged across all vegetation classes occurred after the Deluz fire (0.65 ± 0.21), and the lowest occurred after the Canyon 2 fire (0.16 ± 0.08). Upland burn severity was statistically different from invasive burn severity across seven of the eight fires (Canyon 2, Bernardo, Rye, Deluz, Lilac, Lincoln and Vuelta). Upland burn severity was also significantly different from riparian burn severity for seven fires (Canyon 2, Bernardo, Rye, Deluz, Del Cerro, Lincoln and Vuelta Fire). Four fires (Bernardo, Rye, Deluz and Vuelta Fire) had statistically different burn severities for invasive vegetation compared to native riparian and upland vegetation classifications.

To test if the immediate (0-year) green canopy loss was different between all vegetation classes and between control and invasive fires, dNDVI was averaged by upland, riparian, and invasive classes immediately following each fire (Figure 2B). For the control fires, the average 0-year green canopy loss for the upland and riparian classes for both the Lyon and West fires was over 50% higher than the riparian and upland classes in the Colina fire. Similar to burn severity, the 0-year riparian and upland canopy losses were significantly different for only the Lyon and West fires, where the riparian cover was 28% and 40% higher than the upland vegetation, respectively. In addition, the average green riparian canopy loss for the Lyon and West fires (0.277 ± 0.100 and 0.271 ± 0.142, respectively) was significantly higher than all the fires with non-native vegetation present, except for the Deluz fire (0.285 ± 0.142).

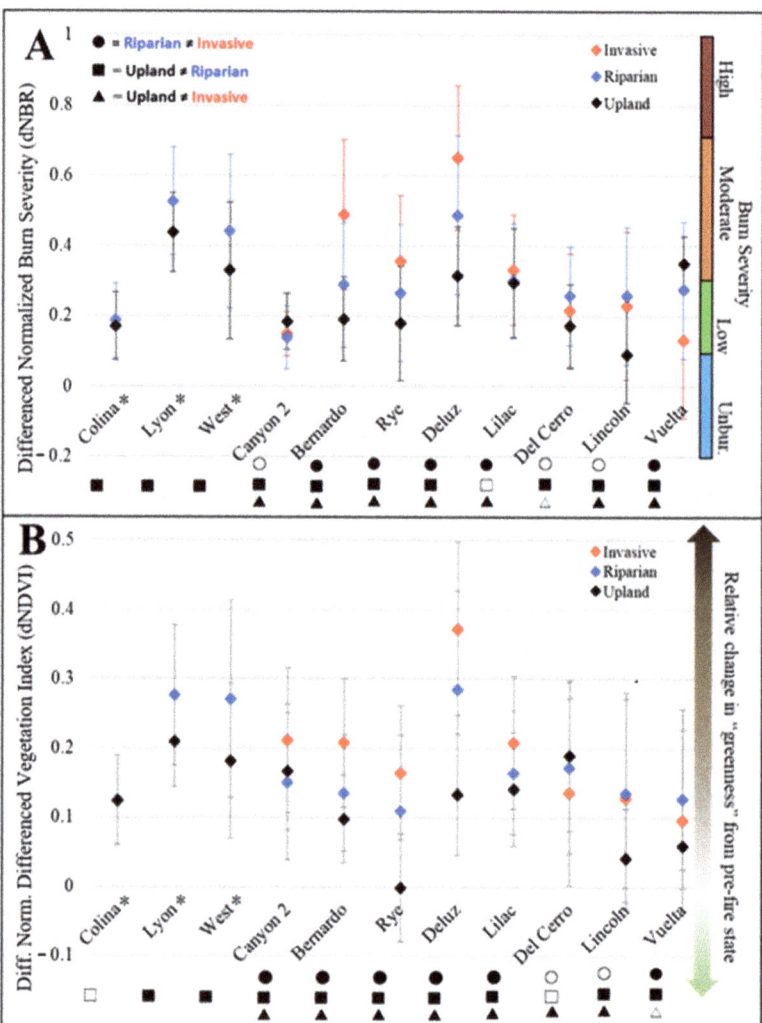

Figure 2. The dNBR and categorized burn severity by vegetation class immediately after each fire (**A**). The dNDVI and relative change in "greenness," by vegetation class immediately after each fire (**B**), where positive values above zero indicate a loss of greenness and negative values indicate an increase in greenness. Fires are in order of ascending percent invasive cover. The control fires have zero percent invasive cover and are denoted by *. The results of the unpaired *t*-tests are denoted for each fire by shapes to represent each null hypothesis. If the null hypothesis was rejected, the results were statistically significant and symbolized by a solid shape. Statistically insignificant results are symbolized by open shapes.

We hypothesized that riparian green canopy loss in the dNDVI would be different from the upland class. This was true for nine of the eleven fires evaluated, including seven of the eight fires that burned invasive vegetation (Figure 2B). The average green canopy loss was significantly different between all vegetation classes for five out of the eight fires (Canyon 2, Bernardo, Rye, Deluz and Lilac). In the five fires that supported hypotheses 1–3, the highest green canopy loss occurred in the invasive class where the canopy losses were, on average, 33% and 86% higher than the riparian and the upland areas, respectively. For six of the seven fires, the riparian vegetation class loss was, on average, 74% higher than the upland class, and was significantly different. The Canyon 2 fire was the only exception, where the upland and riparian burn severities were not significantly different, but the upland green canopy loss was 9.4% higher. The Del Cerro fire was the only case wherein the invasive region had significantly lower green canopy loss than the upland region. The highest 0-year green canopy loss as the dNDVI averaged across all vegetation classes occurred after the Deluz fire (0.263 ± 0.118), and the lowest occurred during the Vuelta fire (0.094 ± 0.098).

3.2. Average Burn Severity and Canopy Loss Patterns over Time

To evaluate the patterns in burn severity between immediately after fire (0-year) and 1-year following fire, dNBR for each vegetation class was averaged across the eight invasive fires and three control fires (Figure 3A; Table 3). Across the eight fires with invasive vegetation, 0-year burn severity was higher for invasive cover (moderate, 0.362 ± 0.240) compared to riparian (moderate, 0.254 ± 0.193) and upland cover (low severity, 0.205 ± 0.134). This was statistically significant ($p < 0.05$). However, the average burn severity for the riparian vegetation in the control fires (moderate, 0.360 ± 0.222) was 35% higher than the burn severity observed in riparian areas of invasive fires. The riparian burn severity for control fires was similar to the burn severity of the invasive class for invasive fires, which was statistically significant at $p < 0.05$.

Table 3. Mean dNBR and dNDVI values and standard deviations by vegetation class for each point in time. Control fire vegetation classes are denoted by *.

	0-Year Post-Fire		1-Year Post-Fire		3-Years Post-Fire	
Veg. Class	dNBR	dNDVI	dNBR	dNDVI	dNBR	dNDVI
Invasive	0.362 ± 0.240	0.210 ± 0.138	0.211 ± 0.181	0.166 ± 0.166	N/A	0.029 ± 0.090
Riparian	0.255 ± 0.193	0.161 ± 0.121	0.109 ± 0.173	0.203 ± 0.134	N/A	0.007 ± 0.113
Riparian *	0.360 ± 0.222	0.218 ± 0.134	0.158 ± 0.172	0.057 ± 0.088	N/A	-0.041 ± 0.093
Upland	0.205 ± 0.135	0.106 ± 0.109	0.035 ± 0.119	0.160 ± 0.107	N/A	0.015 ± 0.107
Upland *	0.298 ± 0.176	0.167 ± 0.087	0.133 ± 0.161	0.057 ± 0.082	N/A	-0.033 ± 0.085

The burn severity was statistically different between 0- and 1-year for all classes, where 1-year burn severity was on average 63% lower than 0-year. The largest percent change in burn severity between 0-year and 1-year occurred in the upland areas (−83%), and the smallest percent change in burn severity occurred in the invasive class (−42%). By 1-year, the average burn severities for all vegetation classes were statistically different from 0-year.

To compare the progressive change in green canopy loss following the fire, average dNDVI was evaluated as a proxy of "greenness" or green canopy loss for invasive and control fires by vegetation class for three points in time (Figure 3B). The highest average green canopy loss values occurred immediately following fire (0-year) and were recorded for riparian control and invasive classes (0.360 ± 0.222 and 0.362 ± 0.240, respectively). The invasive and riparian controls were not statistically similar for 0-year, however, all other classes were statistically different. This included the upland control (0.167 ± 0.087) and riparian class (0.160 ± 0.121), which were similar for 0-year (Figure 3B). The upland class had the lowest immediate loss in the green canopy (0.106 ± 0.109) and was over 20% lower than all other vegetation classes.

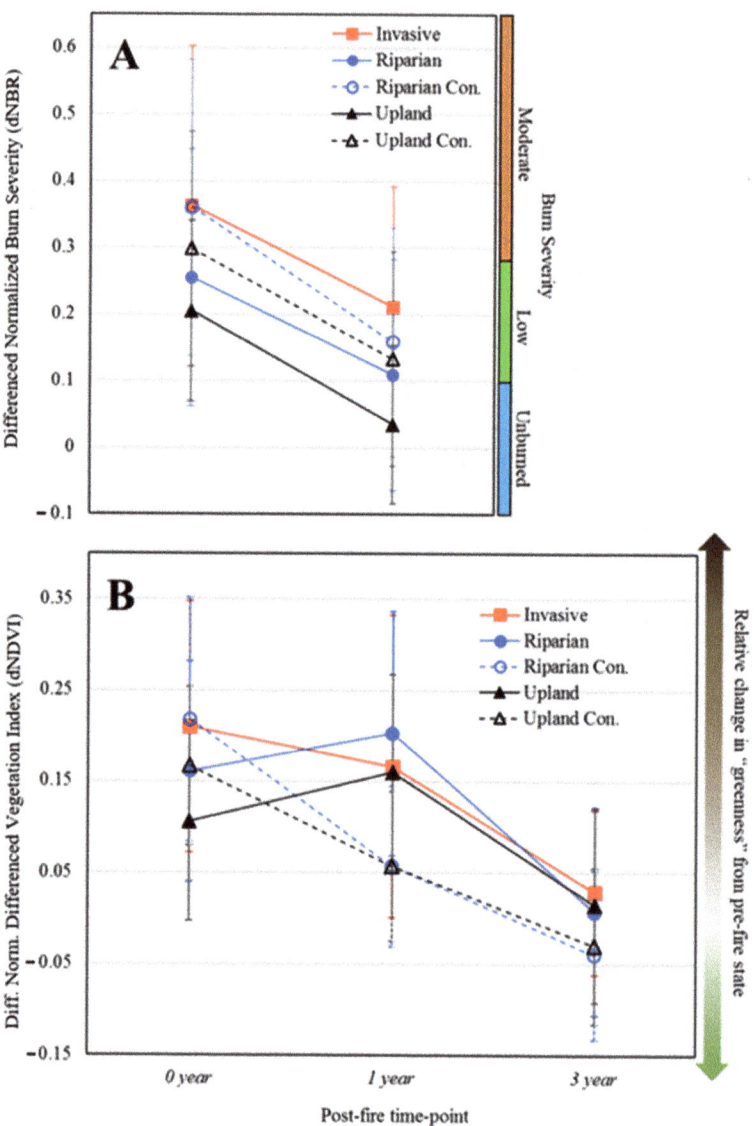

Figure 3. Average dNBR and burn severity across all fires by vegetation class for 0-year post-fire and 1-year post-fire (**A**). Burn severity is categorized by dNBR. Average dNDVI and relative change in "greenness" for invasive and control fires by vegetation class for 0-year post-fire, 1-year post-fire, and 3-year post-fire (**B**). Positive values above zero indicate a loss of greenness, and negative values indicate an increase in greenness.

One year following the fire, the largest green canopy recovery occurred for control fires, where the upland control class decreased by −66%, and the riparian control decreased by −74%. After 1-year, the average dNDVI for the upland and riparian classes in control fires converged and remained statistically similar to each other through the 3-year. The lowest recovery in green canopy 1-year post-fire occurred in the riparian class, where the 1-year dNDVI value (0.203 ± 0.133) was 27% higher than the 0-year value. The upland followed a similar pattern to the riparian but exhibited an even larger relative loss of green canopy with a 50% increase in dNDVI (0.159 ± 0.107). Although an increase of green canopy was detected for the invasive class for the 1-year, it exhibited the smallest absolute change in dNDVI out of all the classes (−21%). Further, the overall 1-year recovery of green canopy was over 20% higher in the invasive class (0.166 ± 0.165) than the riparian class and 4.3% higher than the upland class.

By 3-year, the green canopy loss generally diverged by control and non-control fires. The control upland and riparian classes increased in the green canopy by over 150% from 1-year to 3-year and by 3-year both had 18% more green canopy cover than the pre-fire state. Conversely, the 3-year dNDVI for the riparian, upland, and invasive classes were positive, indicating less green canopy cover than the pre-fire state. The lowest green canopy recovery (dNDVI) by 3-year occurred in the invasive class (0.028 ± 0.089), followed by the upland and riparian classes, respectively (0.015 ± 0.106 and 0.006 ± 0.112). Although the 3-year values appeared similar, all classes were statistically different except for the riparian and upland control classes (Figure 4).

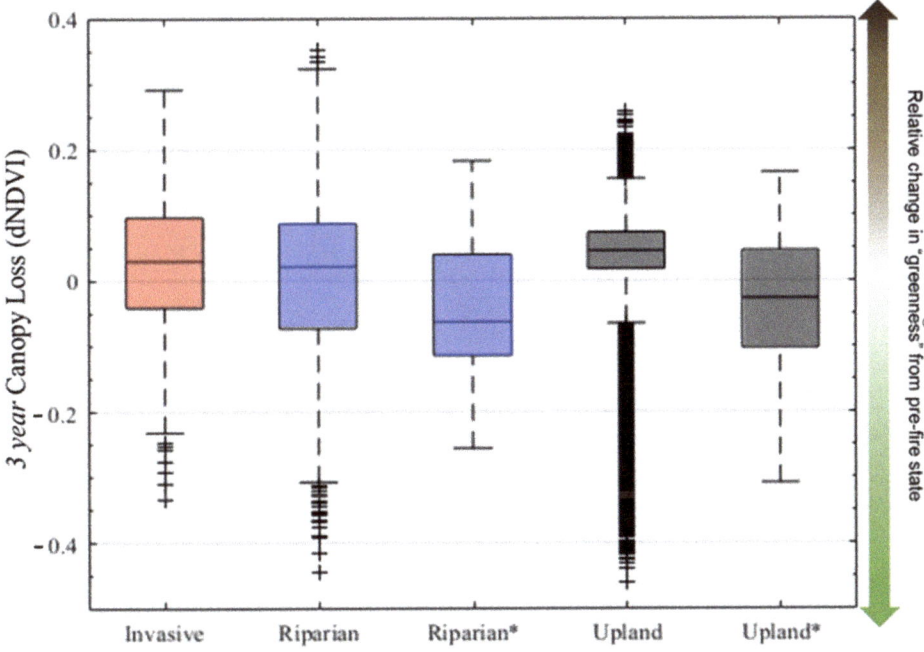

Figure 4. Canopy loss proxied by differenced Normalized Difference Vegetation Index (dNDVI) 3 years post-fire for each vegetation class. Control fire vegetation classes (no invasive species) are denoted by *. Box and whisker plots indicate the median, 25th, and 75th percentiles. The whiskers extend to the most extreme data points not considered outliers, and the outliers are denoted by '+'.

4. Discussion

Dwire and Kauffman [4] noted that there is much to learn about riparian fire, and proposed interesting relations worth evaluating further, such as the compounded impacts of land-use change, invasive plant species, and other anthropogenic pressures on fire ecology and ecological recovery processes. It was recognized that senescence could impact estimates of burn severity and vegetation estimates. Traditionally, when vegetation indices are calculated using imagery from one year after the pre-fire date, it is assumed that error from seasonal senescence is removed due to the assumption that vegetation conditions are similar at the same time of year. Compared to woody annuals in upland environments, riparian perennials grow extremely quickly [34–36]. In particular, grasses and forbs sprout quickly after winter rains and desiccate by early summer [37]. As a result, there are significant seasonal changes in vegetation cover and health within a rapid time frame that may affect the accuracy of remotely sensed plant health and burn severity [38]. Based on an unpaired two-tailed *t*-test, the pre- and post-fire unburned riparian areas were statistically similar. In an auxiliary analysis, we confirmed that reflectance changes were not due to senescence in any of our eleven fires.

4.1. Influence of Invasive Plants on Burn Severity and Green Canopy Loss after Fire

Generally, across all the fires, including the control fires, the 0-year burn severity and green canopy loss was statistically different between the upland and riparian regions (hypothesis 1). This was reflected in the averaged results (Figure 3), where the riparian region experienced higher burn severity and canopy loss immediately following the fire, supporting previous observations by Dwire and Kauffman [4] and Petit and Naiman [17]. Contrary to hypothesis 3, invasive fires did not burn more severely or experience more canopy loss on average in the riparian and upland class than control fires (Figure 3). Instead, invasive fires experienced depressed burn severity and green canopy loss in the riparian and upland classes. However, the average severity and green canopy loss of the invasive class was as high as the control fires. In fact, when the trends of the vegetation indices of each class are examined by fire, it was only fires with low to moderate invasive cover (Table 1) that had the highest burn severity and green canopy loss within the invasive classification of vegetation (Figure 2). Although these trends were statistically significant, our results highlight that the large variability between each fire suggests the need for more field investigations that capture a range of environmental factors to further our understanding of the impact of invasive vegetation on fire behavior in urban Med-sys.

4.2. Evidence of the Grass–Fire Cycle

Hypothesis 2, the highest burn severity and green canopy loss would occur in fires with a higher percentage of highly invasive vegetation, was rejected for Del Cerro, Lincoln, and Vuelta (Table 1 and Figure 2). In accordance with grass–fire cycle theories, our results suggest that the type of vegetation growth forms that make up the fuel loads in the urban Med-sys strongly affect the severity of the fire and the corresponding vegetation loss. A high proportion of grass, herbaceous, and fan palm cover, which is typical in the urban Med-sys, with extreme invasive cover preceding the fire [9,10,15] contrasts with the abundance of woody fuels and deep green characteristics that are typical of chaparral and riparian tree species [39].

Low-severity fires within invasive grass stands were observed by Keeley [40]. This process favors the survival and propagation of alien vegetation species [40]. The homogenous grass and herbaceous cover found in Med-sys that are severely impacted by invasive species colonization generally have lower burn severity. This is due to the lower presence of woody biomass (fuel), which leads to quicker fire movement [41]. Riparian ecosystems with more complex vegetation communities and vertical fuel structures, such as the control fires and fires with low invasive cover (Canyon 2, Bernardo, Rye, and Deluz fires), may burn with a higher intensity and also have a more heterogeneous burn severity or green canopy loss pattern [25].

The most prominent invasive vegetation observed in all burn perimeters investigated was *Arundo donax*, a tall bamboo-like member of the grass family (Poaceae) commonly known as giant reed [11,42]. *Arundo donax* is widely associated with fire in the southern California Med-sys. It thrives and rapidly regrows in monocultures or stands following fire [11,15,43]. In addition to outcompeting native riparian species, *Arundo donax* can use three times more water resources, contributing to the aridification of Med-sys riparian communities [11,44]. The compounded effects of rapid regrowth, the aridification of riparian systems, and increased fuel loads following fire make *Arundo donax* an archetypal plant for grass–fire feedback.

One year after fire, the results averaged over all invasive fires showed that only areas dominated with invasive vegetation (invasive class) saw canopy health recovery (Figure 3B). However, this is influenced by the four fires with the highest percentage of invasive cover (Lilac, Del Cerro, Lincoln, and Vuelta; Figure 5). In contrast, the four fires with the lowest percentage of invasive cover showed an even greater loss of canopy health immediately following the fire, which is similar to the trend exhibited by riparian and upland classes. These results support the grass–fire cycle hypothesis [45] and the promotion of secondary ecological impacts such as invasive vegetation infestations in riparian Med-sys ecosystems [17]. This effect, while already meriting consideration, could be exacerbated by climate change and anthropogenic activities. These disturbances could create favorable conditions for opportunistic invasive species. Since areas colonized by invasive species are more prone to wildfires, a self-perpetuating loop exists that selects for quickly regenerating invasive vegetation [46].

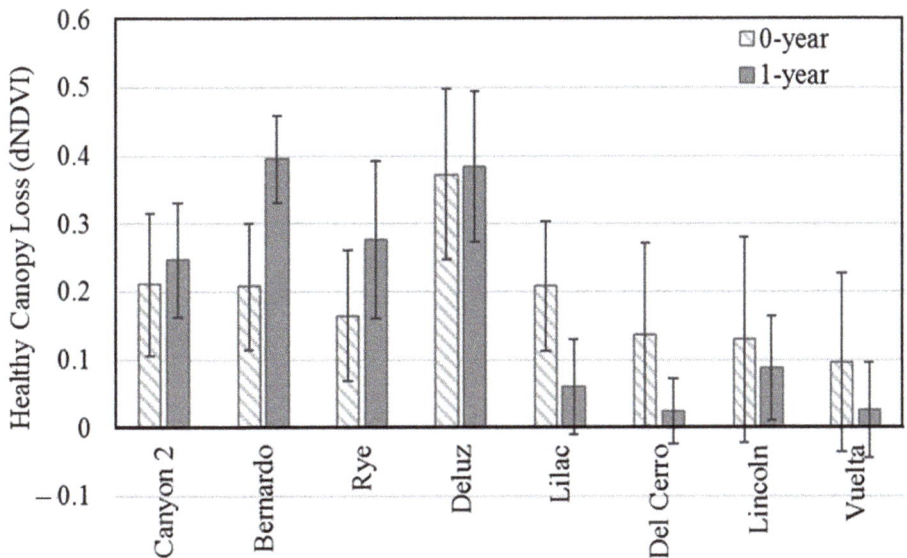

Figure 5. The comparison of green canopy loss (dNDVI) within the invasive class at 0-year and 1-year post-fire for each fire.

4.3. Highlighting the Uncertainty of Burn Severity and Canopy Loss Metrics in Urban Riparian Environments

Studies such as those by Key and Benson [26] and Tran et al. [31] showed that dNDVI may be more effective in identifying fire severity in woodland riparian environments, but less effective in chaparral-dominated ecosystems. When immediate (0-year) dNBR is compared against the corresponding dNDVI values, the relationship is almost linear across all vegetation classes and fires (Figure 6A). The R^2 for the control fires is much greater than the R^2 for the invasive fires (0.92 and 0.39, respectively). However, the agreement between dNBR and dDNVI for invasive fires increased as burn severity increased. In other words, the highest R^2 occurred for invasive fires that experienced both the greatest burn severity as well as canopy cover (Figure 6B). The agreement in dNBR and dNDVI for high-severity fires has been shown in previous work, such as Franco et al. [12].

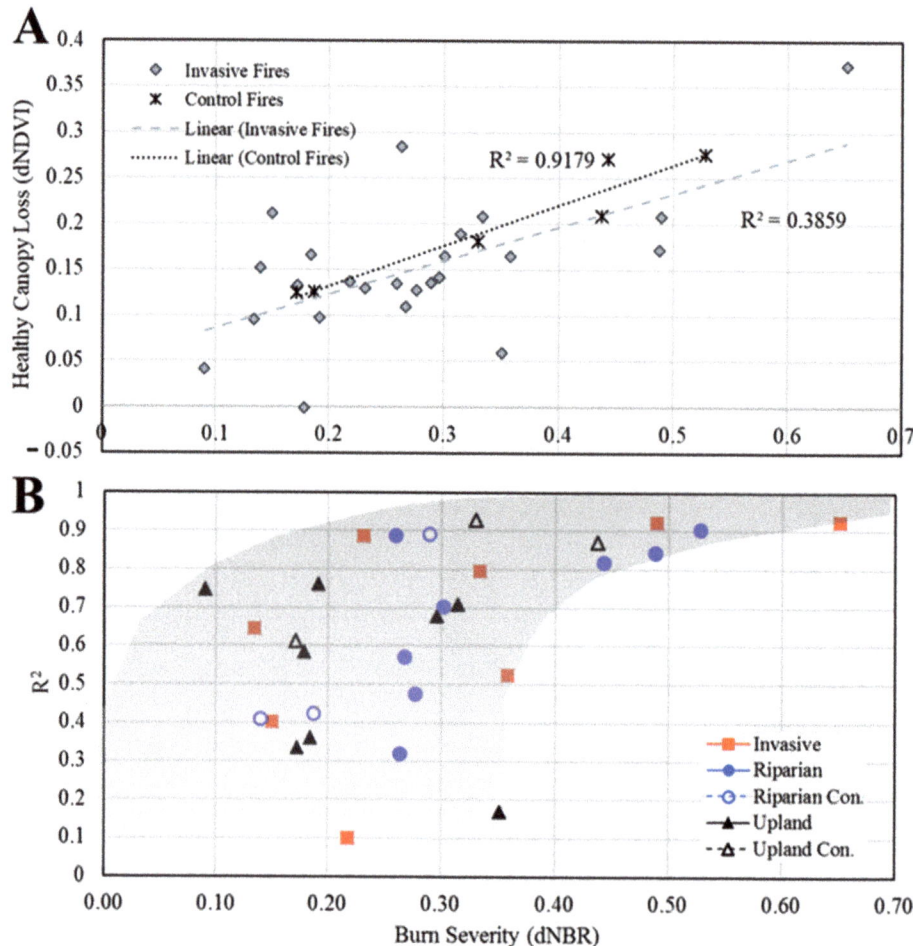

Figure 6. (**A**) The relation between average dNBR and dNDVI by class with respect to post-fire condition immediately after the fire (0-year). The regression equation and R^2 are denoted for invasive and control fires. (**B**) The average dNBR and R^2 by vegetation class with increasing burn severity. The grey shading highlights the approximate horizontal range of burn severity values.

Due to the aggressive nature of non-native vegetation, the rapid greening that occurs in the time between a wildfire and measurements (field-based or remote sensing) can often belie the true severity of a wildfire [15]. This is especially true in southern California, where both native and non-native forbs and grasses re-establish quickly after a wildfire [25]. Some of these species grow at incredible rates, such as *Arundo Donax*, which can grow over 3 cm per day [15]. It is not always possible to obtain satellite images within such a narrow timeframe [25,38]. Thus, the metrics such as those used in this study are susceptible to large errors due to the growth patterns of invasive vegetation.

The remote sensing results revealed that the immediate burn severity (dNBR) in the riparian area was higher than in the upland area. We also noted that the immediate burn severity of the invasive vegetation was not statistically different from in the upland or riparian areas (Figure 3A). In contrast, the smallest immediate loss in the green canopy (dNDVI) occurred in the areas with invasive vegetation, and the greatest loss in canopy occurred in the upland (Figure 3B). The lowest green canopy cover (NDVI) detected immediately post-fire (0-year) occurred in the uplands, which was also the only region that reflected bare ground levels at any time-point measured following a fire (0.19 ± 0.09). These patterns are expected given the typical distribution of chaparral near urban areas [2,3,46,47] compared to the denser canopy cover of riparian areas [4]. These results also highlight the general bias in dNDVI towards green new regrowth, such that the canopy loss signal is dampened in the riparian corridor dominated by invasive vegetation, despite the bare ground being exposed due to the fire.

The concentration of canopy loss on the upland slope nearest the riparian zone suggests that the riparian–upland interfaces in urban areas are vulnerable during and after urban fires. The combination of extra fuel loads, woody debris and abundant grasses accumulating in the riparian zone [48,49], and the high flammability of chaparral vegetation due to volatile compounds [6,25,47,50,51], may encourage intense and sustained fires that can severely impact vegetation health [25,52]. We encourage future field studies to confirm the remotely sensed spatial patterns observed in the ten urban fires we presented.

Overall, the discrepancy between dNDVI and dNBR immediately following fire is attributed to the rapid regeneration of non-native vegetation within the invasive and riparian regions and the inherent limitations of remote sensing metrics. We hypothesize that directly after the low-severity fire, the biomass and water content were low, but the greenness was higher as a result of the presence of small and new invasive grass shoots. Thus, the immediate post-fire measurement of dNDVI has a higher likelihood of underestimating the green canopy loss [53]. This is observed by the pattern found across all fires in our study (Figure 6), where the linearity between low-severity fires and green canopy loss deteriorated. Further, we observed that the accuracy of remotely sensed immediate canopy loss (dNDVI) in the urban Med-sys is highly reliant on the availability of observations less than a week following the fire. If satellite images are unavailable, metrics derived to guide management in post-fire areas should be avoided or used cautiously.

5. Conclusions

This research highlighted the different burn severity and green canopy loss patterns of riparian and upland regions in urban Mediterranean riverine systems across a spectrum of invasive vegetation cover, from zero to over 30% invasive. In all fires observed, with or without invasive vegetation, the riparian class burned more severely and experienced a greater loss of green canopy than the upland class. However, the presence of invasive vegetation did affect the magnitude of burn severity and canopy loss within both the riparian and upland regions of burned Med-sys. On average, fires with no invasive cover (the control fires) burned at an overall higher severity and resulted in a higher immediate loss of green canopy in the upland and riparian areas than invasive fires. Some invasive fires (generally with low to moderate invasive vegetation cover) and control fires had the same magnitude of burn severity, except in different vegetation classes. In fires with low to moderate invasive cover, the burn severity was just as great as that recorded in the riparian area of control fires, but was concentrated in the invasive class. Fires with high invasive cover generally exhibited low burn severity across all vegetation classes (upland, riparian and invasive). These patterns of burn severity across the spectrum of invasive vegetation cover may indicate the current degree of disturbance each landscape has succumbed to within the grass–fire cycle. The low burn severity observed in already invaded landscapes favored the life cycle of opportunistic and invasive grasses, potentially perpetuating and promoting the spread of invasive cover across vegetation classes in urban riverine systems.

There are the significant implications of rapid vegetation regeneration for the accuracy of remote sensing techniques. Contrary to previous literature, satellite imagery of invasive fires did not capture a substantial increase in green canopy in the invasive or riparian regions after fire. However, our results suggest a need for more frequent imagery after the fire to capture the regrowth that occurs between the extinguish date of the fire and the first available image post-fire, especially in areas prone to aggressive non-native vegetation growth. Observations from the field documented the presence of invasive vegetation regrowth prior to the first available satellite image. This demonstrated that satellite-based techniques can miss the initial green canopy loss for rapidly re-sprouting and invasive grass species, such as *Arundo donax*, therefore underestimating the relative increase in green canopy. This study highlighted the role of invasive vegetation in urban fire regimes, and the need for improved monitoring strategies to accurately assess the initial biophysical impact of fire as well as the lasting ecological impacts in urban fluvial systems that are vulnerable to invasive vegetation infestation.

Author Contributions: L.E.H.M. and A.M.K. conceived and designed the work. L.E.H.M. acquired the data. L.E.H.M. analyzed and interpreted the data. L.E.H.M. and A.M.K. contributed substantial writing of the manuscript. All authors have read and agreed to the published version of the manuscript.

Funding: This material is based upon work supported by the National Science Foundation CAREER Program under Grant No. 1848577 and the California State University (CSU) CSU Council on Ocean Affairs, Science & Technology (COAST) Rapid Response Funding Program Award #COAST-RR-2017-04 and Graduate Student Research Award #CSUCOAST-MATLAU-SDSU-AY1718.

Acknowledgments: This work was made possible by undying support of the SDSU Disturbance Hydrology Lab and excellent SDSU IT group. We also thank S. Zorn for his assistance.

Conflicts of Interest: The authors declare no conflict of interest.

References

1. Kinoshita, A.M.; Chin, A.; Simon, G.L.; Briles, C.; Hogue, T.S.; O'Dowd, A.P.; Gerlak, A.K.; Albornoz, A.U. Wildfire, water, and society: Toward integrative research in the "Anthropocene". *Anthropocene* **2016**, *16*, 16–27. [CrossRef]
2. Syphard, A.D.; Radeloff, V.C.; Keeley, J.E.; Hawbaker, T.J.; Clayton, M.K.; Stewart, S.I.; Hammer, R.B. Human Influence on California Fire Regimes. *Ecol. Appl.* **2007**, *17*, 1388–1402. [CrossRef] [PubMed]
3. Syphard, A.D.; Brennan, T.J.; Keeley, J.E. Drivers of chaparral type conversion to herbaceous vegetation in coastal Southern California. *Divers. Distrib.* **2019**, *25*, 90–101. [CrossRef]
4. Dwire, K.A.; Kauffman, J.B. Fire and riparian ecosystems in landscapes of the western USA. *For. Ecol. Manag.* **2003**, *178*, 61–74. [CrossRef]
5. Verkaik, I.; Rieradevall, M.; Cooper, S.D.; Melack, J.M.; Dudley, T.L.; Prat, N. Fire as a disturbance in mediterranean climate streams. *Hydrobiologia* **2013**, *719*, 353–382. [CrossRef]
6. Keeley, J.E.; Fotheringham, C.J. Historic Fire Regime in Southern California Shrublands. *Conserv. Biol.* **2001**, *15*, 1536–1548. [CrossRef]
7. White, M.D.; Greer, K.A. The effects of watershed urbanization on the stream hydrology and riparian vegetation of Los Peñasquitos Creek, California. *Landsc. Urban Plan.* **2006**, *74*, 125–138. [CrossRef]
8. D'Antonio, C.M. Fire, plant invasions, and global changes. In *Invasive Species in a Changing World*, 2nd ed.; Island Press: Washington, DC, USA, 2000; pp. 65–93. ISBN 78-1559637824.
9. Godefroid, S.; Koedam, N. Urban plant species patterns are highly driven by density and function of built-up areas. *Landsc. Ecol.* **2007**, *22*, 1227–1239. [CrossRef]
10. Aronson, M.F.; Lepczyk, C.A.; Evans, K.L.; Goddard, M.A.; Lerman, S.B.; MacIvor, J.S.; Nilon, C.H.; Vargo, T. Biodiversity in the city: Key challenges for urban green space management. *Front. Ecol. Environ.* **2017**, *15*, 189–196. [CrossRef]
11. Coffman, G.C. Factors Influencing Invasion of Giant Reed (*Arundo donax*) in Riparian Ecosystems of Mediterranean-Type Climate Regions. Ph.D. Thesis, University of California, Los Angeles, CA, USA, 2007.
12. Franco, M.G.; Mundo, I.A.; Veblen, T.T. Field-Validated Burn-Severity Mapping in North Patagonian Forests. *Remote Sens.* **2020**, *12*, 214. [CrossRef]
13. Collins, K.M.; Penman, T.D.; Price, O.F. Some Wildfire Ignition Causes Pose More Risk of Destroying Houses than Others. *PLoS ONE* **2016**, *11*, e0162083. [CrossRef] [PubMed]
14. Giessow, J.; Casanova, J.; Leclerc, R.; MacArthur, R.; Fleming, G. *Arundo Donax (Giant Reed): Distribution and Impact Report*; State Water Resources Control Board California Invasive Plant Council (Cal-IPC): Berkeley, CA, USA, 2011; pp. 1–240.
15. Coffman, G.C.; Ambrose, R.F.; Rundel, P.W. Wildfire promotes dominance of invasive giant reed (*Arundo donax*) in riparian ecosystems. *Biol. Invasions* **2010**, *12*, 2723–2734. [CrossRef]
16. Cushman, J.H.; Gaffney, K.A. Community-level consequences of invasion: Impacts of exotic clonal plants on riparian vegetation. *Biol. Invasions* **2010**, *12*, 2765–2776. [CrossRef]
17. Pettit, N.E.; Naiman, R.J. Fire in the Riparian Zone: Characteristics and Ecological Consequences. *Ecosystems* **2007**, *10*, 673–687. [CrossRef]

18. Lutes, D.C.; Keane, R.E.; Caratti, J.F.; Key, C.H.; Benson, N.C.; Sutherland, S.; Gangi, L.J. *FIREMON: Fire Effects Monitoring and Inventory System*; RMRS-GTR-164; U.S. Department of Agriculture, Forest Service, Rocky Mountain Research Station: Ft. Collins, CO, USA, 2006.
19. Halofsky, J.E.; Hibbs, D.E. Controls on early post-fire woody plant colonization in riparian areas. *For. Ecol. Manag.* **2009**, *258*, 1350–1358. [CrossRef]
20. Swain, D.L.; Langenbrunner, B.; Neelin, J.D.; Hall, A. Increasing precipitation volatility in twenty-first-century California. *Nat. Clim. Chang.* **2018**, *8*, 427–433. [CrossRef]
21. US Geological Survey and US Department of Agriculture, Natural Resources Conservation Service. National Hydrography Dataset. Available online: https://www.usgs.gov/core-science-systems/ngp/national-hydrography/nhdplus-high-resolution (accessed on 1 January 2020).
22. California Department of Forestry and Fire Protection. Fire Perimeters Data. In *Fire and Resources Assessment Program (FRAP)*; California Department of Forestry and Fire Protection's Fire and Resource Assessment Program (FRAP): Sacramento, CA, USA, 2018.
23. California Department of Forestry and Fire Protection. Vegetation (FVEG) Database. In *Fire and Resources Assessment Program (FRAP)*; California Department of Forestry and Fire Protection's Fire and Resource Assessment Program (FRAP): Sacramento, CA, USA, 2015.
24. US Geological Survey. Landsat 8 (L8) Data Users Handbook. Available online: https://prd-wret.s3.us-west-2.amazonaws.com/assets/palladium/production/atoms/files/LSDS-1574_L8_Data_Users_Handbook-v5.0.pdf (accessed on 1 January 2020).
25. Lentile, L.B.; Morgan, P.; Hudak, A.T.; Bobbitt, M.J.; Lewis, S.A.; Smith, A.M.S.; Robichaud, P.R. Post-Fire Burn Severity and Vegetation Response Following Eight Large Wildfires Across the Western United States. *Fire Ecol.* **2007**, *3*, 91–108. [CrossRef]
26. Key, C.H.; Benson, N.C. Landscape Assessment (LA). In *FIREMON: Fire Effects Monitoring and Inventory System*; Lutes, D.C., Keane, R.E., Caratti, J.F., Key, C.H., Benson, N.C., Sutherland, S., Gangi, L.J., Eds.; General Technical Report RMRS-GTR-164-CD; U.S. Department of Agriculture, Forest Service, Rocky Mountain Research Station: Fort Collins, CO, USA, 2006; pp. LA-1–LA-55.
27. Keeley, J.E. Fire intensity, fire severity and burn severity: A brief review and suggested usage. *Int. J. Wildland Fire* **2009**, *18*, 116. [CrossRef]
28. Tucker, C.J. Red and photographic infrared linear combinations for monitoring vegetation. *Remote Sens. Environ.* **1979**, *8*, 127–150. [CrossRef]
29. Gamon, J.A.; Field, C.B.; Goulden, M.L.; Griffin, K.L.; Hartley, A.E.; Joel, G.; Peñuelas, J.; Valentini, R. Relationships Between NDVI, Canopy Structure, and Photosynthesis in Three Californian Vegetation Types. *Ecol. Appl.* **1995**, *5*, 28–41. [CrossRef]
30. Hope, A.; Tague, C.; Clark, R. Characterizing post-fire vegetation recovery of California chaparral using TM/ETM+ time-series data. *Int. J. Remote Sens.* **2007**, *28*, 1339–1354. [CrossRef]
31. Tran, B.N.; Tanase, M.A.; Bennett, L.T.; Aponte, C. Evaluation of Spectral Indices for Assessing Fire Severity in Australian Temperate Forests. *Remote Sens.* **2018**, *10*, 1680. [CrossRef]
32. Keeley, J.E.; Keeley, S.C. Post-Fire Regeneration of Southern California Chaparral. *Am. J. Bot.* **1981**, *68*, 524–530. [CrossRef]
33. Sparks, A.M.; Kolden, C.A.; Talhelm, A.F.; Smith, A.M.S.; Apostol, K.G.; Johnson, D.M.; Boschetti, L. Spectral Indices Accurately Quantify Changes in Seedling Physiology Following Fire: Towards Mechanistic Assessments of Post-Fire Carbon Cycling. *Remote Sens.* **2016**, *8*, 572. [CrossRef]
34. Holstein, G. Deciduous Islands in an Evergreen Sea. In *California Riparian Forests: Ecology, Conservation and Productive Management*; University of California Press: Berkeley, CA, USA, 1984.
35. Davis, F.W.; Keller, E.A.; Parikh, A.; Florsheim, J. Recovery of the Chaparral Riparian Zone After Wildfire. In Proceedings of the California Riparian Systems Conference: Protection, Management, and Restoration for the 1990s, Davis, CA, USA, 22–24 September 1988; Gen. Tech. Rep. PSW-GTR-110. Pacific Southwest Forest and Range Experiment Station, Forest Service, U.S. Department of Agriculture: Berkeley, CA, USA, 1989; pp. 194–203.
36. Bendix, J.; Cowell, C.M. Impacts of Wildfire on the Composition and Structure of Riparian Forests in Southern California. *Ecosystems* **2010**, *13*, 99–107. [CrossRef]
37. Minnich, R.A. Fire Mosaics in Southern California and Northern Baja California. *Science* **1983**, *219*, 1287–1294. [CrossRef] [PubMed]
38. Veraverbeke, S.; Lhermitte, S.; Verstraeten, W.W.; Goossens, R. The temporal dimension of differenced Normalized Burn Ratio (dNBR) fire/burn severity studies: The case of the large 2007 Peloponnese wildfires in Greece. *Remote Sens. Environ.* **2010**, *114*, 2548–2563. [CrossRef]
39. Park, I.W.; Hooper, J.; Flegal, J.M.; Jenerette, G.D. Impacts of climate, disturbance and topography on distribution of herbaceous cover in Southern California chaparral: Insights from a remote-sensing method. *Divers. Distrib.* **2018**, *24*, 497–508. [CrossRef]
40. Keeley, J.E. Fire Management Impacts on Invasive Plants in the Western United States. *Conserv. Biol.* **2006**, *20*, 375–384. [CrossRef]
41. Bowman, D.M.J.S.; MacDermott, H.J.; Nichols, S.C.; Murphy, B.P. A grass–fire cycle eliminates an obligate-seeding tree in a tropical savanna. *Ecol. Evol.* **2014**, *4*, 4185–4194. [CrossRef]
42. Perdue, R.E. Arundo donax—Source of musical reeds and industrial cellulose. *Econ. Bot.* **1958**, *12*, 368–404. [CrossRef]
43. Bell, G.P. Ecology and Management of Arundo donax, and Approaches to Riparian Habitat Restoration in Southern California. Available online: https://ic.arc.losrios.edu/~{}veiszep/05spr2001/stelmok/g26files/attach5.html (accessed on 13 May 2020).

44. Iverson, M.E.; Jackson, N.E.; Frandsen, P.; Douthit, S. The impact of Arundo donax on water resources. In *Arundo Donax Workshop Proceedings, California Exotic Pest Plant Council, Riverside*; Jackson, N.E., Frandsen, P., Duthoit, S., Eds.; Team Arundo Riverside: Ontario, CA, USA, 1994; pp. 19–26.
45. D'Antonio, C.M.; Vitousek, P.M. Biological Invasions by Exotic Grasses, the Grass/Fire Cycle, and Global Change. *Annu. Rev. Ecol. Syst.* **1992**, *23*, 63–87. [CrossRef]
46. Keeley, J.E.; Baer-Keeley, M.; Fotheringham, C.J. Alien Plant Dynamics Following Fire in Mediterranean-Climate California Shrublands. *Ecol. Appl.* **2005**, *15*, 2109–2125. [CrossRef]
47. Keeley, J.E. Chaparral. In *North American Terrestrial Vegetation*; Cambridge University Press: Cambridge, UK, 2000; pp. 203–254.
48. Keeley, J.E.; Brennan, T.; Pfaff, A.H. Fire severity and ecosytem responses following crown fires in California shrublands. *Ecol. Appl.* **2008**, *18*, 1530–1546. [CrossRef]
49. Kobziar, L.N.; McBride, J.R. Wildfire burn patterns and riparian vegetation response along two northern Sierra Nevada streams. *For. Ecol. Manag.* **2006**, *222*, 254–265. [CrossRef]
50. Park, I.W.; Jenerette, G.D. Causes and feedbacks to widespread grass invasion into chaparral shrub dominated landscapes. *Landsc. Ecol.* **2019**, *34*, 459–471. [CrossRef]
51. Barro, S.C.; Conard, S.G. Fire effects on California chaparral systems: An overview. *Environ. Int.* **1991**, *17*, 135–149. [CrossRef]
52. Coppoletta, M.; Merriam, K.E.; Collins, B.M. Post-fire vegetation and fuel development influences fire severity patterns in reburns. *Ecol. Appl.* **2016**, *26*, 686–699. [CrossRef]
53. Morgan, P.; Keane, R.E.; Dillon, G.K.; Jain, T.B.; Hudak, A.T.; Karau, E.C.; Sikkink, P.G.; Holden, Z.A.; Strand, E.K. Challenges of assessing fire and burn severity using field measures, remote sensing and modelling. *Int. J. Wildland Fire* **2014**, *23*, 1045. [CrossRef]

Article

Evaluation of the MODIS LAI/FPAR Algorithm Based on 3D-RTM Simulations: A Case Study of Grassland

Jiabin Pu [1], Kai Yan [1,2,*], Guohuan Zhou [1], Yongqiao Lei [1], Yingxin Zhu [1], Donghou Guo [1], Hanliang Li [1], Linlin Xu [1,3], Yuri Knyazikhin [2] and Ranga B. Myneni [2]

[1] School of Land Science and Techniques, China University of Geosciences, Beijing 100083, China; pujiabin@cugb.edu.cn (J.P.); zgh@cugb.edu.cn (G.Z.); leiyongqiao@cugb.edu.cn (Y.L.); zhuyingxin@cugb.edu.cn (Y.Z.); guodonghou@cugb.edu.cn (D.G.); lihanlaing1998@cugb.edu.cn (H.L.); xulinlin@cugb.edu.cn (L.X.)
[2] Department of Earth and Environment, Boston University, Boston, MA 02215, USA; jknjazi@bu.edu (Y.K.); ranga.myneni@bu.edu (R.B.M.)
[3] Department of Systems Design Engineering, University of Waterloo, Waterloo, ON N2L 3G1, Canada
* Correspondence: kaiyan@cugb.edu.cn

Received: 12 August 2020; Accepted: 14 October 2020; Published: 16 October 2020

Abstract: Uncertainty assessment of the moderate resolution imaging spectroradiometer (MODIS) leaf area index (LAI) and the fraction of photosynthetically active radiation absorbed by vegetation (FPAR) retrieval algorithm can provide a scientific basis for the usage and improvement of this widely-used product. Previous evaluations generally depended on the intercomparison with other datasets as well as direct validation using ground measurements, which mix the uncertainties from the model, inputs, and assessment method. In this study, we adopted the evaluation method based on three-dimensional radiative transfer model (3D RTM) simulations, which helps to separate model uncertainty and other factors. We used the well-validated 3D RTM LESS (large-scale remote sensing data and image simulation framework) for a grassland scene simulation and calculated bidirectional reflectance factors (BRFs) as inputs for the LAI/FPAR retrieval. The dependency between LAI/FPAR truth and model estimation serves as the algorithm uncertainty indicator. This paper analyzed the LAI/FPAR uncertainty caused by inherent model uncertainty, input uncertainty (BRF and biome classification), clumping effect, and scale dependency. We found that the uncertainties of different algorithm paths vary greatly (−6.61% and +84.85% bias for main and backup algorithm, respectively) and the "hotspot" geometry results in greatest retrieval uncertainty. For the input uncertainty, the BRF of the near-infrared (NIR) band has greater impacts than that of the red band, and the biome misclassification also leads to nonnegligible LAI/FPAR bias. Moreover, the clumping effect leads to a significant LAI underestimation (−0.846 and −0.525 LAI difference for two clumping types), but the scale dependency (pixel size ranges from 100 m to 1000 m) has little impact on LAI/FPAR uncertainty. Overall, this study provides a new perspective on the evaluation of LAI/FPAR retrieval algorithms.

Keywords: MODIS; leaf area index (LAI); fraction of photosynthetically active radiation absorbed by vegetation (FPAR); three-dimensional radiative transfer model (3D RTM); uncertainty assessment

1. Introduction

Leaf area index (LAI), defined as half of the total green leaf area of per unit horizontal ground area, is a basic parameter for measuring the vegetation canopies [1,2]. This variable plays a key roles in hydrology, biogeochemistry, and ecosystem models that connect vegetation to the climate observing system through the carbon, water cycles, and radiation [3]. Fraction of photosynthetically active radiation (0.4–0.7 μm) absorbed by vegetation (FPAR) measures the proportion of the solar radiation entering at the top of the plant canopy that contributes to the photosynthetic activity [3–6]. LAI/FPAR retrieved from remote sensing observations in the reflective solar domain, are used as input parameters for models monitoring the Earth's surface continuously and are key parameters recognized by the global climate observing system (GCOS) to describe climatic characteristics [3,7]. LAI/FPAR products, derived from atmospherically corrected surface reflectances, have entered a new era since the moderate resolution imaging spectroradiometer (MODIS) became operational in 1999 [8–10]. The MODIS LAI/FPAR products (MOD15), based on the radiative transfer (RT) model [11], have been widely used to corroborate global climate change [12], to serve as key inputs for terrestrial carbon cycle models [13], and to support the research of both phenomena and possible reasons of large scale vegetation dynamics [14–16]. Moreover, the generation of MODIS LAI/FPAR products does not depend on other LAI/FPAR datasets and they are commonly used as input and reference data for the generation and intercomparison of other products [17,18].

Intensive evaluation and validation efforts have been carried out to examine the uncertainty of MODIS LAI/FPAR products and the corresponding retrieval algorithm. These works mainly included: (1) theoretical derivation based on model mechanisms and error propagation [19]; (2) intercomparison with other LAI/FPAR products or related variables (e.g., GLASS, CYCLOPES, VIIRS) [20–25]; (3) direct validation using ground LAI/FPAR measurements [25,26]. The theoretical derivation has an explicit mathematical basis and does not require other datasets; however, this approach is highly correlated with the algorithm itself and is easily affected by model limitations and uncertainties [27,28]. Intercomparison with other LAI/FPAR products is an approach that can effectively analyze the spatio-temporal consistency of long-term LAI/FPAR, but the results cannot meet the requirement of product usage and algorithm refinement. Ground-based validations are essential as the basis of all validations, but the accuracy of this validation method includes the uncertainty of the ground measurements, the spatial heterogeneity-caused uncertainty [29] in the upscaling process from the point measurement to the pixel scale, and the product uncertainty. Above all, the previous studies mainly focused on the evaluation of product uncertainty, which introduces the coupled uncertainties from the model, inputs, and assessment method. Therefore, it would hinder the process of evaluating the uncertainty of the algorithm itself and understanding the deficiencies of the algorithm, thus hampering future improvements to the algorithm.

In the above context, real scene computer simulations provide a new approach for remote sensing evaluation and validation [30]. As computing power improves, several 3D RT models have been developed for scene simulation [31,32], such as DART (discrete anisotropic radiative transfer) [33], RAPID (radiosity applicable to porous individual objects) [34], and LESS (large-scale remote sensing data and image simulation framework) [35]. These models have become an important tool in the field of quantitative remote sensing, particularly for studying the radiometric properties of the Earth's surface [31,36]. 3D RT models can analyze the detailed interactions between solar radiation and vegetation canopies [37], analyze the radiative properties of specific biome types [38], and help the science team define the characteristics of optical sensors through model simulation [39]. Data from simulations based on 3D RT models are widely used for model validation and evaluation. The DART model has been used in studies on the surface energy budget [40], the impact of canopy structure on satellite image texture [41], the 3D distribution of photosynthesis and primary production rates of vegetation canopies [42], and forest biophysical parameter retrieval [43,44]. The LESS model can synergistically use spectral and angular information to simulate the radiation properties of complex realistic landscapes, which can be used for simulating datasets of 3D landscapes [45]. The outputs of LESS can serve as benchmarks for retrieval algorithm evaluation since it has a solid theoretical foundation and its accuracy has already been well-assessed by comparison with other models of radiation transfer model intercomparison (RAMI) [35] and field measurements [46].

This study aimed to provide a new perspective on the evaluation of MODIS LAI/FPAR retrieval algorithms, which differs from previous research by evaluating the algorithm itself rather than the product. In this paper, a computer simulation of a real grassland scene is performed using the ray-tracing LESS model to analyze the uncertainty of the MODIS LAI/FPAR retrieval algorithm. The advantage of simulation-based model evaluation is that the uncertainty caused by a single variable can be analyzed to avoid the effects caused by the mixing of multiple factors. The uncertainty of the MODIS LAI/FPAR algorithm was evaluated by separating the model and input uncertainties. In addition, further analysis was conducted to understand the impact of scale dependency and clumping. The results can serve as guidance for improving this algorithm continuously.

The structure of this paper is organized as follows. Section 2 briefly describes the MODIS LAI/FPAR retrieval algorithm, how we use LESS to analyze the retrieval algorithm for uncertainty, and the methodologies for uncertainty evaluation. Section 3 details the results of LAI/FPAR uncertainty caused by inherent model, reflectance, and biome type uncertainties as well as the clumping effect, and scale dependency. The discussions, including the analysis of the experiment results, are detailed in Section 4. Finally, Section 5 provides some concluding remarks.

2. Materials and Methods

2.1. MODIS LAI/FPAR Retrieval Algorithm

The MODIS LAI/FPAR retrieval algorithm consists of a main algorithm based on the radiative transfer equation (RTE) and a backup algorithm using the relationship between vegetation index and LAI/FPAR. The retrieval algorithm exploits the spectral information content of MODIS surface reflectances at up to 7 spectral bands (band 1: 620–670 nm; band 2: 841–876 nm; band 3: 459–479 nm; band 4: 545–565 nm; band 5: 1230–1250 nm; band 6: 1628–1652 nm; band 7: 2105–2155 nm) [4,8]. Inputs of this algorithm include BRFs at red and near-infrared (NIR) bands (band 1 and 2), their uncertainties, sun–sensor geometry (SZA: solar zenith angle, SAA: solar azimuth angle, VZA: view zenith angle, VAA: view azimuth angle), and a biome classification map. Note that in the current algorithm version, different biome types use different RT models. Herbaceous biomes (B1: grasses and cereal crops; B2: shrubs; B3: broadleaf crops;) were modelled using 1D RT due to the good continuity of the grass distribution and in consideration of the computational efficiency. Savannas (B4) were modelled by a stationary Poisson germ-grain stochastic process (so called stochastic radiative transfer (SRT) model) [47,48]. Forest biomes (B5: evergreen broadleaf forests; B6: deciduous broadleaf forests; B7: evergreen needleleaf forests; and B8: deciduous needleleaf forests) were based on a 3D RTM (3D structures were represented by columns uniformly (deterministically) spaced on the ground). With these RTMs, the science team constructed an LAI/FPAR main algorithm based on angular information, biome type, and spectral information in which the mean and standard deviation values of the LAI and FPAR selected in the spectral retrieval space are reported for retrieval value and its uncertainty. The main look up table (LUT)-based algorithm was designed as follows. Firstly, the main algorithm evaluates a weight coefficient as a function of sun–sensor geometry, wavelength, and LAI by using a field-tested canopy reflectance model. Then it calculates the BRFs by using the weight coefficient and the same model [4,8]. The algorithm tests the eligibility of a canopy radiation model to generate the LUT file where a subset of coefficients is satisfied within a given accuracy [9]. The given atmosphere-corrected BRFs are then compared with the modeled BRFs, which are stored in the biome-specific LUT files. Finally, all candidates of LAI/FPAR are used to calculate the mean values and uncertainty of the retrieval [9]. In the case of highly dense canopies, reflectance will be saturated and insensitive to changes in canopy properties. Therefore, LAI and FPAR values acquired under saturated conditions are less reliable than those generated by unsaturated BRFs. When the main algorithm fails to localize a solution, the backup algorithm is used to retrieve values through an empirical relationship between the normalized difference vegetation index (NDVI) and the canopy LAI/FPAR [11,21]. Such retrievals are flagged in the algorithm path quality assessment (QA) variable [8], which consists of two values for the main algorithm and two values for the backup algorithm (from high quality to low): the main algorithm without saturation (QA = 0), the main algorithm with saturation (QA = 1), the backup algorithm due to sun–sensor geometry (QA = 3), and the backup algorithm due to other reasons (QA = 4) [9,14,49].

2.2. Three-Dimensional Grassland Scene Simulation

We used the newly proposed but well validated 3D RT model LESS to simulate the interaction between the solar radiation and landscape elements based on the spectral response functions (SRFs) (from ENVI software) of MODIS and calculated the scene BRFs [35,45]. LESS simulates BRFs by a weighted forward photon tracing method as well as simulated energy transfer and generates images by a backward path tracing method [35]. Qi et al. [35] described the comparison between BRFs simulated by LESS and average BRF results from other models (e.g., SPRINT3, RAYTRAN, and RAYSPREAD) over several different homogeneous and heterogeneous canopies from the RAMI website to evaluate the accuracy of LESS.

The input parameters of LESS include 3D landscape elements, optical properties, and sun–sensor geometries. The simulated scenes are covered by grass (Johnson grass) and its component spectra were obtained from the LOPEX93 dataset on the OPTICLEAF website [50]. The soil (grayish brown loam) spectra were selected from the soil spectral library in ENVI software, and the transmittance of the soil is 0 (Figure 1). Then we calculated the two MODIS bands (red: band 1 and NIR: band 2) reflectance and transmittance (see Table 1) using SRFs (the shaded part of Figure 1) of the MODIS sensor by the following equation [51]:

$$\begin{cases} R = \sum_{\lambda=\lambda_{min}}^{\lambda_{max}} S_\lambda R_\lambda / \sum_{\lambda=\lambda_{min}}^{\lambda_{max}} S_\lambda \\ T = \sum_{\lambda=\lambda_{min}}^{\lambda_{max}} S_\lambda T_\lambda / \sum_{\lambda=\lambda_{min}}^{\lambda_{max}} S_\lambda \end{cases} \quad (1)$$

where, R and T are MODIS band reflectance and transmittance, respectively. The R_λ and T_λ are mean narrow-band reflectance and transmittance derived from the spectral curves. The S_λ is the SRF value of the MODIS sensor. λ is the value of wavelength, which has a specific upper (λ_{max}: red = 670 μm, NIR = 876 μm) and lower (λ_{min}: red = 620 μm, NIR = 841 μm) limit for each band.

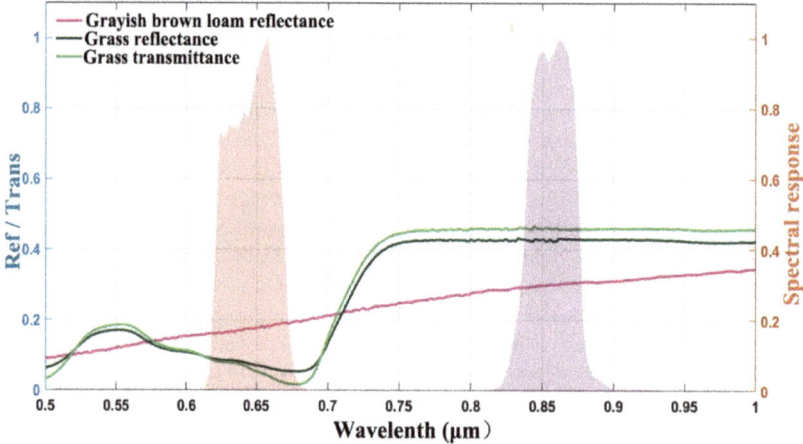

Figure 1. Variation of reflectance (Ref), transmittance (Trans), and spectral response function (SRF) values at different wavelengths. The dark green and magenta curves represent the grass and soil reflectances, and the light green represents the grass transmittance. The shades of red and purple represent the SRFs of the MODIS sensor in the red (620–670 μm) and NIR (841–876 μm) bands, respectively.

Table 1. Broad-band reflectance and transmittance of grass and soil used in this study. R and T are abbreviations for broad-band reflectance and transmittance, respectively.

	R (Red)	T (Red)	R (NIR)	T (NIR)
Johnson grass	0.0738	0.0577	0.4276	0.4607
Grayish brown loam	0.1755	0	0.3021	0

The 3D landscape elements were created with the third-party software OnyxTree, which uses the calculated reflectance and transmittance (Table 1) to make a grass 3D model (*obj* format file). As shown in Figure 2, we created nine randomly distributed grasslands with different LAIs (0.25, 0.50, 0.75, 1.0, 1.25, 1.5, 2.5, 3.5, and 4.5) using the LESS and grass 3D model. Moreover, LESS calculates FPAR by performing a band integration of the PAR between 380 nm and 710 nm and dividing by the incident radiation (slightly different from MODIS for which the wavelength interval is 400–800 nm) based on the LESS simulation of the collision of photons and the transfer of energy. In addition, to match the canopy structure of grasses in the MODIS LAI/FPAR retrieval algorithm (all organs other than leaves are ignored), only foliage is present in the scene. There is also only direct radiation in these scenes. The size of these scenes is 500 m × 500 m, which matches the spatial resolution of the MODIS LAI/FPAR products.

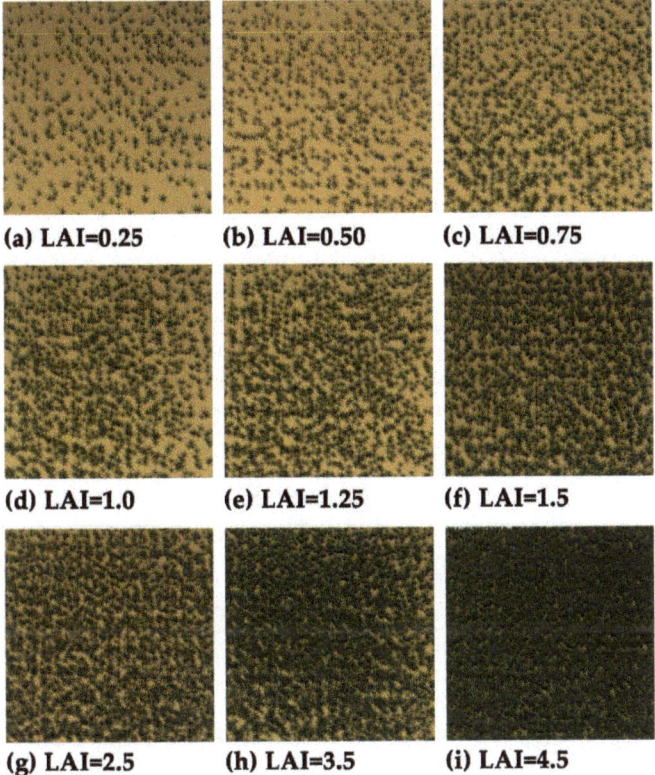

Figure 2. Simulated scenes with nine different LAI values using the LESS 3D RT model. Panels (a)–(i) are with LAI = 0.25, 0.50, 0.75, 1.0, 1.25, 1.5, 2.5, 3.5, and 4.5, respectively. The plots represent a smaller portion (5 m × 5 m) of a 500 m × 500 m scene. The grasses are randomly distributed in these scenes.

2.3. Experimental Design

We utilized the standard deviation of all LAI/FPAR candidates (StdLAI and StdFPAR), the retrieval index (RI), and the relative and absolute LAI/FPAR differences as the indicators of LAI/FPAR uncertainty. According to the uncertainty theory, StdLAI and StdFPAR are the standard deviations of all acceptable LAI/FPAR solutions in the LUT, which are the function of both the input uncertainty (biome type and BRF uncertainty) and model uncertainty [4,8]. StdLAI and StdFPAR have been proven and evaluated as quality metrics for MODIS LAI/FPAR products [28,52]. However, these two metrics have limitations due to the regularization introduced by the LUT algorithm and are artificially lowered at large LAIs [8]. Therefore, in this paper, we have also selected the RI (see Equation (2)) as an uncertainty metric, which is defined as the percentage of pixels for which the main RTE-based algorithm generates retrieval results. We note that the RI is used to characterize the overall uncertainty of all pixels [21,25,53], while StdLAI and StdFPAR are used to characterize individual pixel uncertainty.

$$RI = \frac{Number\ of\ pixels\ retrieved\ by\ the\ main\ algorithm}{Total\ number\ of\ processed\ pixels} \quad (2)$$

To evaluate the consistency between true LAI/FPAR and MODIS retrievals, the difference between the simulation results of LAI (input to the LESS)/FPAR (output from the LESS) and the LAI/FPAR retrieved by the MODIS retrieval algorithm were used. The relative difference (RD, see Equation (3)) and absolute difference (AD, see Equation (4)), were utilized to quantify any differences.

$$RD = (Retrieval - Truth)/Truth \quad (3)$$

$$AD = Retrieval - Truth \quad (4)$$

Based on the uncertainty theory, the retrieval uncertainty is a function of both model and input uncertainty and is embedded in the MODIS algorithm. In this study, we explored the relationship between retrieval uncertainty and the retrieval space, sun–sensor geometry, surface reflectance uncertainty, and biome type uncertainty using the variable-controlling approach (see Table 2). We analyzed the inherent model uncertainty in two steps: 1) analysis of the retrieval space; 2) uncertainty changed with sun–sensor geometry. We obtained 4000 red-NIR BRF pairs by adding normally distributed errors (errors with 5% and 15% standard deviation) to the LESS simulated red and NIR band BRFs (1000: red without uncertainties and NIR with 5% standard deviation, 1000: red without uncertainties and NIR with 15% standard deviation, 1000: NIR without uncertainties and red with 5% standard deviation, and 1000: NIR without uncertainties and red with 15% standard deviation). Then we analyzed the LAI/FPAR uncertainty caused by BRF uncertainty within the 4 groups of samples. In addition, we analyzed uncertainties due to biome type misclassification, which is one of the main factors affecting the LAI/FPAR retrieval accuracy [4,54]. Each red-NIR BRF pair was sequentially combined with each biome type as the inputs for the MODIS LAI/FPAR algorithm. In this experiment, only B1 (grasses and cereal crops) was correct while the remaining seven combinations represented the biome type misclassification cases. Finally, we analyzed the influence of scale dependency and clumping effect ("tree groups") [55] on the uncertainty of LAI/FPAR retrievals. We simulated a randomly distributed 1 km × 1 km scene (Figure 9a-1) and two clumping 1 km × 1 km scenes. Clumping type 1 (CT1, Figure 9a-2) had random clumping and Clumping type 2 (CT2, Figure 9a-3) was half bare ground and half grass. The LAI of the three scenes remained constant and these scenes were downscaled into four 500 m × 500 m scenes, sixteen 250 m × 250 m scenes, and one hundred 100 m × 100 m scenes for the discussion of scale dependency and clumping effect.

Table 2. Parameter configuration for designed experiments. SZA, SAA, VZA, VAA means solar zenith angle, solar azimuth angle, view zenith angle, and view azimuth angle, respectively.

Experiment	LAI	SZA	SAA	VZA	VAA	Uncertainty Metrics
Retrieval Space	/	0°	0°	0°	0°	StdLAI, StdFPAR
Sun–Sensor Geometry	0.50, 1.5, 3.5	30°/ 0°:10°:60°	90°/ 0°:30°:330°	0°:10°:60°/30°	0°:30°:330°/ 90°	RD, StdLAI, StdFPAR
BRF Uncertainty	1.5	0°	0°	−60°:10°:60°	0°	RD, StdLAI, StdFPAR
Biome Type Uncertainty	0.25, 0.50, 0.75, 1.0, 1.25, 1.5, 2.5, 3.5, 4.5	0°	0°	0°:10°:60°	0°:30°:330°	RI, AD
Clumping and Scale Effect	1.5	30°	0°	0°:30°:60°	0°:60°:300°	RI, StdLAI, StdFPAR

3. Results

3.1. Inherent Model Uncertainty

To evaluate the inherent model uncertainty of the MODIS LAI/FPAR retrieval algorithm, we analyzed the effect of the retrieval space and sun–sensor uncertainty, separately. In performing the evaluation of the retrieval space, we paid more attention to the changes in the uncertainty of algorithm paths. While the difference between the LAI/FPAR retrieval and LESS simulations were analyzed when evaluating of the sun–sensor geometry.

3.1.1. Analysis of Retrieval Space

Figure 3 indicates the variation of LAI/FPAR and its uncertainty in the retrieval space. As we can see, LAI/FPAR is nonlinearly related to surface reflectance (Figure 3a,b), and FPAR is also nonlinearly related to LAI. Moreover, the relationship between LAI/FPAR and its uncertainty (StdLAI and StdFPAR) is also nonlinear. The StdLAI and StdFPAR are very low for lower LAI/FPAR and then increase to the highest values, and then steadily decrease (from the bottom right to the top left of Figure 3d,e) as the LAI/FPAR gets progressively larger (from the bottom right to the top left of Figure 3a,b). It is also obvious that there is a clear division between the saturated (QA = 1) and unsaturated (QA = 0) parts where the LAI/FPAR values are higher in the saturated part (Figure 3c). Compared to the unsaturated part, the bias of LAI (+4.64) and FPAR (+0.631) are high, but the bias of StdLAI (−0.052) and StdFPAR (−0.169) are low in the saturated part. Figure 3d,e show that StdLAI and StdFPAR are relatively small at the boundaries of the area retrieved by the main algorithm due to the regularization of the algorithm [4,8].

3.1.2. Retrieval Uncertainty as a Function of Sun–Sensor Geometry

The relationship between LAI/FPAR uncertainty and sun–sensor geometry are presented in Figures 4 and 5. In the high LAI scene (Figure 4a, LAI = 3.50), the retrieval results of LAI/FPAR show low consistency with the truth (it yields to an overall uncertainty of 20.01% for RD of LAI and 13.96% for RD of FPAR). The main algorithm shows an averaged 6.61% underestimation of LAI, while the backup algorithm results in an averaged 84.85% overestimation of LAI. In this scene, the backup algorithm appears at the "hotspot" geometry and where the difference between SAA and VAA is large. It can also be seen that the large VZA will lead to saturation. Nevertheless, for the low LAI scene (LAI = 0.50), the retrieved LAI/FPAR showed a significant overestimation (+111.86% RD for LAI, +162.50% RD for FPAR) and large uncertainty (StdLAI = 0.285, and StdFPAR = 0.238). Figure 5 shows the same analysis as above, but we controlled the view position and varied the sun position. Comparing Figures 4 and 5, the distribution of LAI and its uncertainty (Figure 4a,c, and Figure 5a,c) show higher consistency, while FPAR and its uncertainty (Figure 4b,d and Figure 5b,d) are slightly different.

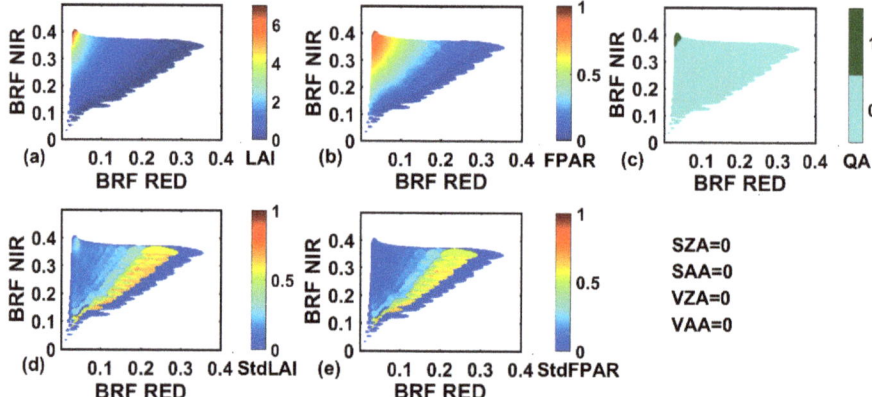

Figure 3. Distribution of LAI/FPAR values and associated uncertainty derived from the main RT-based algorithm in the red-NIR space. The SZA, SAA, VZA, and VAA were all fixed at 0°. Panel (**a–e**) represent the retrieved LAI, FPAR, algorithm path (QA = 0: main algorithm without saturation, QA = 1: main algorithm with saturation), StdLAI, and StdFPAR, respectively.

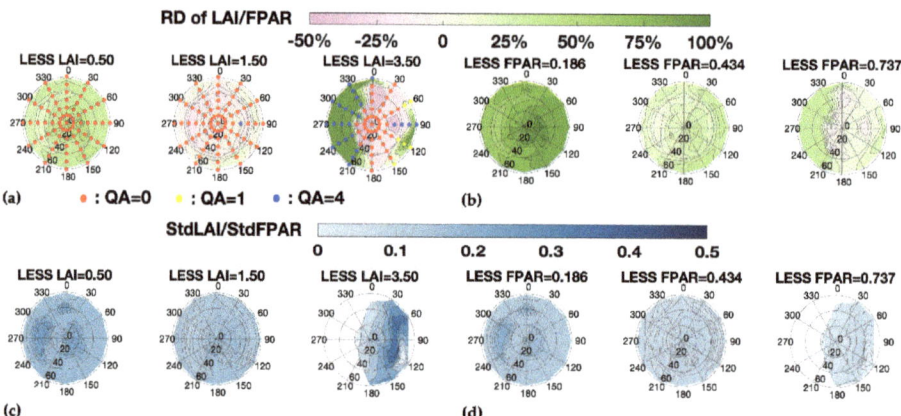

Figure 4. The uncertainty (LAI/FPAR RD, StdLAI, and StdFPAR) as a function of sensor geometry when the SZA is 30° and SAA is 90°. (**a**)–(**d**) are RD of LAI, RD of FPAR, StdLAI, and StdFPAR in three different scenes (Scene 1: LAI = 0.50 and FPAR = 0.186, Scene 2: LAI = 1.50 and FPAR = 0.434, Scene 3: LAI = 3.50 and FPAR = 0.737), respectively. The colored dots in panel (**a**) represent different algorithm paths (main without saturated: QA = 0, main with saturated: QA = 1, backup: QA = 4).

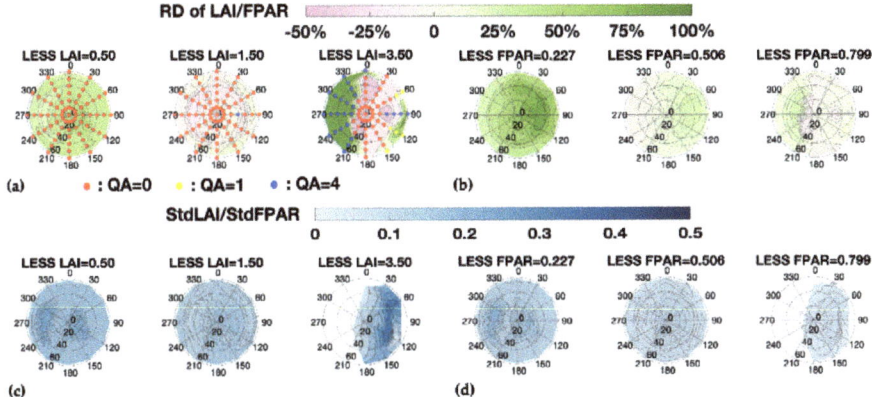

Figure 5. The LAI/FPAR uncertainty as a function of sun geometry. Same as Figure 4 but for sun geometry and the VZA is 30° and VAA is 90°. The FPAR values of the scenes are calculated as the mean of different solar angles.

3.2. Input BRF Uncertainty

Here, we calculated the effects of input BRF uncertainty on the LAI/FPAR retrieval. Figure 6 shows that the uncertainty in the LAI/FPAR in the shadow area of the 15% BRF uncertainty is much larger, which means that larger BRF uncertainty will result in larger LAI/FPAR uncertainty. The StdLAI and StdFPAR due to a 5% BRF uncertainty are close to the StdLAI and StdFPAR due to a 15% BRF in the red band. This is because both 5% and 15% BRF uncertainty in the red band will trigger the backup algorithm with no StdLAI and StdFPAR. Comparing the shadow area in panels (a) and (b), we found that the same level of uncertainty in the NIR band BRF has a greater impact on the retrieval than the red band BRF. The main algorithm was not used in the hotspot (VZA = 0) geometry leading to the absence of both StdLAI and StdFPAR in panel (a).

3.3. Input Biome Type Uncertainty

Different biome types have different canopy structures, and the MODIS retrieval algorithm uses photon transport theory and the corresponding RT model for different biome types to parameterize the canopy structures (e.g., reflectance and transmittance of leaves, crown shadowing), which form the LUTs of the MODIS retrieval algorithm. To check the sensitivity of the algorithm to biome type, we modified the input biome type for the retrieval algorithm from the correct type (B1: grasses and cereal crops) to incorrect types (B2: shrubs; B3: broadleaf crops; B4: savannas; B5: evergreen broadleaf forests; B6: deciduous broadleaf forests; B7: evergreen needleleaf forests; and B8: deciduous needleleaf forests). As seen from Figures 7 and 8, the retrieval uncertainty is similar when the input biome types are non-forest biomes (B1–B4) with a greater than 59.5% RI for all four biome types except for B2 in the LAI = 4.5 scene. However, the RI gets much lower when the grassland pixel is misclassified into forest biomes. As shown in Figures 7c and 8c, the uncertainties of the retrieved LAI are high at the scene with high LAI (e.g., LAI = 3.5, 4.5). A significant overestimation (+0.727, +1.434 for AD of LAI) in B2, and a significant underestimation (−1.608, −2.344 for AD of LAI) in B3 is also evident. For B5, the RI is high (>69%) but AD of LAI (>1.656) is also high when LAI is relatively high (e.g., LAI = 2.5, 3.5, 4.5). As shown in Figure 8c, the FPAR calculated from the MODIS algorithm is significantly overestimated for all cases except for B4 high LAI scenes, which appear to be underestimated.

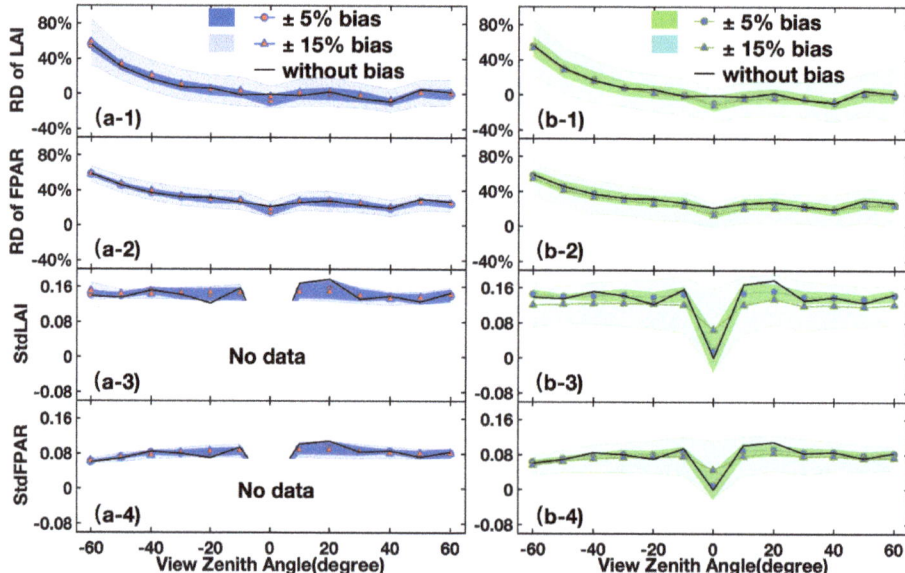

Figure 6. LAI/FPAR uncertainty caused by input BRF uncertainty as a function of the view zenith angle (VZA). Panel (**a**) and panel (**b**) represent the red and NIR band, respectively. The LAI value of the scene is 1.5, and the SZA, SAA and VAA are all set to 0. The upper two panels show the RD of LAI/FPAR and the lower two panels show the StdLAI and StdFPAR, respectively. Dots are the mean values of LAI/FPAR calculated by 1000 different BRFs and shadow indicates the standard deviation of these retrievals. "No data" means that RI is equal to 0 and neither StdLAI nor StdFPAR exists in this VZA condition.

3.4. Impact of Clumping Effect and Scale Dependency

The model scale dependency and clumping effect have attracted much attention from the community in the development of quantitative remote sensing. In this experiment, the model scale dependency refers to the discrepancy between LAI/FPAR uncertainties that are derived from the same algorithm but at different spatial resolutions. The scale dependency determines the adaptive capacity of an algorithm for different pixel size. The clumping effect refers to the discrepancy between retrieved LAI/FPARs with same LAI/FPAR truth but different vegetation spatial distributions. The model nonlinearly and surface heterogeneity together result in the well-known phenomenon called "Inversion first and aggregation later is different from aggregation first and inversion later" [11].

Comparing the algorithm performance at different scales, we found that the MODIS algorithm is nearly scale-invariant from 100 m to 1000 m. Both LAI/FPAR and their uncertainty nearly remain unchanged with increasing pixel size except for the CT2, which shows that the StdLAI and StdFPAR are lower than other scales (Table 3). The retrieved LAIs for all scenes are less than the LAI truth at 1000 m scale, and the underestimations for Uniform, CT1, and CT2 vegetation distributions are −0.005, −0.846, and −0.525, respectively. Comparing the three clumping scenes, we found that the LAI of a uniform scene is very close to the LAI truth (Figure 9b-1). CT1 shows a significant underestimation, while CT2 shows a significant overestimation except in the 1000 m scale. For FPAR, there is a significant overestimation in all three scenes (Figure 9b-2). At the same spatial resolution, the RI of CT1 is the highest, followed by Uniform, and the lowest is CT2 (Table 3). While the values of StdLAI and StdFPAR are as follows (from small to large): Uniform, CT1, and CT2. The standard deviations of StdLAI and StdFPAR also get larger in this order.

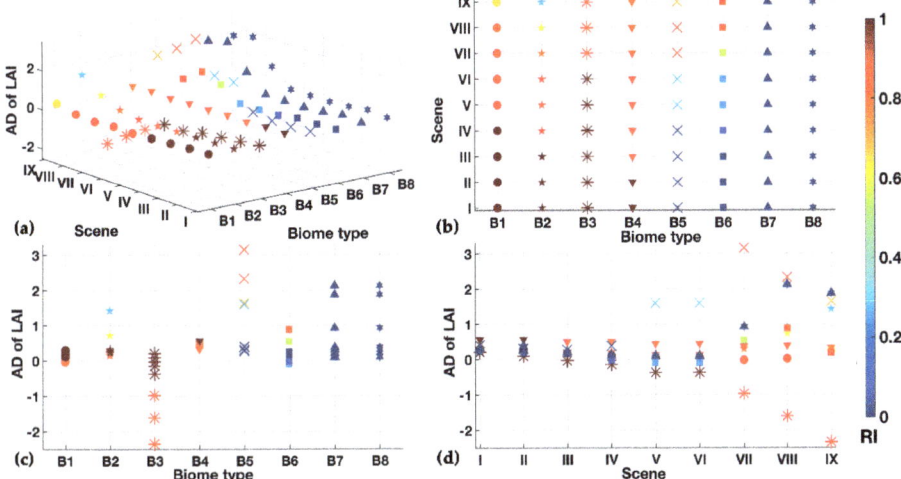

Figure 7. Illustration of the retrieval index (RI, indicated by different colors) and the absolute difference (AD) of LAI as a function of biome type and different scenes. Panel (**b**) and (**d**) are the x–z (x means biome type and z means AD) sections of the panel (**a**) and (**c**), respectively, which show the approximate range of AD of LAI in different biome types. Scenes I to IX represent the LAI truth being equal to 0.25, 0.5, 0.75, 1.00, 1.25, 1.50, 2.50, 3.50, and 4.50, respectively. The colors in the figure are the values of RI. The eight biome types are: grasses and cereal crops (B1); shrubs (B2); broadleaf crops (B3); savannas (B4); evergreen broadleaf forests (B5); deciduous broadleaf forests (B6); evergreen needleleaf forests (B7); and deciduous needleleaf forests (B8), where B1 is the correct input, and B2–B8 all represent misclassification. The shapes of the different symbols correspond to different biome types (one by one in panel (**b**) and (**c**)).

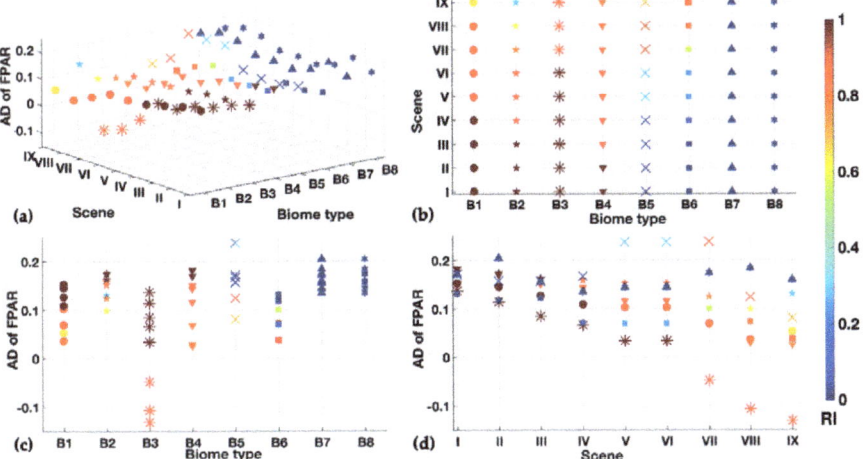

Figure 8. Illustration of the RI and the AD of FPAR as a function of biome type and different scenes. Same as Figure 7 but for FPAR.

Table 3. The uncertainty metrics of three different clumping scenes and four scales.

	Scene	100 m	250 m	500 m	1000 m
RI (N. of main/N. of all)	Uniform	1700/1800	272/288	68/72	17/18
	CT1	1743/1800	279/288	70/72	18/18
	CT2	1565/1800	251/288	62/72	17/18
StdLAI (mean ± Std)	Uniform	0.147± 0.019	0.148± 0.019	0.149± 0.020	0.150± 0.021
	CT1	0.251± 0.108	0.218± 0.074	0.225± 0.066	0.179± 0.074
	CT2	0.340± 0.160	0.339± 0.160	0.342± 0.159	0.181± 0.027
StdFPAR (mean ± Std)	Uniform	0.088± 0.012	0.088± 0.012	0.089± 0.012	0.089± 0.013
	CT1	0.208± 0.109	0.177± 0.070	0.180± 0.058	0.144± 0.061
	CT2	0.269± 0.188	0.269± 0.188	0.271± 0.187	0.130± 0.022

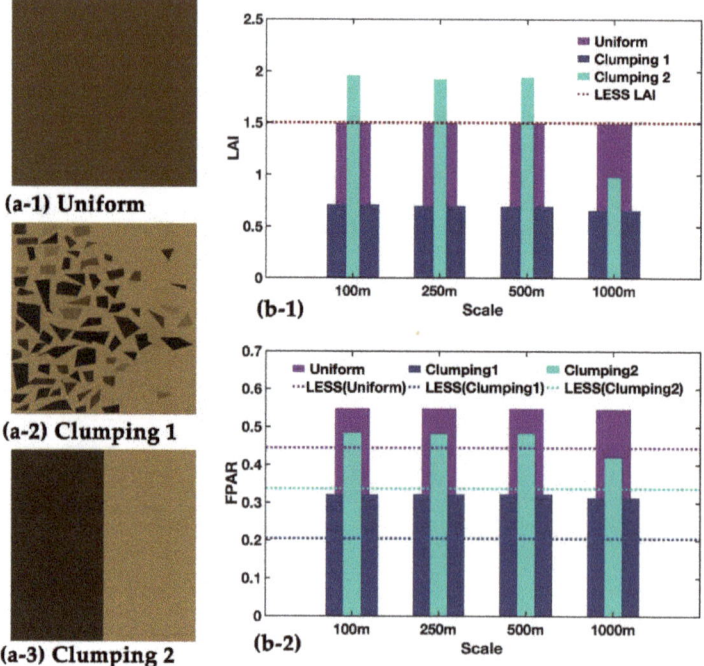

Figure 9. Comparison of LAI/FPAR retrievals over different clumping scenes and scales. Panel (**a**) is for three 1 km² scenes (**a-1**: uniform, **a-2**: randomly generated clumping (CT1), **a-3**: half-and-half clumping (CT2)) and panel (**b**) shows the retrievals over three different scenes and four different scales (100 m, 250 m, 500 m, and 1000 m) where the dashed line represents the LAI/FPAR truth.

4. Discussion

Because of the different sensitivities of LAI/FPAR to surface reflectances, we note that there would be a gap of uncertainty between the saturated part and the unsaturated part [4,8]. However, Figure 3 indicates that for large LAI/FPAR, their theoretical uncertainty is artificially reduced by the method of regularization, which causes the retrieval to have varying degrees of confidence and leads to a problematic evaluation of high LAI/FPAR scenes using the provided StdLAI and StdFPAR. This also places new requirements on future algorithm refinement that the LUT algorithm should be consistent in the saturated case as in the unsaturated case. The LAI/FPAR values estimated by the backup algorithm and calculated by the main algorithm also show significant discontinuity [4,8–10] (Figure 4a). Based on this, we point out that future algorithm refinement should increase the coverage of the main algorithm usage, which will greatly improve the overall accuracy of the product. In addition, according to the 3D RT model, the hotspot means that the radiation field tends to peak around the retro-illumination direction. The results of this study indicate that the uncertainty of the MODIS algorithm in the hotspots is quite large (Figures 4–6), due to which the science team decided not to include additional hotspot parameters since their inclusion would make algorithm calibration difficult [56,57]. We note that this will not cause large problems in the MODIS LAI/FPAR production because of the fact that observations near the hotspot are rare for MODIS. However, this points out a new refinement direction of this algorithm to improve the accuracy of hotspot modeling for other sensors.

As is known, the uncertainty of the inputting BRFs has some influence on the uncertainty of the retrieval algorithm. In particular, our results show that the uncertainty of NIR BRFs has a larger effect on LAI/FPAR uncertainty compared to the red BRFs (Figure 6). We know that insufficient input information will lead to the "ill-posed" retrieval problem [11]; however, the inputting BRFs of the MODIS operational algorithm are currently only for the red and NIR bands. Therefore, in the future we may try to make use of BRFs in other bands to improve the retrieval accuracy. The MODIS algorithm depends on a priori information about the land surface given by biome type representing the pattern of the architecture of vegetation, as well as patterns of spectral reflectance and transmittance of vegetation [8]. Figures 7 and 8 confirm that the misclassification of biome types with similar structures will result in smaller LAI/FPAR uncertainty, and vice versa [11,58]. This means that the improvement of biome classification accuracy is an efficient way to improve the LAI/FPAR products. Moreover, different biome types also lead to different clumping types. As Figure 9 shows, the underestimation of LAI is significant for two clumping scenes at 1000 m scale. For the other three scales, however, the overestimation of CT2 is due to the backup algorithm retrievals. As our results show, the algorithm only considers the clumping effect at one scale (e.g., B1 is minimal leaf clumping) [4], which can result in large differences in the retrievals; therefore, we suggest that future algorithms consider the clumping effect at more scales (e.g., leaf, branch, and crown).

We note that there are some problems with the way we use LESS to simulate specific scenes and evaluate the MODIS algorithm. First, according to the algorithm, the retrieved LAI/FPAR is a weighted average of the probability values within the error range. Therefore, the probability distribution of LAI/FPAR within the error range based on a great number of realizations has more statistical significance thus may differ from the specific realization (scene) that was used. Secondly, although the LESS model has been well validated, the confidence of our evaluation results depends on the accuracy of the LESS simulation.

In short, validation in the field of remote sensing utilizing computer simulations has proved feasible. In future studies, we will analyze the other seven biome types, which will provide a more comprehensive evaluation of the MODIS LAI/FPAR retrieval algorithm. In addition, we will change the mode of a single specific scene to obtain retrieval results by simulating multiple scenes. Moreover, evaluation of the algorithm at different levels of vegetation clumping will be the focus of our future research.

5. Conclusions

This paper presents an uncertainty assessment of the MODIS LAI/FPAR retrieval algorithm over B1 (grassland) based on computer simulation. To accomplish this assessment, we first analyzed the theoretical uncertainty caused by inherent model uncertainty, then we calculated the uncertainty caused by input parameters (BRF and biome type) over simulated 3D grass scenes. Finally, we analyzed the effects of vegetation clumping and scale dependency of the MODIS algorithm. The 3D grass scenes were simulated by a well validated 3D RT model (LESS), which helps to separate the model uncertainty and other uncertainties. We found that the uncertainty of the main and backup algorithm varies considerably. In the same scene, there is a −6.61% bias for the main algorithm retrieval, while the backup algorithm retrieval has a +84.85% bias. We noted that the uncertainty of the saturated retrievals is artificially reduced compared with unsaturated retrievals. At the same time, MODIS showed significant overestimation at low LAI scenes, with a maximum bias of +111.86% for LAI and +162.50% for FPAR. In the high LAI scenes, the "hotspot" geometry results in greater retrieval uncertainty from the backup algorithm. Moreover, input uncertainties further increased the uncertainty of LAI/FPAR retrieval. We found that the uncertainties in BRF in the NIR band has a greater impact than in the red band. The biome type uncertainty also leads to great retrieval uncertainty. Large uncertainties occurred when grassland was misclassified into forest biomes, while smaller uncertainties occurred when the misclassification was within the non-forest biomes. In addition, the clumping effect results in underestimation (−0.846 and −0.525 for the two clumping types, respectively) and we found that the MODIS algorithm is nearly scale-invariant from 100 m to 1000 m pixel sizes. Overall, these results, based on novel computer simulation experiments, can guide the future refinements of the MODIS LAI/FPAR algorithm.

Author Contributions: J.P.: formal analysis, writing—original draft preparation, investigation. K.Y.: conceptualization, methodology, writing—review and editing, funding acquisition, supervision, project administration. G.Z., Y.L., Y.Z., D.G., and H.L.: software, formal analysis. L.X.: writing—original draft preparation. Y.K. and R.B.M.: conceptualization, methodology. All authors have read and agreed to the published version of the manuscript.

Funding: This work was supported by the National Natural Science Foundation of China (41901298), the open fund of the State Key Laboratory of Remote Sensing Science (OFSLRSS201924), the open fund of the Key Laboratory of Digital Earth Science, the Institute of Remote Sensing and Digital Earth, the Chinese Academy of Sciences (2018LDE002), the Fundamental Research Funds for the Central Universities (2652018031) and the open fund of Shanxi Key Laboratory of Resources, Environment and Disaster Monitoring(2019-04).

Acknowledgments: We thank the MODIS LAI&FPAR team for all of their help and Jianbo Qi for support with the LESS 3D RT model. We also appreciate the fruitful suggestions from the anonymous reviewers which made the work better.

Conflicts of Interest: The authors declare no conflict of interest. The funders had no role in the design of the study; in the collection, analyses, or interpretation of data; in the writing of the manuscript, or in the decision to publish the results.

References

1. Jacquemoud, S.; Baret, F.; Hanocq, J. Modeling spectral and bidirectional soil reflectance. *Remote Sens. Environ.* **1992**, *41*, 123–132. [CrossRef]
2. Chen, J.M.; Black, T.A. Defining leaf area index for non-flat leaves. *Plant Cell Environ.* **1992**, *15*, 421–429. [CrossRef]
3. GCOS. Systematic observation requirements for satellite-based products for climate. 2011 update supplemetnatl details to the satellite 39 based component og the implementation plan for the global observing system for climate in support of the unfccc (2010 update). In *Technical Report*; World Meteorological Organisation (WMO): Geneva, Switzerland, 2011.
4. Knyazikhin, Y.; Martonchik, J.; Myneni, R.B.; Diner, D.; Running, S.W. Synergistic algorithm for estimating vegetation canopy leaf area index and fraction of absorbed photosynthetically active radiation from MODIS and MISR data. *J. Geophys. Res. Atmos.* **1998**, *103*, 32257–32275. [CrossRef]
5. Sellers, P.; Dickinson, R.E.; Randall, D.; Betts, A.; Hall, F.; Berry, J.; Collatz, G.; Denning, A.; Mooney, H.; Nobre, C. Modeling the exchanges of energy, water, and carbon between continents and the atmosphere. *Science* **1997**, *275*, 502–509. [CrossRef] [PubMed]
6. Zhu, Z.; Bi, J.; Pan, Y.; Ganguly, S.; Anav, A.; Xu, L.; Samanta, A.; Piao, S.; Nemani, R.R.; Myneni, R.B. Global data sets of vegetation leaf area index (LAI) 3g and fraction of photosynthetically active radiation (FPAR) 3g derived from global inventory modeling and mapping studies (GIMMS) normalized difference vegetation index (NDVI3g) for the period 1981 to 2011. *Remote Sens.* **2013**, *5*, 927–948.
7. Mason, P.; Zillman, J.; Simmons, A.; Lindstrom, E.; Harrison, D.; Dolman, H.; Bojinski, S.; Fischer, A.; Latham, J.; Rasmussen, J. *Implementation Plan for the Global Observing System for Climate in Support of the UNFCCC (2010 Update)*; World Meteorological Organization: Geneva, Switzerland, 2010; p. 180.

8. Knyazikhin, Y. MODIS Leaf Area Index (LAI) and Fraction of Photosynthetically Active Radiation Absorbed by Vegetation (FPAR) Product (MOD 15) Algorithm Theoretical Basis Document. Available online: https://modis.gsfc.nasa.gov/data/atbd/atbd_mod15.pdf (accessed on 2 February 2017).
9. Yan, K.; Park, T.; Yan, G.; Chen, C.; Yang, B.; Liu, Z.; Nemani, R.; Knyazikhin, Y.; Myneni, R. Evaluation of MODIS LAI/FPAR Product Collection 6. Part 1: Consistency and Improvements. *Remote Sens.* **2016**, *8*, 359. [CrossRef]
10. Myneni, R.; Park, Y. MODIS Collection 6 (C6) LAI/FPAR Product User's Guide. Available online: https://lpdaac.usgs.gov/sites/default/files/public/product_documentation/mod15_user_guide.pdf (accessed on 1 January 2016).
11. Myneni, R.B.; Hoffman, S.; Knyazikhin, Y.; Privette, J.; Glassy, J.; Tian, Y.; Wang, Y.; Song, X.; Zhang, Y.; Smith, G. Global products of vegetation leaf area and fraction absorbed PAR from year one of MODIS data. *Remote Sens. Environ.* **2002**, *83*, 214–231. [CrossRef]
12. Chen, L.; Dirmeyer, P.A. Adapting observationally based metrics of biogeophysical feedbacks from land cover/land use change to climate modeling. *Environ. Res. Lett.* **2016**, *11*, 034002. [CrossRef]
13. Kala, J.; Decker, M.; Exbrayat, J.-F.; Pitman, A.J.; Carouge, C.; Evans, J.P.; Abramowitz, G.; Mocko, D. Influence of leaf area index prescriptions on simulations of heat, moisture, and carbon fluxes. *J. Hydrometeorol.* **2014**, *15*, 489–503. [CrossRef]
14. Chen, C.; Park, T.; Wang, X.; Piao, S.; Xu, B.; Chaturvedi, R.K.; Fuchs, R.; Brovkin, V.; Ciais, P.; Fensholt, R.; et al. China and India lead in greening of the world through land-use management. *Nat. Sustain.* **2019**, *2*, 122–129. [CrossRef]
15. Zhu, Z.; Piao, S.; Myneni, R.B.; Huang, M.; Zeng, Z.; Canadell, J.G.; Ciais, P.; Sitch, S.; Friedlingstein, P.; Arneth, A.; et al. Greening of the Earth and its drivers. *Nat. Clim. Chang.* **2016**, *6*, 791–795. [CrossRef]
16. Zhang, Y.; Song, C.; Band, L.E.; Sun, G.; Li, J. Reanalysis of global terrestrial vegetation trends from MODIS products: Browning or greening? *Remote Sens. Environ.* **2017**, *191*, 145–155. [CrossRef]
17. Baret, F.; Weiss, M.; Lacaze, R.; Camacho, F.; Makhmara, H.; Pacholcyzk, P.; Smets, B. GEOV1: LAI and FAPAR essential climate variables and FCOVER global time series capitalizing over existing products. Part1: Principles of development and production. *Remote Sens. Environ.* **2013**, *137*, 299–309. [CrossRef]
18. Xiao, Z.; Liang, S.; Wang, J.; Chen, P.; Yin, X.; Zhang, L.; Song, J. Use of General Regression Neural Networks for Generating the GLASS Leaf Area Index Product From Time-Series MODIS Surface Reflectance. *IEEE Trans. Geosci. Remote Sens.* **2014**, *52*, 209–223. [CrossRef]
19. Baret, F.; Buis, S. Estimating canopy characteristics from remote sensing observations: Review of methods and associated problems. In *Advances in Land Remote Sensing*; Springer: Berlin/Heidelberg, Germany, 2008; pp. 173–201.
20. Claverie, M.; Vermote, E.F.; Weiss, M.; Baret, F.; Hagolle, O.; Demarez, V. Validation of coarse spatial resolution LAI and FAPAR time series over cropland in southwest France. *Remote Sens. Environ.* **2013**, *139*, 216–230. [CrossRef]
21. Yan, K.; Park, T.; Chen, C.; Xu, B.; Song, W.; Yang, B.; Zeng, Y.; Liu, Z.; Yan, G.; Knyazikhin, Y.J.I.T.o.G.; et al. Generating global products of lai and fpar from snpp-viirs data: Theoretical background and implementation. *IEEE Trans. Geosci. Remote Sens.* **2018**, *56*, 2119–2137. [CrossRef]
22. Serbin, S.P.; Ahl, D.E.; Gower, S.T. Spatial and temporal validation of the MODIS LAI and FPAR products across a boreal forest wildfire chronosequence. *Remote Sens. Environ.* **2013**, *133*, 71–84. [CrossRef]
23. Fuster, B.; Sánchez-Zapero, J.; Camacho, F.; García-Santos, V.; Verger, A.; Lacaze, R.; Weiss, M.; Baret, F.; Smets, B. Quality Assessment of PROBA-V LAI, fAPAR and fCOVER Collection 300 m Products of Copernicus Global Land Service. *Remote Sens.* **2020**, *12*, 1017. [CrossRef]
24. Weiss, M.; Baret, F.; Block, T.; Koetz, B.; Burini, A.; Scholze, B.; Lecharpentier, P.; Brockmann, C.; Fernandes, R.; Plummer, S. On Line Validation Exercise (OLIVE): a web based service for the validation of medium resolution land products. Application to FAPAR products. *Remote Sens.* **2014**, *6*, 4190–4216. [CrossRef]
25. Yan, K.; Park, T.; Yan, G.; Liu, Z.; Yang, B.; Chen, C.; Nemani, R.; Knyazikhin, Y.; Myneni, R. Evaluation of MODIS LAI/FPAR Product Collection 6. Part 2: Validation and Intercomparison. *Remote Sens.* **2016**, *8*, 460. [CrossRef]
26. De Kauwe, M.G.; Disney, M.; Quaife, T.; Lewis, P.; Williams, M. An assessment of the MODIS collection 5 leaf area index product for a region of mixed coniferous forest. *Remote Sens. Environ.* **2011**, *115*, 767–780. [CrossRef]

27. Loew, A.; Bell, W.; Brocca, L.; Bulgin, C.E.; Burdanowitz, J.; Calbet, X.; Donner, R.V.; Ghent, D.; Gruber, A.; Kaminski, T. Validation practices for satellite based earth observation data across communities. *Rev. Geophys.* **2017**, *55*, 779–817. [CrossRef]
28. Fang, H.; Baret, F.; Plummer, S.; Schaepman-Strub, G. An overview of global leaf area index (LAI): Methods, products, validation, and applications. *Rev. Geophys.* **2019**, *57*, 739–799. [CrossRef]
29. Fang, H.; Wei, S.; Liang, S. Validation of MODIS and CYCLOPES LAI products using global field measurement data. *Remote Sens. Environ.* **2012**, *119*, 43–54. [CrossRef]
30. Somers, B.; Tits, L.; Coppin, P. Quantifying Nonlinear Spectral Mixing in Vegetated Areas: Computer Simulation Model Validation and First Results. *IEEE J. Sel. Top. Appl. Earth Observ. Remote Sens.* **2014**, *7*, 1956–1965. [CrossRef]
31. Schneider, F.D.; Leiterer, R.; Morsdorf, F.; Gastelluetchegorry, J.P.; Lauret, N.; Pfeifer, N.; Schaepman, M.E. Simulating imaging spectrometer data: 3D forest modeling based on LiDAR and in situ data. *Remote Sens. Environ.* **2014**, *152*, 235–250. [CrossRef]
32. Lanconelli, C.; Gobron, N.; Adams, J.; Danne, O.; Blessing, S.; Robustelli, M.; Kharbouche, S.; Muller, J. *Report on the Quality Assessment of Land ECV Retrieval Algorithms*; Scientific and Technical Report JRC109764; European Commission, Joint Research Centre: Ispra, Italy, 2018.
33. Gastellu-Etchegorry, J.-P.; Yin, T.; Lauret, N.; Cajgfinger, T.; Gregoire, T.; Grau, E.; Feret, J.-B.; Lopes, M.; Guilleux, J.; Dedieu, G. Discrete anisotropic radiative transfer (DART 5) for modeling airborne and satellite spectroradiometer and LIDAR acquisitions of natural and urban landscapes. *Remote Sens.* **2015**, *7*, 1667–1701. [CrossRef]
34. Huang, H.; Qin, W.; Liu, Q. RAPID: a Radiosity Applicable to Porous IndiviDual Objects for directional reflectance over complex vegetated scenes. *Remote Sens. Environ.* **2013**, *132*, 221–237. [CrossRef]
35. Qi, J.; Xie, D.; Yin, T.; Yan, G.; Gastellu-Etchegorry, J.-P.; Li, L.; Zhang, W.; Mu, X.; Norford, L.K. LESS: LargE-Scale remote sensing data and image simulation framework over heterogeneous 3D scenes. *Remote Sens. Environ.* **2019**, *221*, 695–706. [CrossRef]
36. Widlowski, J.L.; Pinty, B.; Lopatka, M.; Atzberger, C.; Buzica, D.; Chelle, M.; Disney, M.; Gastelluetchegorry, J.; Gerboles, M.; Gobron, N. The fourth radiation transfer model intercomparison (RAMI-IV): Proficiency testing of canopy reflectance models with ISO-13528. *J. Geophys. Res.* **2013**, *118*, 6869–6890. [CrossRef]
37. Disney, M.; Lewis, P.; Saich, P. 3D modelling of forest canopy structure for remote sensing simulations in the optical and microwave domains. *Remote Sens. Environ.* **2006**, *100*, 114–132. [CrossRef]
38. Widlowski, J.-L.; Côté, J.-F.; Béland, M. Abstract tree crowns in 3D radiative transfer models: Impact on simulated open-canopy reflectances. *Remote Sens. Environ.* **2014**, *142*, 155–175. [CrossRef]
39. Kuusk, A. 3.03—Canopy Radiative Transfer Modeling. In *Comprehensive Remote Sensing*; Liang, S., Ed.; Elsevier: Oxford, UK, 2018; pp. 9–22. [CrossRef]
40. Gastelluetchegorry, J.P.; Martin, E.; Gascon, F. DART: a 3D model for simulating satellite images and studying surface radiation budget. *Int. J. Remote Sens.* **2004**, *25*, 73–96. [CrossRef]
41. Bruniquelpinel, V.; Gastelluetchegorry, J.P. Sensitivity of Texture of High Resolution Images of Forest to Biophysical and Acquisition Parameters. *Remote Sens. Environ.* **1998**, *65*, 61–85. [CrossRef]
42. Guillevic, P.; Gastellu-Etchegorry, J. Modeling BRF and radiative regime of tropical and boreal forests—PART II: PAR regime. *Remote Sens. Environ.* **1999**, *68*, 317–340. [CrossRef]
43. Demarez, V.; Gastelluetchegorry, J.P. A Modeling Approach for Studying Forest Chlorophyll Content. *Remote Sens. Environ.* **2000**, *71*, 226–238. [CrossRef]
44. Malenovsky, Z.; Homolova, L.; Zuritamilla, R.; Lukes, P.; Kaplan, V.; Hanus, J.; Gastelluetchegorry, J.P.; Schaepman, M.E. Retrieval of spruce leaf chlorophyll content from airborne image data using continuum removal and radiative transfer. *Remote Sens. Environ.* **2013**, *131*, 85–102. [CrossRef]
45. Qi, J.; Xie, D.; Yan, G.; Gastelluetchegorry, J.P. Simulating Spectral Images with Less Model Through a Voxel-Based Parameterization of Airborne Lidar Data. In Proceedings of the International Geoscience and Remote Sensing Symposium, Yokohama, Japan, 28 July–2 August 2019; pp. 6043–6046.
46. Qi, J.; Xie, D.; Guo, D.; Yan, G. A Large-Scale Emulation System for Realistic Three-Dimensional (3-D) Forest Simulation. *IEEE J. Sel. Top. Appl. Earth Observ. Remote Sens.* **2017**, *10*, 4834–4843. [CrossRef]
47. Huang, D.; Knyazikhin, Y.; Wang, W.; Deering, D.W.; Stenberg, P.; Shabanov, N.V.; Tan, B.; Myneni, R.B. Stochastic transport theory for investigating the three-dimensional canopy structure from space measurements. *Remote Sens. Environ.* **2008**, *112*, 35–50. [CrossRef]

48. Yang, B.; Knyazikhin, Y.; Mottus, M.; Rautiainen, M.; Stenberg, P.; Yan, L.; Chen, C.; Yan, K.; Choi, S.; Park, T. Estimation of leaf area index and its sunlit portion from DSCOVR EPIC data: Theoretical basis. *Remote Sens. Environ.* **2017**, *198*, 69–84. [CrossRef]
49. Yang, W.; Tan, B.; Huang, D.; Rautiainen, M.; Shabanov, N.V.; Wang, Y.; Privette, J.L.; Huemmrich, K.F.; Fensholt, R.; Sandholt, I. MODIS leaf area index products: From validation to algorithm improvement. *IEEE Trans. Geosci. Remote Sens.* **2006**, *44*, 1885–1898. [CrossRef]
50. Hosgood, B.; Jacquemoud, S.; Andreoli, G.; Verdebout, J.; Pedrini, G.; Schmuck, G. Leaf optical properties experiment 93 (LOPEX93). *Rep. Eur.* **1995**, *16095*.
51. Trigg, S.; Flasse, S. Characterizing the spectral-temporal response of burned savannah using in situ spectroradiometry and infrared thermometry. *Int. J. Remote Sens.* **2000**, *21*, 3161–3168. [CrossRef]
52. Fang, H.; Jiang, C.; Li, W.; Wei, S.; Baret, F.; Chen, J.M.; Garcia-Haro, J.; Liang, S.; Liu, R.; Myneni, R.B.; et al. Characterization and intercomparison of global moderate resolution leaf area index (LAI) products: Analysis of climatologies and theoretical uncertainties. *J. Geophys. Res.* **2013**, *118*, 529–548. [CrossRef]
53. Xu, B.; Park, T.; Yan, K.; Chen, C.; Zeng, Y.; Song, W.; Yin, G.; Li, J.; Liu, Q.; Knyazikhin, Y.; et al. Analysis of Global LAI/FPAR Products from VIIRS and MODIS Sensors for Spatio-Temporal Consistency and Uncertainty from 2012–2016. *Forests* **2018**, *9*, 73. [CrossRef]
54. Knyazikhin, Y.; Martonchik, J.V.; Diner, D.J.; Myneni, R.B.; Verstraete, M.M.; Pinty, B.; Gobron, N. Estimation of vegetation canopy leaf area index and fraction of absorbed photosynthetically active radiation from atmosphere-corrected MISR data. *J. Geophys. Res.* **1998**, *103*, 32239–32256. [CrossRef]
55. Chen, J.M.; Leblanc, S.G. A four-scale bidirectional reflectance model based on canopy architecture. *IEEE Trans. Geosci. Remote Sens.* **1997**, *35*, 1316–1337. [CrossRef]
56. Kuusk, A. The hot spot effect on a uniform vegetative cover. *Sov. J. Remote Sens* **1985**, *3*, 645–658.
57. Roujean, J.-L. A parametric hot spot model for optical remote sensing applications. *Remote Sens. Environ.* **2000**, *71*, 197–206. [CrossRef]
58. Myneni, R.B.; Ramakrishna, R.; Nemani, R.R.; Running, S.W. Estimation of global leaf area index and absorbed par using radiative transfer models. *IEEE Trans. Geosci. Remote Sens.* **1997**, *35*, 1380–1393. [CrossRef]

Publisher's Note: MDPI stays neutral with regard to jurisdictional claims in published maps and institutional affiliations.

© 2020 by the authors. Licensee MDPI, Basel, Switzerland. This article is an open access article distributed under the terms and conditions of the Creative Commons Attribution (CC BY) license (http://creativecommons.org/licenses/by/4.0/).

Article

Assessment of Workflow Feature Selection on Forest LAI Prediction with Sentinel-2A MSI, Landsat 7 ETM+ and Landsat 8 OLI

Benjamin Brede [1,*], Jochem Verrelst [2], Jean-Philippe Gastellu-Etchegorry [3], Jan G. P. W. Clevers [1], Leo Goudzwaard [4], Jan den Ouden [4], Jan Verbesselt [1] and Martin Herold [1]

1. Laboratory of Geo-Information Science and Remote Sensing, Wageningen University & Research, Droevendaalsesteeg 3, 6708 PB Wageningen, The Netherlands; jan.clevers@wur.nl (J.G.P.W.C.); jan.verbesselt@wur.nl (J.V.); martin.herold@wur.nl (M.H.)
2. Image Processing Laboratory (IPL), Parc Cientific, Universitat de València, 46980 Paterna, València, Spain; jochem.verrelst@uv.es
3. Centre d'Etudes Spatiales de la BIOsphere, Paul Sabatier University, CNES-CNRS, 18 avenue Edouard Belin, CEDEX 4, BPi 2801-31401 Toulouse, France; jean-philippe.gastellu-etchegorry@cesbio.cnes.fr
4. Forest Ecology and Forest Management Group, Wageningen University & Research, Droevendaalsesteeg 3, 6708 PB Wageningen, The Netherlands; leo.goudzwaard@wur.nl (L.G.); jan.denouden@wur.nl (J.d.O.)
* Correspondence: benjamin.brede@wur.nl

Received: 13 January 2020; Accepted: 8 March 2020; Published: 12 March 2020

Abstract: The European Space Agency (ESA)'s Sentinel-2A (S2A) mission is providing time series that allow the characterisation of dynamic vegetation, especially when combined with the National Aeronautics and Space Administration (NASA)/United States Geological Survey (USGS) Landsat 7 (L7) and Landsat 8 (L8) missions. Hybrid retrieval workflows combining non-parametric Machine Learning Regression Algorithms (MLRAs) and vegetation Radiative Transfer Models (RTMs) were proposed as fast and accurate methods to infer biophysical parameters such as Leaf Area Index (LAI) from these data streams. However, the exact design of optimal retrieval workflows is rarely discussed. In this study, the impact of five retrieval workflow features on LAI prediction performance of MultiSpectral Instrument (MSI), Enhanced Thematic Mapper Plus (ETM+) and Operational Land Imager (OLI) observations was analysed over a Dutch beech forest site for a one-year period. The retrieval workflow features were the (1) addition of prior knowledge of leaf chemistry (two alternatives), (2) the choice of RTM (two alternatives), (3) the addition of Gaussian noise to RTM produced training data (four and five alternatives), (4) possibility of using Sun Zenith Angle (SZA) as an additional MLRA training feature (two alternatives), and (5) the choice of MLRA (six alternatives). The features were varied in a full grid resulting in 960 inversion models in order to find the overall impact on performance as well as possible interactions among the features. A combination of a Terrestrial Laser Scanning (TLS) time series with litter-trap derived LAI served as independent validation. The addition of absolute noise had the most significant impact on prediction performance. It improved the median prediction Root Mean Square Error (RMSE) by $1.08\,m^2\,m^{-2}$ when 5% noise was added compared to inversions with 0% absolute noise. The choice of the MLRA was second most important in terms of median prediction performance, which differed by $0.52\,m^2\,m^{-2}$ between the best and worst model. The best inversion model achieved an RMSE of $0.91\,m^2\,m^{-2}$ and explained 84.9% of the variance of the reference time series. The results underline the need to explicitly describe the used noise model in future studies. Similar studies should be conducted in other study areas, both forest and crop systems, in order to test the noise model as an integral part of hybrid retrieval workflows.

Keywords: leaf area index (LAI); Sentinel-2; forest; machine learning; vegetation radiative transfer model; Discrete Anisotropic Radiative Transfer (DART) model

1. Introduction

Vegetation represents a primary component in Earth's terrestrial carbon cycle, with respect to its role both as a source of CO_2 through respiration and as a sink of CO_2 through photosynthesis [1]. Its photosynthetic capacity is a function of available leaf area, which can be quantified in terms of Leaf Area Index (LAI). LAI is the leaf area per horizontally projected ground area [2]. It was acknowledged as an Essential Climate Variable (ECV) with high priority in the European Space Agency (ESA)'s Copernicus program [3] and is a focus area of the Committee on Earth Observing Satellites (CEOS) Working Group on Calibration and Validation (WGCV) Land Product Validation (LPV) subgroup [4].

Multi-spectral Earth observation data are sensitive to the amount of canopy foliage, primarily in the NIR and SWIR spectral region [5]. In the NIR, the leaf mesophyll with its enclosed air and air–tissue interfaces results in high reflectance. In the SWIR, the water contained in the leaves leads to high reflectance. Denser leaf layers increase these effects on the canopy level. In this context, the multi-spectral Landsat missions and especially its latest satellite Landsat 8 [6] have shown potential to estimate crop LAI [7–9], forest LAI [10] and in combination with Radiative Transfer Models (RTMs)' generic LAI [11]. However, the missions' orbits defines the revisit time as 16 days, which may be insufficient to track fast vegetation changes such as spring leaf flush. The Sentinel-2 mission was awaited for the purpose of estimation of biophysical parameters such as LAI. Its higher spatial resolution, higher revisit frequency and additional red edge bands are emphasised as potentials for performance advances, when compared to the Landsat missions. Agricultural testing campaigns such as SEN3Exp and SicilyS2EVAL offered measurements to explore opportunities for estimating vegetation parameters with spectral observations from Sentinel-2 [12–15]. The SPOT5 Take5 satellite campaign delivered a dataset with a 5-day revisit time to explore the temporal domain [16]. However, the domain of forest remote sensing has not seen this extent of targeted preparation campaigns. Nevertheless, approaches are advancing to make use of the Sentinel-2 open data in terms of biophysical parameter estimation [17,18]. All these campaigns and connected studies underlined that Sentinel-2 has high potential for estimation of LAI and Chlorophyll a and b (C_{ab}) in diverse canopies.

In parallel with the advances in sensor technology and data availability, Machine Learning Regression Algorithms (MLRAs) were introduced as retrieval techniques [14,19–23]. Their main advantages are their ability to map the nonlinear relationship between canopy parameters and the reflectance signal, and their fast mapping speed, compared to look-up tables [23]. Especially Gaussian Process Regression (GPR)—a kernel-based MLRA—showed good results in terms of prediction accuracy and could achieve the 10% precision for C_{ab} retrieval required by the Global Climate Observing System (GCOS) [14]. Like traditional retrieval techniques, MLRAs require a database of biophysical parameters, and their associated reflectance signal to learn the mapping between spectral bands and biophysical parameters. This database can originate from field observations, but also from vegetation RTMs. RTMs simulate the spectral properties of vegetation based on a limited set of biophysical parameters and the laws of radiative transfer, which makes them universally applicable. Verrelst et al. [23] conclude that MLRAs and RTMs combined in training workflows have a potential for implementation in operational processing chains for retrieving biophysical parameters. These workflows are referred to as hybrid training workflows.

Turbid medium RTMs, which approximate the vegetation canopy as one homogeneous layer, dominate inversion workflows (e.g., [19,24]). Among these RTMs, PROSAIL—a combination of the PROSPECT leaf and the SAIL canopy model [5]—is one of the most widely used. The main reasons for the use of turbid medium models are their fast processing speed and the low number of input parameters. However, PROSAIL does not agree well with geometrically explicit RTMs in the case of heterogeneous scenes, i.e., scenes where natural objects such as trees are explicitly modelled [25]. As modelling capabilities have outrun means to collect ground truth, a final evaluation of the differences remains open [26]. Apart from this, so-called emulators make it possible to build fast surrogates for complex models [27]. For that, a complex physical model is replaced with a statistical learning model that was trained on input–output combinations of the physical model. This makes it practically possible to exploit heterogeneous RTMs in operational contexts.

In the context of hybrid training workflows, addition of noise to the RTM generated spectral data has been a common practice [19,28–30]. This is typically achieved by adding an absolute term or multiplying with a multiplicative term across the RTM generated spectral bands following a Gaussian distribution. The addition of noise has multiple purposes: it simulates errors of radiometric calibration, atmospheric noise and residuals from the atmospheric correction, but to some extent also bridges between the simplified representation of the RTM and the actual radiometric behaviour of the canopy [19]. Generally, noise prevents the inverse model from over-fitting on the training database. However, an accurate quantification of all error terms in the sensing process remains difficult [19]. Additionally, the quantification of the optimal noise level to be added has not been attempted for hybrid workflows yet.

Considering these developments together—decametric observations from Sentinel-2 and Landsat, fast mapping with MLRA algorithms and fast radiative transfer modelling with emulators—a decametric LAI product would be possible. Such a product would offer better opportunities to monitor ecosystems in fragmented landscapes than comparable products on hectometric scale such as the MODerate-resolution Imaging Spectroradiometer (MODIS) [31] and CYCLOPES [19] LAI products. However, the discrete implementation of hybrid learning processing chains requires many decisions to be taken on—for example, the used RTM and MLRA. Often implemented, but rarely discussed in depth so far, is the addition of noise to the synthetic spectral data.

The aim of this study was to compare different hybrid retrieval workflows that make combined use of Sentinel-2A (S2A) MultiSpectral Instrument (MSI), Landsat 7 (L7) Enhanced Thematic Mapper Plus (ETM+) and Landsat 8 (L8) Operational Land Imager (OLI) observations for forest LAI retrieval. Features in the retrieval workflow were altered in a fully-factorised way to explore their effects. The tested features were: (1) addition of prior knowledge of leaf chemistry, (2) the choice of RTM, (3) the addition of Gaussian noise to RTM produced training data, (4) possibility of using Sun Zenith Angle (SZA) as an additional MLRA training feature, and (5) the choice of MLRA. Performance statistics for the realisations were derived to identify the relevance of the features with respect to prediction performance. Apart from this, features were analysed in terms of their first degree interactions with other features.

2. Data

2.1. Study Site

This study focussed on the Speulderbos Fiducial Reference site in the Veluwe forest area (N 52°15.15′ E 5°42.00′), The Netherlands [32,33] (www.wur.eu/fbprv). In an earlier forest inventory, the site was marked with a wooden pole grid with 40 m spacing, which was geo-located with a Leica™ Robotic Total Station (Leica Geosystems AG, Heerbrugg, Switzerland) and triangulation. This grid served to define the five plots A to E used as validation for this study (Figure 1). The plot locations were chosen with a distance of at least 80 m between them. This distance is four times the image registration error according to S2A mission definition [34].

The stand is predominantly composed of European beech (*Fagus sylvatica*). A few specimens of pedunculate oak (*Quercus robur*) and sessile oak (*Quercus petraea*) can be found as well. In the understorey, few specimens of evergreen European holly (*Ilex aquifolium*) can be found with heights of <7 m in plots A, B, and E. The stand was created as a plantation in 1835 and left unmanaged from then on, so that dominant trees are of even age. Recruitment took only place in canopy openings caused by falling trees as was the case in plot D. This was reflected by the total number of stems, which was 1059 ha^{-1} in plot D compared to 280, 250, 280 and 202 ha^{-1} in plots A, B, C and E, respectively, as determined in a forest inventory in 2013/2014. Basal area was 43.0, 42.5, 31.4, 34.8 and 37.2 m ha^{-1} for plots A, B, C, D, and E, respectively.

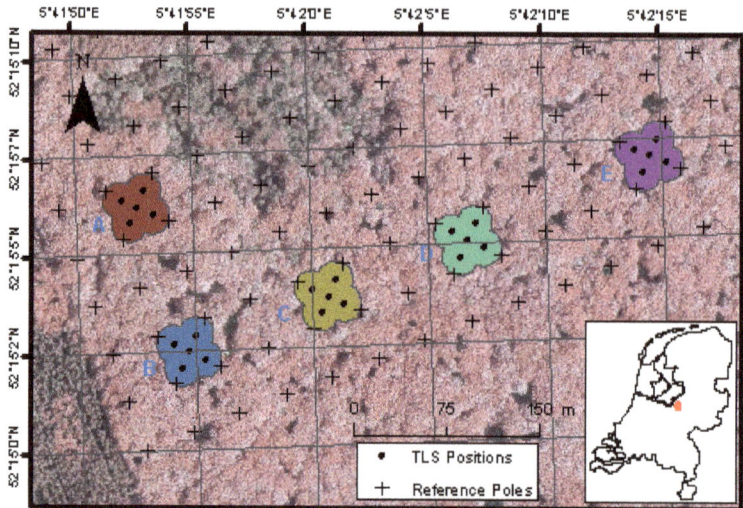

Figure 1. Speulderbos study site with Terrestrial Laser Scanning (TLS) scan positions and polygons representing the five plots. Background image is an airborne false-colour composite of 2013. The inset shows the location of the study site within the Netherlands.

2.2. Field Data

2.2.1. Terrestrial Laser Scanning (TLS) Data

During a field campaign in 2016, a TLS time series was acquired that followed the phenological development of the trees in the study site. The five plots were revisited 28 times with a RIEGL VZ-400 scanner (RIEGL LMS GmbH, Horn, Austria). Samples were taken during the growing season, with an increased intensity during leaf flush and senescence. Rain events and wet canopy conditions were avoided as wet canopy elements tend to absorb the laser pulses, thereby introducing a bias in the estimation of gap fraction. A comprehensive list of all sampling dates can be found in Table A1. In each plot, five scan positions were established that have been visited on each field visit, resulting in a total of 25 positions per field visit. The positions were arranged in a star shape with a centre position determined according to the wooden poles and four positions at the corners of imaginary squares with 20 m edge length.

The individual point clouds were processed with the PyLidar package (http://pylidar.org) based on the methodology developed by Calders et al. [35]. This method basically treats the TLS laser pulses as virtual probes similar to point quadrats, which have been proposed for measuring LAI [36]. The underlying assumption is that the canopy is a random medium with a density proportional to its LAI. The canopy density is proportional to the TLS hit probability, while the TLS hit probability is the inverse of the gap fraction. In this aspect, the TLS gap fraction is similar to the gap fraction derived from digital hemispherical photography [37], which is a standard method to measure LAI [38]. The gap fraction was derived by counting pulses that exited the canopy without return versus all pulses fired in the *hinge angle* region, which was approximated by the zenith angle region between 50° and 60°. Terrain correction for vertical profiles as proposed by Calders et al. [39] was not used because only total canopy Plant Area Index (PAI) values were required, which is indifferent to the terrain. Finally, PAI was derived as follows:

$$PAI = -1.1 \log(P_{gap}(57.5°)) \qquad (1)$$

where $P_{gap}(57.5°)$ is the gap fraction at the hinge angle. PAI is defined as the one sided surface area of all plant material per unit area ground surface [35]. Alternatively, it can be defined as $PAI = $

$LAI + WAI$, where WAI is the wood area index. This is different from LAI, which only includes foliage material [35].

2.2.2. Littertrap Data

Additionally, 25 litter traps were installed in the area to directly measure LAI per season. In each plot, five litter traps were positioned close to the TLS sampling points. Their construction was based on recommendations of the Center for Tropical Forest Science (CTFS) Global Forest Carbon Research Initiative Litterfall Monitoring Protocol, version March 2010 (https://forestgeo.si.edu/sites/default/files/litterfall_protocol_2010.pdf). Each trap consisted of a PVC pipe bent to a circle with an area of $0.7\,m^2$ and holding a plastic net. The net allows water to drain and prevent decomposition of the litter content. For each trap, the pipe circles were levelled to assure correct surface area and are held in place 1 m over the ground with four wooden poles.

Litter was collected seven times in 2016 over the course of the season on 1 June, 12 August, 7 and 21 October, 11 November, 2 December, as well as on 10 January in 2017 to collect the last fallen leaves of 2016. Sampling dates were chosen to account for the increased litter-fall in autumn. Litter was collected with paper bags; sorted by species and components, i.e., leaves, twigs, and husks; dried for at least 24 h at 65 °C; and weighted. For each litter collection, 100 leaves were randomly sampled, and their area determined with a leaf area meter (Licor Area Meter 3100, LI-COR, Inc., Lincoln, NE, USA). Specific Leaf Area (SLA)—the unit area of leaf per unit mass—was estimated based on these sub-samples [40]. The total LAI per trap for the whole season was then inferred from the total collected dry leaf weights taking into account the litter trap surface area of $0.79\,m^2$. Plot level LAI was estimated as the mean of the single traps.

2.2.3. Leaf Sampling Data

Apart from these canopy structural measurements, the leaf chemistry was monitored over the course of the year by inversion of leaf spectral samples. For this purpose, two beech trees between plot A and B were rigged with ropes and climbed four times at different points of leaf development. On each tree, five branches were cut off at each sampling event from near the crown top. From each branch, five leaves were sampled randomly when no differences in development stage were visible. Especially during the last sampling event, the leaves showed different stages of senescence. In that case, the leaves were sampled to represent the abundance of the respective senescence stage on the branch. Leaf reflectance spectra were acquired with a Fieldspec Pro 3 (ASD Incorporated, Boulder, CO, USA) equipped with an integration sphere. Additionally, the same leaves were sampled with a Minolta SPAD-500 chlorophyll meter (Spectrum Technologies, Inc., Plainfield, IL, USA). For this, four SPAD measurements per leaf were averaged. The sampling resulted in a total of 173 reflectance spectra.

2.3. Satellite Data

2.3.1. Sentinel-2A MSI

S2A MSI was primarily designed for land cover and disaster monitoring, but also for retrieval of biophysical parameters such as Fraction Absorbed Photosynthetically Active Radiation (FAPAR), LAI and Fractional vegetation cover (FCover) [34]. Operational products incorporate the Top Of Atmosphere (TOA) Bidirectional Reflectance Factor (BRF) and recently the Surface Reflectance (SR) BRF. Table 1 gives an overview of the MSI spectral bands.

Table 1. Spectral band specifications for bands and missions used in this study (band centres and widths in nm), bands used for atmospheric correction were omitted [34,41].

Domain	Landsat 7 ETM+			Landsat 8 OLI			Sentinel-2A MSI		
	Name	Center	Width	Name	Center	Width	Name	Center	Width
VIS	B1	485	70	B2	482	60	B2	490	65
	B2	560	80	B3	561	57	B3	560	35
	B3	660	60	B4	654	37	B4	665	30
NIR	B4	835	130	B5	864	30	B5	705	15
							B6	740	15
							B7	783	20
							B8	842	115
							B8A	865	20
SWIR	B5	1650	200	B6	1608	84	B11	1610	90
	B7	2220	260	B7	2200	187	B12	2190	180

S2A MSI TOA BRF products for tile T31UFT were downloaded from the Copernicus Open Access Hub (https://scihub.copernicus.eu/) for the period of January 2016 until December 2016. The relative orbits R008 and R051 both include the Speulderbos site, thereby doubling the number of observations compared to single orbit observation. TOA BRF products were further processed with sen2cor 2.4.0 (http://step.esa.int/main/third-party-plugins-2/sen2cor/) to derive SR BRF products. During further processing, 60 m bands (B01, B09, B10) were excluded because they are heavily affected by atmospheric conditions and, therefore, surface reflectance products were not provided for these bands. Cloud and quality screening was performed manually under consideration of the scene classification delivered with sen2cor. For this, Normalised Difference Vegetation Index (NDVI) time series for the extracted observations were inspected. Potentially cloud free scenes were identified as high NDVI values in the time series and, therefore, those images were accepted for further processing.

Of the 64 dates in 2016 when MSI observations of the Speulderbos site were available, 21 dates yielded usable observations, which is 32.8% of all. Figure 2 shows the bands B04 and B8A, which are the red and NIR spectral bands (Table 1) and hold most information on change in canopy characteristics. The automatic scene classification could identify most of the cloud affected conditions with an accuracy of 91.6%. It should be noted that all observations that were not clouds were further processed, no matter their assigned Scene Classification (SCL) class. Overall, the time series depicted the start of season in April and May: the red reflectance decreased over all plots from 0.059 ± 0.003 on May 1 to 0.022 ± 0.001 on May 11 due to absorption by chlorophyll. At the same time, reflectance in band B8A increased from 0.188 ± 0.008 to 0.406 ± 0.014, which can be attributed to the leaves' characteristic scattering behaviour. During late summer, the overall temporal course remained stable with a slightly decreasing trend. This trend could also be observed in the validation time series (Figure 6). An exception was 17 July, when B8A reflectance jumped to 0.526 ± 0.023. There were no clouds over the plots on that day, but clouds close by probably caused adjacency effects.

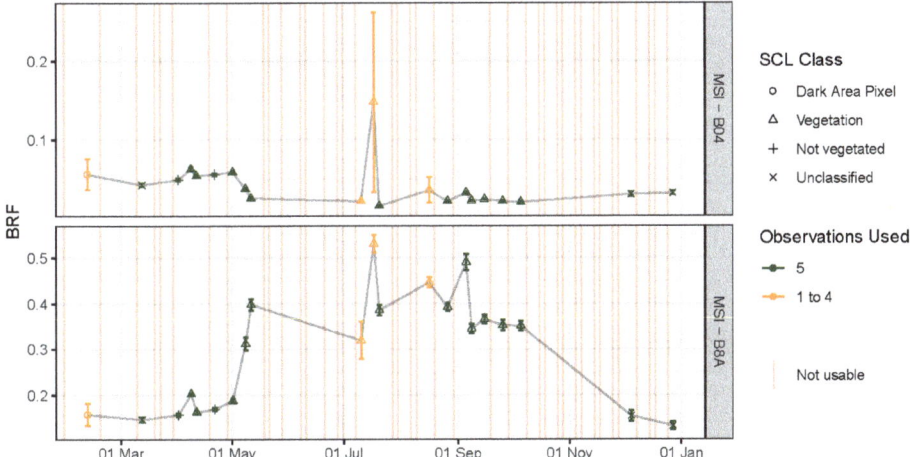

Figure 2. Observed BRFs for S2A MSI red and NIR band over the year 2016 for the five Speulderbos plots (see Figure 1). Points represent average BRF over the five plots, error bars one standard deviation. Colour codes the number of plots for which the observations were useful (clear sky). SCL class refers to the mode of all *Scene Classification* (SCL) given by sen2cor over the five plots. Observations used refers to how many of the BRF of the five plots were not affected by clouds and used for analysis.

2.3.2. Landsat ETM+ and OLI

Similar to S2A, atmospherically corrected data are available from the Landsat Archive. For this study, L7 ETM+ and L8 OLI SR BRF products at Worldwide Reference System (WRS) row 24 and WRS path 197 and 198 were obtained as on-demand download products provided by the United States Geological Survey (USGS) Earth Resources Observation and Science (EROS) Center Science Processing Architecture (ESPA) On Demand Interface (https://landsat.usgs.gov/landsat-surface-reflectance-high-level-data-products). Both Landsat time series profited from two orbits from which Speulderbos can be observed. Clouds and cloud shadows were identified in the same manner as for MSI and with the support of the pixel quality layer delivered with Landsat Collection 1 products.

In order to extract SR from the satellite SR products, the plots needed to be defined in terms of geo-referenced polygons. For this, the circular field of view of the TLS at the five scan positions within each plot (Figure 1) was considered. Each position was buffered with a circle of 14.4 m radius, which corresponds to the top of canopy of the approximated canopy height of 25 m and maximum observation angle 60° of the TLS. The combined area of the circles represented the plots. The square SR product pixels did not correspond exactly to the polygonal definitions of our plots. Therefore, pixels overlaying each plot were weighted according to their overlap area with the respective plot polygon in order to estimate the SR of the plot.

In case of the two Landsat missions, 41 and 42 observation dates were available of which 8 (17.1%) and 9 (15.8%) were usable for ETM+ and OLI, respectively (Figure 3). The pixel quality bits for cloud occurence indicated at least medium confidence, which appeared to indicate clear sky conditions after checking the corresponding images. As for MSI, these could mostly be found during the spring green-up and late summer periods. Cloud conditions, represented by bits set to high confidence cloud, were identified with an accuracy of 85.9%.

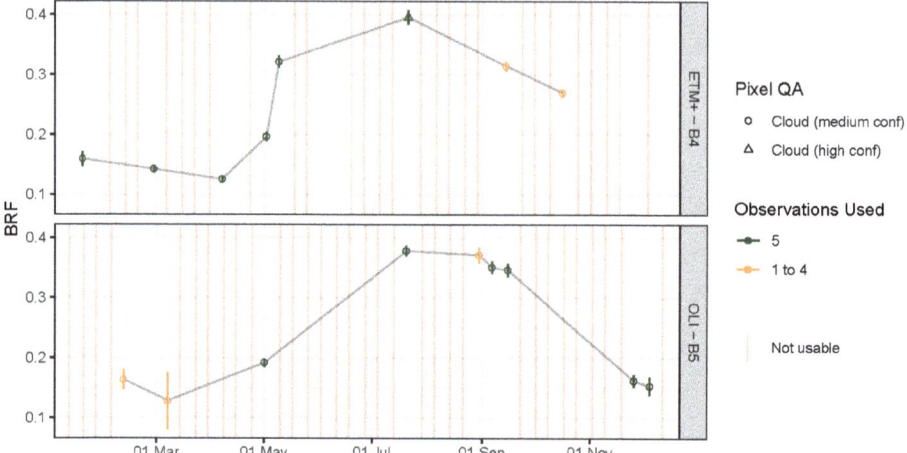

Figure 3. Observed BRFs for Landsat 7 ETM+ and Landsat 8 OLI MSI NIR bands over the year 2016 for the five Speulderbos plots (see Figure 1). Points represent average BRF over the five sites, error bars one standard deviation. Colour codes the number of plots for which the observations were useful (clear sky). Pixel QA refers to the pixel quality bits. Six observations were discarded because they exceeded a reflectance of 1 (undetected clouds). Observations used refer to how many of the BRF of the five plots were not affected by clouds and used for analysis.

In September, when multiple observations of MSI and OLI were available, they produced comparable SR BRFs when excluding September 5: average per band BRFs showed differences of less than 5%, which is the S2A mission requirement for SR products. Only L7 ETM+ and S2A MSI showed differences in the NIR of 7.4%, which can be explained by the wider spectral response curve of the ETM+ NIR band 4. These similarities give confidence in comparable behaviour of SR BRFs and to combine observations from these sensors. However, for a detailed comparison the spectral response functions of the sensors and the used atmospheric processors need to be taken into account (e.g., [42]). In fact, for optimal inter-operability, the S2A and Landsat products should be harmonised before combined processing [43].

3. Methods

The aim of this study was to modify features—or elements—of a hybrid retrieval workflow for LAI in order to test their impact on the prediction performance. The general order for a hybrid retrieval workflow is as follows [23,44]:

1. A vegetation RTM is run in forward mode to create a database of training samples, i.e., biophysical parameters serve as input for the RTM to predict spectral BRFs. The parameter values are altered to cover multiple canopy conditions.
2. Gaussian noise is added to the spectra to prevent the MLRA from over-fitting and simulate observation noise.
3. Multiple MLRAs are trained on the database to learn the inverse mapping, i.e., from spectral bands to biophysical parameter. Model hyperparameter tuning is performed on a part of the generated database, while the rest is used for testing the trained model.
4. The MLRAs are applied to the observed spectra to predict the biophysical parameter of interest. MLRAs performance is compared.

The following list gives an overview of the modified features of this study and introduces reference terms under which the feature domains were treated. Figure 4 summarises the workflow and features visually.

- Biochemical Prior: Using leaf biochemical parameters inferred from field spectroscopy observations to restrict the RTM input parameter space (label: *prior knowledge*) versus using a free range (label: *free*) (two alternatives).
- RTM: Two underlying, structurally contrasting RTMs were tested: turbid medium PROSAIL (SAIL 4 coupled with PROSPECT 5) and structurally-explicit Discrete Anisotropic Radiative Transfer (DART) (with PROSPECT 5) (two alternatives).
- Noise scenario: Using multiple noise levels for two types of noise (four and five alternatives).
- SZA: Using the SZA as an additional learning feature (label: *SZA*) or not (label: *no SZA*) (two alternatives).
- MLRA: Using multiple MLRAs: Ordinary Least Squares (OLS), Multi-Layer Perceptron (MLP), Regression Tree (RT), Support Vector Regression (SVR), Kernel Ridge Regression (KRR), GPR (six alternatives).

Each unique combination of these features is referred to as a *realisation* in the following, while realisations with the same feature were summarised as *ensembles*. For example, a realisation may have used *biochemical prior knowledge*, DART, specific levels of noise, SZA, and OLS. All realisations that implement DART make up the ensemble. All possible combinations of the list above were tested, resulting in 960 realisations. Model training was performed independently for each satellite mission, and model predictions were combined later to form one time series per realisation. A separate per mission performance assessment was not conducted because the effective number of observations varied strongly between the missions and thus would not allow fair comparison.

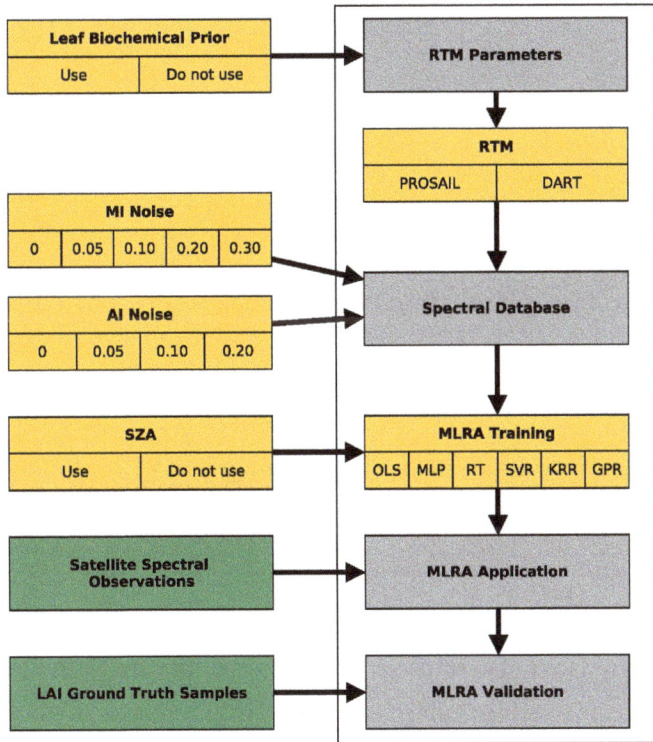

Figure 4. Flowchart of the applied workflow. Varied feature domains have orange background, LAI ground truth green background. For abbreviations, the reader is referred to the text.

3.1. Leaf Biochemical Parameter Estimation

The spectral signature of vegetation canopies is sensitive to LAI mostly in the NIR, but to a certain extent also in the VIS and red edge, and SWIR [5]. However, the VIS and red edge regions are also strongly affected by C_{ab}, and the SWIR by leaf water content (C_w) and leaf dry matter content (C_m). For LAI retrieval, this means having knowledge about leaf biochemistry allows restricting the parameter range that the training database needs to cover, thereby improving the learning of the variations that are caused by LAI. For this reason, a leaf biochemical prior was tested as a possible feature of the training workflow, even though the retrieval of leaf biochemical compounds was not the aim of this study.

For the retrieval of leaf chemical properties, the collected field spectral samples (Section 2.2) were inverted with a gradient descent approach utilising the PROSPECT 5 model as implemented in the R package hsdar (https://cran.r-project.org/web/packages/hsdar) [45,46]. The quasi-Newton method after Byrd et al. [47] that allows box constraints was chosen. Box constraints allow for limiting the searched parameter space to hyper-cubes according to given parameter ranges. The parameter ranges given in Table 2 were chosen as constraints. Results were inspected to identify biochemical compounds whose abundance was stable over the season and could be assumed fixed over the course of the year, or otherwise needed to be treated as variable.

Table 2. RTM parameters with their symbols, units, ranges (in case of *free* realisations) and best-estimate values (in case of *prior knowledge* realisations; based on PROSPECT inversion of sampled leaf spectra, Section 3.1). The best-estimate values were used for the *prior knowledge* realisations.

	Model Parameter	Unit	Free	Best Estimate
Leaf parameters: PROSPECT-5B				
N	Leaf structure index	-	1–2.5	1.27
C_{ab}	Leaf chlorophyll content	µg cm^{-2}	0–80	-
Car	Leaf carotenoid content	µg cm^{-2}	0–20	8.60
C_m	Leaf dry matter content	g cm^{-2}	0.001–0.025	0.00263
C_w	Leaf equivalent water thickness	cm	0.002–0.025	0.0053
C_{brown}	Brown pigment fraction	-	0–1	-
Canopy parameters: SAIL4 and DART				
LAI	Leaf area index	m^2 m^{-2}	0–8	0–8
θ_s	Sun zenith angle	°	27.5–80	27.5–80
θ_o	View zenith angle	°	0	0
ϕ	Sun-sensor azimuth angle	°	0	0
LAD	Leaf angle distribution	-	Plagiophile	Plagiophile
Canopy parameters: SAIL4				
α_{soil}	Soil wet/dry factor	-	0	0
hspot	Hot spot parameter	-	0	0
Canopy parameters: DART				
TreeHeight	Tree height	m	20	-
CrownDiameter	Tree crown diameter	m	5–9	-
CrownHeight	Tree crown height	m	7	-

3.2. RTMs and Training Database Generation

Two contrasting canopy RTMs were used to represent different levels of canopy complexity. One was PROSAIL, which is a turbid medium model, i.e., it treats the canopy as a homogeneous medium. It is a combination of the PROSPECT leaf and the SAIL canopy bidirectional reflectance model [5]. It has been widely used in the fields of agriculture, plant physiology, and ecology, including estimation of biophysical parameters [16,19,48,49]. In this study, PROSAIL 5B, as implemented in the R package hsdar, was used.

On the other hand, the DART model is a voxel-based flux-tracing model that allows building complex 3D scenes, including vegetation canopies [50,51]. DART contains a PROSPECT module to simulate leaf reflectance and transmission. Applications of DART can be found in the fields of surface energy budget studies [52] and forest biophysical parameter retrieval [53–55], where its advantages of explicitly modelling 3D structure were exploited. DART can be obtained from CESBIO with free licences for publicly funded research and teaching (https://dart.omp.eu).

As both RTMs use PROSPECT 5 as the underlying leaf model, they have many common parameters. Table 2 gives an overview of the used parameter ranges. In the case of *free* realisations, parameters were allowed to vary within the ranges that adopted from the literature [44]. In case of *prior knowledge* realisations, the values were estimated with field spectroscopy as described in Section 3.1. The range of sun zenith angles is based on the geographic location of the Speulderbos site. Since ETM+, OLI and MSI have narrow fields of view of 15°, 15° and 21°, respectively, view zenith angle and relative azimuth angles were assumed to be 0. This was found to be a reasonable assumption for mid and high latitudes [56]. Soil spectra were estimated from MSI barren and winter observations. BRFs were extracted as described above and averaged over all sites. Sensor spectral response curves were approximated as Gaussian with centre wavelength and Full Width at Half Maximum (FWHM) according to published specifications (Table 1). In summary, three models were trained, one for each sensor.

The DART scene was built up of five trees as squared, repetitive scene with 10 m edge length and grid size of 1 m horizontal and 0.5 m vertical cell size (Figure 5). This means that the scene was duplicated along the edges. The trees' heights and diameters were roughly approximated with TLS point clouds (Table 2, Section 2.2). However, the trees were based on generic forms consisting of 8-faceted stems and ellipsoidal crowns. The stems had a diameter of 0.5 m below and 0.25 m within the crown. The crown leaf volume was simulated as turbid medium cells. TLS was not used to build explicit 3D tree models in order to keep the number of input parameters minimal. For fast computation, an emulator was built to replace actual DART simulations [27]. For this, 2500 samples of DART input parameters were drawn with Latin hypercube sampling [57] according to the *free* option in Table 2. Then, the same MLRAs as for the inversion were trained to predict the single spectral bands of ETM+, OLI and MSI. The best performing MLRA was identified according to the lowest Root Mean Square Error (RMSE) in a five-fold cross-validation.

In total, 2500 parameter samples were drawn with Latin hypercube sampling using uniform distributions to evenly cover the parameter space for all parameters with range specifications. Of these 2500, 30% were modified to represent barren, winter conditions. This means that all leaf chemical parameters and LAI were set to 0.

Figure 5. Nadir view of a DART sample scene that is used to represent the heterogeneous canopy. The centre scene with five trees is replicated along the edges. Colour is reflectance with low values black and high values white.

3.3. RTM Sensitivity Analysis

In order to assess the importance of the RTM input parameters (Table 2) on the single spectral outputs, a global sensitivity analysis of the RTMs was conducted. Such an approach also helps to gauge how good input parameters can be estimated from spectral outputs. The approach here generally followed the approach of Verrelst et al. [27] based on Sobol sensitivity indices [58,59] modified by Saltelli et al. [60] and as implemented in the R package sensitivity (https://cran.r-project.org/web/packages/sensitivity). Here, only the total effect indices were considered that describe the sensitivity of the model output to an input parameter and its interactions with other parameters. The sum of the sensitivity indices with respect to all input parameters varies per spectral band output. Therefore, indices were normalised to sum up to 1 to ease comparison across spectral band outputs. Furthermore, only bands of MSI were taken into account because they cover the same spectral domains as ETM+ and OLI (Table 1).

3.4. Noise Scenarios

In this study, two types of noise were tested: multiplicative wavelength-independent (MI) and additive wavelength-independent (AI) noise [61]. MI is dependent on the BRF. Its term is larger for NIR compared to red spectral bands for typical vegetation spectral responses. MI and AI were added to the RTM spectral bands:

$$\rho' = \rho_{RTM} + \rho_{RTM} \cdot \epsilon_{MI} + \epsilon_{AI} \qquad (2)$$

where ρ' is the noise contaminated spectral band, ρ_{RTM} is the RTM spectral band output, ϵ_{MI} the MI noise term with $\epsilon_{MI} \sim \mathcal{N}(0, \sigma_{MI})$ and ϵ_{AI} the AI noise term with $\epsilon_{AI} \sim \mathcal{N}(0, \sigma_{AI})$. Apart from noise free, realisations with MI noise of 0.05, 0.1, 0.2 and 0.3, and AI noise of 0.05, 0.1 and 0.2 were tested. The 0.05 noise level was motivated by the Sentinel-2 mission requirement of 5% error on SR [34]. The other noise levels were pessimistic variations. However, as mentioned before, the noise term has multiple purposes, so that the mission requirements can only be an indication.

3.5. Solar Zenith Angle

Illumination conditions greatly affect the reflectance of canopies [5,62,63]. As the SZA changes over the course of the year, the internal canopy shadowing varies. Furthermore, SZA is an easy to obtain feature, as it is solely a function of location and time. Therefore, SZA was incorporated in the training workflow to test if it improves LAI prediction. SZA was calculated for local overpass times of the respective missions with the R package RAtmosphere ([64], https://cran.r-project.org/web/packages/RAtmosphere). For the respective realisations, it was treated as an extra training feature next to spectral bands.

3.6. Machine Learning Regression Algorithms

Studies on MLRA typically test a range of algorithms to explore their respective (dis-)advantages and cross-comparison results. This was adopted in this study as well. All models were trained to predict LAI, while the independent variables depended on the learning realisation. Multi-variate OLS regression was chosen as a benchmark method. For neural networks, the classic MLP was used (e.g., [19]). In particular, this was the implementation of the Stuttgart Neural Network Simulator in the R package RSNNS (https://cran.r-project.org/web/packages/RSNNS). Networks with $n+1$ neurons in a single hidden layer were trained, where n corresponded to the number of independent variables [19]. Random Forest was selected as an RT algorithm [65,66] and used as implemented in the R ranger package (https://cran.r-project.org/web/packages/ranger). The forests were grown with 500 trees.

Furthermore, three kernel-based methods were used. This type of regression methods translates the—possibly nonlinear—regression problem from the parameter space into a higher dimensional feature space, where it can be solved linearly. Kernel functions implement a notion of similarity function. SVR [67], KRR [14] and GPR [68] with Radial Basis Function (RBF) kernels were tested here.

The kernel σ hyperparameter was estimated with the `sigest` function from the `kernlab` package (https://cran.r-project.org/web/packages/kernlab/). Although developed to estimate σ for SVR, results in initial tests were promising for KRR and GPR. For further reading on MLRAs in biophysical parameter estimation, the reader is referred to Verrelst et al. [23].

The general workflow for MLRA application typically involves splitting of the feature database into training and validation sets to tune model parameters. Here, five-fold cross-validation was performed during the tuning process. This was based on the training dataset, which held 2500 samples of the RTM-based database. Next, the model performance for the best tuned parameters was evaluated with the test dataset, which held 500 samples of the RTM based database. This set was never seen by the models during training. Finally, the models were applied on the actual sensor observations and compared with the validation dataset (Section 3.7). However, as we inverted observations of three different sensors, in fact, for each realisation, three separate models were trained.

3.7. Ensemble Analysis and Validation

In order to analyse how well the MLRAs were able to learn the RTM-produced band-LAI relationships, the test error was evaluated. However, this error represents only the theoretical performance in case the RTM produces true results for the scene and the induced noise properties correspond to the noise of the actual spectral observations.

For validation purposes, the advantages of the TLS time series—i.e., high precision due to independence of illumination conditions [69,70]—was combined with the direct estimation of LAI with the litter-traps. Litter-traps are considered among the most accurate methods in terms of absolute LAI for forest canopies [71]. This approach follows the suggestion of Woodgate et al. [72] to calibrate TLS with other techniques. Specifically, the TLS time series was scaled with the litter-trap total LAI separately for each plot:

$$LAI_i = \frac{PAI_{TLS,i} - min(PAI_{TLS})}{max(PAI_{TLS})} \cdot LAI_{LT} \quad (3)$$

where LAI_i is the LAI at time i, PAI_{TLS} is the TLS derived PAI time series and LAI_{LT} is the litter-trap LAI. The observations were averaged per plot and linearly interpolated to obtain a continuous time series.

For each realisation, the time series of predicted LAI was compared with this validation time series. The performance metric was the RMSE:

$$RMSE = \sqrt{\frac{1}{t}\sum_{i=1}^{t}(LAI_{realisation,i} - LAI_{valid,i})^2} \quad (4)$$

where t is the length of the time series, and $LAI_{realisation,i}$ and $LAI_{valid,i}$ the realisation and validation LAI at time i, respectively.

4. Results and Discussion

4.1. Validation Time Series

Figure 6 shows the derived validation time series. The seasonal pattern with a fast spring leaf flush in May and autumn leaf falling in November dominated the temporal behaviour. Calders et al. [35] observed this speed in spring leaf flush for a mixed oak forest in the Netherlands. Maximum LAI of $6.1\,\text{m}^2\,\text{m}^{-2}$ was reached on 26 May in plot A. Plots B and E showed LAI values just below $6.0\,\text{m}^2\,\text{m}^{-2}$. Plot C had a larger gap in its centre so that overall LAI was lower there. Measured LAI for plot D was about two units lower than for the other plots with a peak of $3.2\,\text{m}^2\,\text{m}^{-2}$. This was due to the age composition of this plot, which was dominated by younger trees. The maximum LAI compares well with the results of Leuschner et al. [73], who measured LAI by litter-traps in 23 mature Beech stands in Germany. They found an average LAI of $7.4\,\text{m}^2\,\text{m}^{-2}$ with a range between 5.6 to 9.5 $\text{m}^2\,\text{m}^{-2}$.

Another feature is the slow decrease in LAI starting in August that could be observed in all plots. After the 2016 growing season, few brown leaves were still remaining on the trees until new leaves flushed in 2017. Overall, the obtained time series show the expected dynamic behaviour of the canopy during spring and autumn (Figure 6).

With respect to the uncertainty of the validation data, the standard deviation of the mean for the litter-trap samples was calculated as 0.43, 0.25, 0.41, 0.32 and 0.24 $m^2 m^{-2}$ for plots A to E, respectively. GCOS specified an accuracy of $0.5 m^2 m^{-2}$ as a target for LAI products for local and regional applications [38]. However, this is the requirement for the final LAI products, so that the achieved uncertainties are rather high for a validation measurement. This affected especially the realisation evaluations in terms of RMSE.

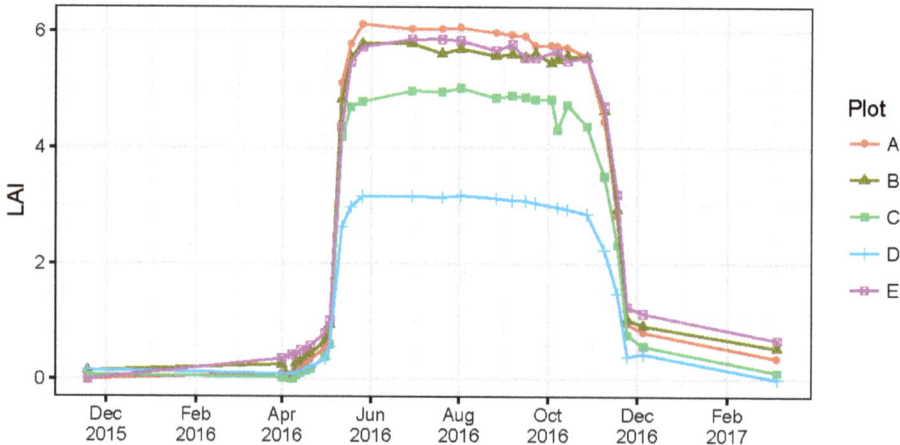

Figure 6. Speulderbos LAI time series for 2016 derived from TLS, points are observations. Lines are linear interpolations. Interpolations for outside of the measurement campaign were performed with the values from the previous and next year, which are added to the graph for clarity.

4.2. Leaf Biochemical Parameters' Retrieval

The leaf chemistry assessment indicated that some leaf components remained stable over the course of the season, while others showed a dynamic behaviour, resulting in multi-modal distributions (Figure 7). The static ones were the leaf structure parameter (N) parameter, the dry matter content C_m, and to some extent the equivalent water thickness C_w. The mean of the carotenoids remained stable, but its variance was increasing at the last sampling day. C_{ab} showed clear dynamics, with a strong decrease during the last sampling day. When compared to the readings of the SPAD meter, C_{ab} retrievals showed a quadratic relationship (Figure A1), which is confirmed by other studies [74,75]. This strong relationship supports the validity of the C_{ab} retrievals, and thereby the retrieval of the other biochemical constituents since they were inverted simultaneously. On the basis of these results, it was decided to constrain the training with fixed, central values of the N parameter, leaf carotenoid content (Car), C_w and C_m, but vary C_{ab} and leaf brown pigment fraction (C_{brown}) as given in Table 2.

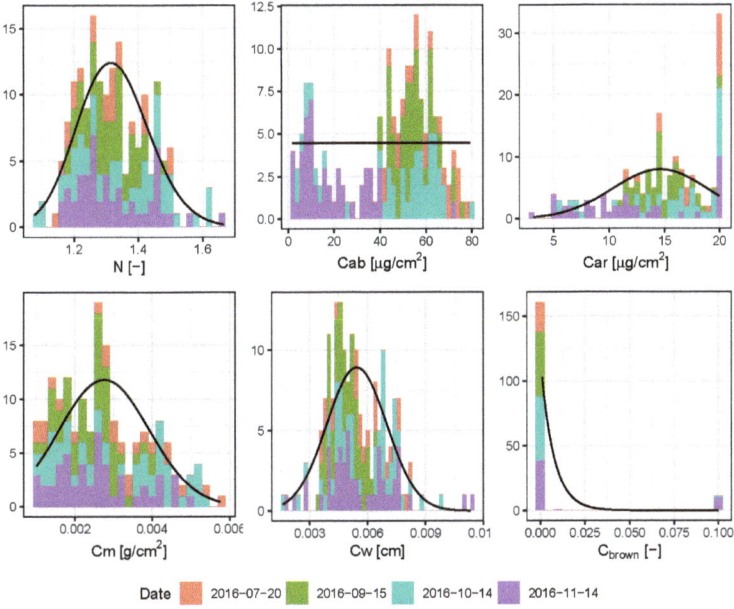

Figure 7. Leaf chemical properties based on PROSPECT inversions of 173 leaf samples. Leaf properties are leaf structure parameter (N), Chlorophyll a and b (C_{ab}), leaf carotenoid content (Car), leaf dry matter content (C_m), leaf water content (C_w), and leaf brown pigment fraction (C_{brown}). Stacks are coloured by acquisition date. Solid lines are fitted density models. Ordinate axis represents observation counts and modelled counts for the observations and the fitted models, respectively.

4.3. RTM Sensitivity

PROSAIL's and DART's sensitivity to their input parameters is depicted in Figure 8. In case of PROSAIL, BRFs in the visible bands were primarily driven by C_{ab} with a contribution of 74.8 and 49.4% in B03 and B04, respectively. This extended into the first red edge band B05, but strongly decreased to 6.1% at B06. In the red edge and NIR bands, LAI was the most important parameter with a relative contribution of 68.5% for bands B06 to B8A on average. This sensitivity is the reason why LAI retrieval relies on the NIR domain. The SWIR bands were mostly dependent on leaf water content C_w, which had 52.8 and 46.2% contribution in B11 and B12, respectively. These results are in line with those of Jacquemoud et al. [5].

On the other hand, DART's output sensitivity was dominated by canopy structural parameters (LAI and crown diameter) in the NIR spectral outputs (Figure 8). The contribution of LAI was maximal in B04 with 64.1%, while that of crown diameter was maximal in B06 with 52.2%. Their combined contribution was minimal in B12 with 12.7%. In contrast to PROSAIL, DART showed some sensitivity towards SZA, which was 5.1% on average in bands B05 to B8A. This reflects the effect of shadowing and DART's vertical heterogeneous character.

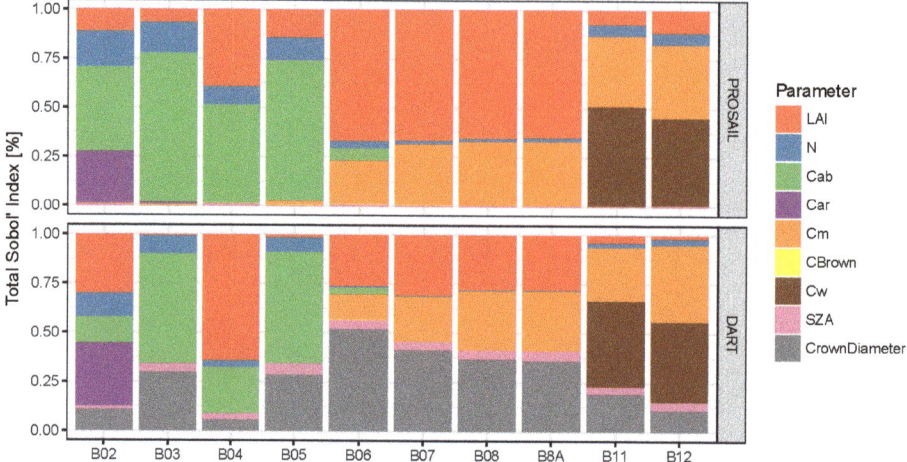

Figure 8. Global sensitivity of S2A MSI spectral bands to PROSAIL and DART input parameters according to Sobol' Indices. Total Sobol' Indices were normalised per band to sum up to 1. For band specifications, see Table 1 and for parameters see Table 2.

4.4. Impact of Training Scheme Features on Prediction Performance

This section presents the single inversion workflow features and their role for predicting LAI, starting with a comparison across domains in the following paragraphs. Since the RMSE results were not consistently normal distributed within ensembles, the median was calculated for all realisations that implemented a specific feature. In this way, the feature's role could be evaluated. It should also be noted that RMSE refers to the validation error. Only in some cases was the training error evaluated, but this is always explicitly mentioned.

Table 3 summarises the effects of all features on the validation RMSE. The 25 and 75% Interquartile Range (IQR) is listed as a quantification for the variability within the ensembles. The feature with the strongest influence on performance in terms of RMSE was the adding of AI noise to the RTM generated spectral outputs. Adding 5% AI noise decreased the median RMSE from $2.33\,\mathrm{m^2\,m^{-2}}$ for no noise to $1.25\,\mathrm{m^2\,m^{-2}}$. Realisations with 5% AI noise also achieved the overall lowest RMSE. The second-most important feature was the choice of MLRA. Here, realisations varied between RMSE of $2.04\,\mathrm{m^2\,m^{-2}}$ for MLP, and RMSE of $1.52\,\mathrm{m^2\,m^{-2}}$ for SVR. Restraining the training database with prior information on leaf biochemical contents generally decreased prediction performance.

The following sections present the performance results of all training features in more detail and elaborate on their first order interactions, whereas typically only the strongest interaction is discussed. Interactions were investigated by comparing the respective groups of realisations that implement the features. For example, if feature A has two realisations A1 and A2, and its strongest first order interaction feature B has B1 and B2, all realisations that implement them (A1/B1, A1/B2, A2/B1, A2/B2) were compared to each other. The strongest interaction (feature B in the previous example) was identified as the one that varies performance in terms of validation RMSE the most after the feature under consideration (feature A in the previous example). For example, when the leaf chemical prior was inspected, the feature that produced the largest differences among the realisations was identified as the strongest interaction, which was the RTM feature.

Table 3. Validation median RMSE and Interquartile Range (IQR) (25 to 75%) for training features. ΔRMSE refers to the difference to the best feature per group.

Feature	Realisation	Median RMSE	ΔRMSE	RMSE IQR
Leaf chemical prior	Free range	1.47	—	0.49
	Prior	2.10	0.63	0.96
RTM	PROSAIL	1.93	0.42	0.95
	DART	1.51	—	0.69
MI Noise	0%	1.73	0.10	1.12
	5%	1.70	0.07	1.08
	10%	1.70	0.07	1.01
	20%	1.63	—	0.84
	30%	1.63	—	0.73
AI Noise	0%	2.33	1.08	0.80
	5%	1.25	—	0.67
	10%	1.38	0.13	0.68
	20%	1.74	0.49	0.51
SZA	Without SZA	1.69	0.06	0.97
	With SZA	1.63	—	0.95
MLRA	OLS	1.72	0.20	0.54
	MLP	2.04	0.52	0.88
	RT	1.57	0.05	1.04
	SVR	1.52	—	1.06
	KRR	1.63	0.11	1.06
	GPR	1.60	0.08	0.98

4.4.1. Leaf Biochemical Prior

Among other information, Figure 9 summarises how the leaf biochemical prior information affected the training results. PROSAIL performance was generally more affected by introducing prior information than DART performance (Figure 9): median RMSE decreased from 1.91 to 1.38 $m^2 \, m^{-2}$ when prior information was included in the test data sets, while it increased from 1.43 to 2.37 $m^2 \, m^{-2}$ in case of the validation data. The former can be explained by the sensitivity of PROSAIL to biochemical parameters C_{ab}, C_m and C_w (Section 4.3). Using prior information effectively decreases the parameter input space that the MLRA has to learn, thereby making it easier for the MLRA. This also made it possible to reach testing RMSE as low as 0.01 $m^2 \, m^{-2}$. However, inversion of actual observations was impaired by the constraint of the leaf chemical parameter space. This may be due to the fact that leave chemical properties were not representative for the whole study area, as leaves were only sampled from two beech trees, while oak was also present and environmental conditions may change the chemical composition throughout the study area. Additionally, the way the constraint was implemented—as mean estimates allowing no deviation—may also play a role.

Contrary to this, the DART inversion performance was less sensitive to leaf biochemical parameters (Section 4.3). Median performance in terms of RMSE was similar at 2.29 and 2.16 $m^2 \, m^{-2}$ for test data and improved slightly with a decrease in RMSE from 1.55 to 1.46 $m^2 \, m^{-2}$ for the validation data set when introducing leaf chemical information. Thus, reducing the input parameter space had a small positive effect during training and validation.

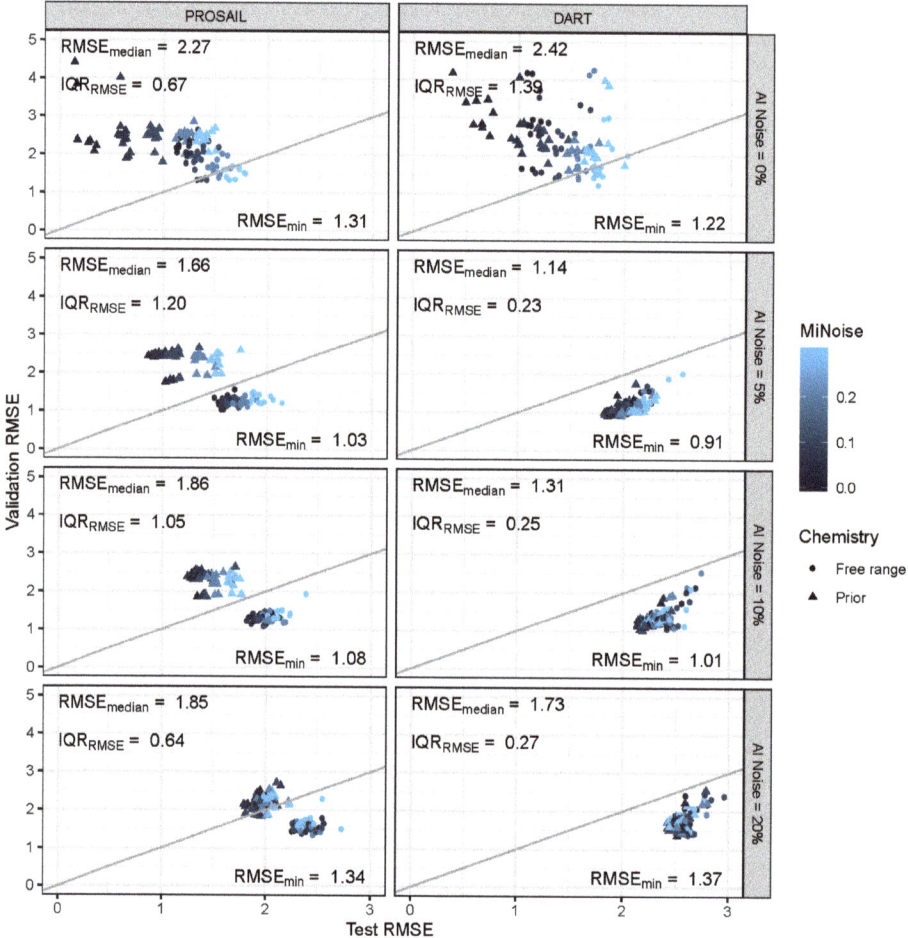

Figure 9. Prediction performance in terms of testing and validation RMSE with dependence on the chosen RTM, AI and MI noise level, and bio-chemical prior. $RMSE_{median}$, IQR_{RMSE} and $RMSE_{min}$ refer to the validation error in the respective panel cell. Grey line is 1:1 line. In case of the validation results, 14 realisations were trimmed with RMSE larger than $10\,m^2\,m^{-2}$ (all with 0 % AI noise) because they prevented proper display.

4.4.2. RTM Choice

Among other features, Figure 9 compares the inverse model realisation in terms of their used RTM. Both testing error, i.e., the error based on RTM produced samples, and validation error, i.e., the error based on the validation time series for the Speulderbos site, are shown. Overall, PROSAIL and DART achieved a median test RMSE of 1.63 and 2.21 $m^2\,m^{-2}$, while the validation RMSE was 1.93 and 1.51 $m^2\,m^{-2}$, respectively. The lower performance of DART on the testing samples can be explained with its additional freely varying parameter, the crown diameter. However, with this parameter came the capability to model crown gaps (Figure 5), which led to the better performance in terms of validation RMSE.

Apart from the overall better median RMSE of DART, using this RTM generally decreased spread of error for realisations that implemented this RTM, at least when some AI noise larger than 0% was chosen. For example, the validation RMSE IQR for DART with 0% AI noise was $0.23\,m^2\,m^{-2}$, while it was $1.20\,\text{m}^2\,\text{m}^{-2}$ for PROSAIL. This means that choosing this DART reduced the importance for a particular choice of the other training features.

There were realisations for which the testing RMSE was disproportionally larger than the validation RMSE. In fact, this was the case for 43.3 and 77.7% of the PROSAIL and DART cases, respectively (Figure 9). Under circumstances where data of the same origin would have been used, this would be unlikely to occur, especially in scenarios with added noise. However, training in this study was based on RTM output and validation on field acquired data.

4.4.3. Noise Scenarios

Both the AI and MI noise had different effects on the testing and validation error. Generally testing errors increased with increasing noise—in the case of AI from $1.29\,\text{m}^2\,\text{m}^{-2}$ for 0% to $2.46\,\text{m}^2\,\text{m}^{-2}$ median RMSE for 20% AI noise, and in the case of MI from $1.89\,\text{m}^2\,\text{m}^{-2}$ for 0% to $2.09\,\text{m}^2\,\text{m}^{-2}$ median RMSE for 20% MI noise. This showed the general effect of the noise to blur the relationship between spectral output and associated LAI, and prevent the MLRA to learn the true RTM produced pattern.

AI noise was generally more successful at reducing validation RMSE (Table 3): it decreased median RMSE by $1.08\,\text{m}^2\,\text{m}^{-2}$ (from $2.33\,\text{m}^2\,\text{m}^{-2}$ for 0% AI noise to $1.25\,\text{m}^2\,\text{m}^{-2}$ for 5% AI noise). MI noise reduced median RMSE only by $0.10\,\text{m}^2\,\text{m}^{-2}$ (from $1.73\,\text{m}^2\,\text{m}^{-2}$ for 0% MI noise to $1.63\,\text{m}^2\,\text{m}^{-2}$ for 20% MI noise).

Considering AI noise alone, the addition of any in comparison to 0% AI noise strongly changed the distribution of realisations in terms of RMSE (Figure 9). This did not only include the best, but also low performing results. More precisely, AI noise prevented occurrence of very bad results. For the 0% AI noise level, worst performance reached up to $29.29\,\text{m}^2\,\text{m}^{-2}$ and the 95% quantile lay at $10.4\,\text{m}^2\,\text{m}^{-2}$, while for 5% these statistics were 2.66 and 2.48 $\text{m}^2\,\text{m}^{-2}$, respectively. Here, the added noise prevented the MLRA from over-fitting on clean RTM simulations.

Noise was also found an important training workflow element in PROSAIL-Look Up Table (LUT)-based inversions of observations from agricultural crops by Rivera et al. [29] and Verrelst et al. [30]. They added up to 50 and 30% noise to PROSAIL spectra, respectively, but did not specify how the error was implemented. However, the required magnitude of noise corresponds to the MI noise in this study, which was optimal at 20 to 30%. Baret et al. [19] added 0.04 absolute Gaussian white (AI) noise in a global retrieval workflow based on PROSAIL. This is in line with the optimal 5% AI noise in this study. Koetz et al. [28] adopted a wavelength-dependent, relative noise term of maximal 10% in the 444 nm band. Both Koetz et al. [28] and Baret et al. [19] based their choice for a specific noise level on experience of observation errors, but do not evaluate other possibilities.

Baret et al. [19] argued that the quantification of the error term is difficult, as it includes errors stemming from the radiometric calibration of the sensor, Bidirectional Reflectance Distribution Function (BRDF) normalisation, atmospheric correction, cloud residuals and the RTM representativeness for the actual canopy. Additionally, the interaction of these single terms plus the properties of the used MLRA in an inversion workflow complicates the choice based on experience of errors of the sub-systems. Surely, it is more practical to conduct a sensitivity analysis over validation samples rather than characterising the sensor-inversion system in detail. Moreover, noise terms need to be defined properly to compare them across studies.

4.4.4. SZA

As shown in Table 3, realisations that made use of SZA as an extra training feature achieved overall $0.06\,\text{m}^2\,\text{m}^{-2}$ smaller median RMSE than realisations that did not include SZA. A Wilcoxon signed-rank test confirmed that the two groups differed significantly ($p < 0.01$). This difference was most prominent with the multiple MLRAs. For example, SVR benefited strongest with a decrease in median RMSE by $0.07\,\text{m}^2\,\text{m}^{-2}$. This can be explained by the richer feature space that the MLRAs had available for learning. Additionally, SZA correlated with the general phenological patterns of the study area with low SZA in summer.

Strategies to include SZA into inversion workflows of multi-temporal observations are not consolidated yet. Koetz et al. [28] computed independent LUTs for different observation dates and consequently SZAs, but they did not investigate the error that would occur if they would not have done so. However, training separate models for multi-temporal time series with multiple observations per year is undesirable due to computational load and the checks that would be necessary to ensure consistent model properties. Campos-Taberner et al. [16] conducted PROSAIL inversions over a full season of L8 OLI, L7 ETM+ and Satellite Pour l'Observation de la Terre (SPOT) High Resolution Geometric (HRG) sensor data and they mention solar-sensor geometry as inputs for PROSAIL, but do not elaborate how these parameters were dealt with in the training database. Baret et al. [19] included SZA as a training feature in their neural network based inversion for the VEGETATION based CYCLOPES LAI product, but again did not evaluate this strategy. However, as the CYCLOPES product has global extents, it spans several degrees of latitude, leading to different regimes of illumination dynamics over the year. Thus, SZA was given importance in past studies and showed some importance here as well, but has not been yet evaluated for global LAI products. At least adding SZA as an additional training feature is an easy implementable option. In addition, SZA is efficient to compute, as only the location and observation time are needed to calculate it. However, interactions with other parameters that change over the time of the year such as soil background and LAI itself also need to be considered.

4.4.5. MLRA

As demonstrated in the discussion of the other training features, the choice for a specific MLRA was not the most important factor affecting validation performance in this study (Table 3). However, studies employing MLRA typically compare various algorithms (e.g., [14,16]). Figure 10 gives an overview over all realisations grouped by their used MLRA. The MLRAs reached best (and median) RMSE of 0.91 (1.57), 0.92 (1.63), 0.93 (1.52), 0.95 (1.60), 1.02 (1.72) and 1.08 (2.04) in case of RT, KRR, SVR, GPR, OLS and MLP, respectively. Hence, even though RT and KRR produced the best, SVR produced the overall best realisations. Maximum differences among the MLRAs in median RMSE of $0.52\,\text{m}^2\,\text{m}^{-2}$ were found between SVR and MLP realisations. There were also 18 realisations using OLS and seven using KRR that exceeded RMSE of $5\,\text{m}^2\,\text{m}^{-2}$, all of which were training without noise applied to the training spectral features.

Apart from the MLRA validation, performance processing time is an important property especially for routine and large scale production. The time required for the training of the described realisations in this study was on average 0.03, 7.18, 2.29, 3.17, 125.37 and 123.81 s for the OLS, RT, MLP, SVR, KRR and GPR, respectively. This is in contrast to Verrelst et al. [14] who found KRR and GPR required around the tenth of the time of a neural network or an SVR. However, their models were implemented with Matlab, while this study used R. Additionally, the implementations of RT, SVR, KRR and GPR in this study could make use of parallelisation. Concerning time required for prediction of 10,000 random samples, the models needed 0.08, 0.09, 0.09, 0.15, 0.30 and 0.19 s in the case of OLS, RT, MLP, SVR, KRR and GPR, respectively. Hence, implementation details and optimisation can play a significant role in processing time.

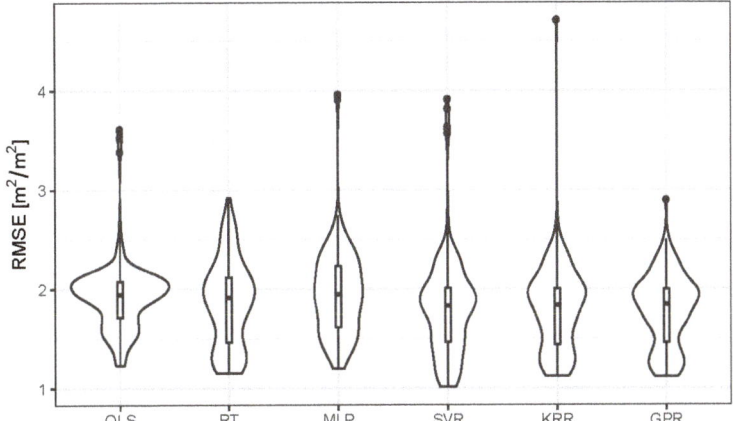

Figure 10. Violin plots [76] of prediction performance for the different MLRAs. It should be noted that 25 realisations were trimmed with RMSE larger than $5.0\,\mathrm{m^2\,m^{-2}}$ (18 for OLS and 7 for KRR) because they prevented proper display.

4.5. Best Performing Feature Combination

Figure 11 shows the best performing realisation that reached RMSE of $0.91\,\mathrm{m^2\,m^{-2}}$. It was built with an RT based on a DART produced database restricted with prior information from the leaf sampling, 5% AI and 20% MI noise, and with SZA as an additional training feature. Maximum LAI of $5.80\,\mathrm{m^2\,m^{-2}}$ was reached in plot A on 17 July. LAI before May 1st was on average $(0.27 \pm 0.31)\,\mathrm{m^2\,m^{-2}}$, and thereafter $(4.81 \pm 0.59)\,\mathrm{m^2\,m^{-2}}$ until end of October. The predicted LAI explained 84.9% of variation of the reference time series.

In general, these results for a single time series are in the range of previously published results. Schlerf and Atzberger [77] achieved $0.66\,\mathrm{m^2\,m^{-2}}$ with a two-band combination chosen from simulated Landsat TM bands over a Norway Spruce site. For beech canopies within the same site, Schlerf and Atzberger [78] report $2.12\,\mathrm{m^2\,m^{-2}}$ RMSE with a multi-spectral, near-nadir viewing set-up. Brown et al. [18] obtained a RMSE of $0.47\,\mathrm{m^2\,m^{-2}}$ for a beech site in Southern England. All three studies used the INFORM RTM to create the training database [79].

However, as can be observed in Figure 11, the predictions showed different biases for the single plots. In fact, summer LAI was underestimated on average by $0.74\,\mathrm{m^2\,m^{-2}}$ in plots A, B, C and E, and overestimated by $1.53\,\mathrm{m^2\,m^{-2}}$ in plot D. The bias in plot D could result from the structure of the plot, which consisted of more young trees compared to the other plots. The position of the litter traps was chosen close to the TLS scan positions, which were possibly not representative for the whole of the plot. Additionally, RMSE as the choice of error metric during retrieval evaluation does not allow the assessment of bias. In fact, additional error metrics would be needed to characterise the bias as well as temporal consistency of the time series.

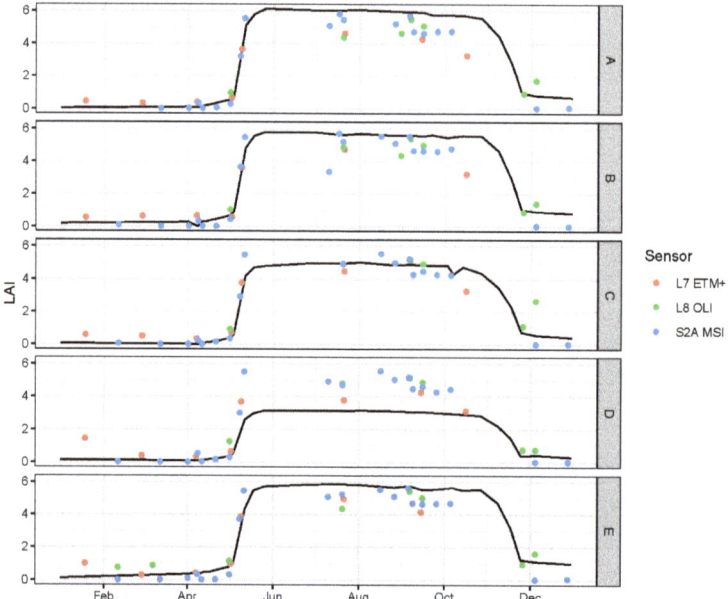

Figure 11. Best performing realisation in terms of RMSE over the 2016 study period. Black solid line is the validation time series.

5. Conclusions

The Sentinel-2 mission provides a new science-grade data stream for monitoring of dynamic vegetation behaviour. Together with other missions such as Landsat 7, 8 and future Landsat 9, a characterisation of temporal dynamics in biophysical parameters becomes possible at decametric resolution. Accurate retrieval and a universal processing workflow for potential Sentinel-2 and Landsat harmonised biophysical products remains a challenge. Previous studies identified MLRAs combined with RTMs in hybrid retrieval workflows as a potential solution. This study investigated the impact of multiple properties of such workflows on the retrieval performance for LAI over a Dutch beech forest site.

Addition of AI noise on the RTM spectral database was found to be most important for prediction performance with a difference of $1.09\,m^2\,m^{-2}$ in median RMSE compared to no noise. A level of 5% was optimal in this study. On the other hand, MI noise showed less improvements. Added noise helps the MLRAs to generalise and prevent over-fitting on the pure RTM output. Previous studies did not investigate the effect of different noise definitions and some did not report precisely how noise was defined. With respect to its importance, a clear definition and careful sensitivity analysis should be paramount for future studies.

The choice for the heterogeneous DART RTM in comparison with the turbid medium PROSAIL model resulted in a median RMSE difference of $0.42\,m^2\,m^{-2}$. An additional advantage of DART in this study was the lower spread of performance of other inversion workflow features, i.e., DART led to more consistent inversion workflows. Apart from this, care must be taken when using the PROSAIL model with biochemical prior information, as PROSAIL is very sensitive to this. The choice of a specific MLRA was found to be less critical in terms of prediction performance. However, MLRAs varied significantly in run-time, also depending on the implementation and code optimisation. When choosing a particular MLRA, these secondary benefits should be weighted together with the expected accuracy.

Author Contributions: Conceptualization, B.B., J.V. (Jochem Verrelst), J.G.P.W.C.; methodology, B.B., J.V. (Jochem Verrelst), J.-P.G.-E., J.G.P.W.C.; formal analysis, B.B., J.-P.G.-E., L.G., J.d.O.; data curation, B.B., L.G.; writing–original draft preparation, B.B., J.G.P.W.C.; writing–review and editing, B.B., J.V. (Jochem Verrelst), J.-P.G.-E., J.G.P.W.C.,

J.V. (Jan Verbesselt); visualization, B.B.; supervision, J.G.P.W.C., J.V. (Jan Verbesselt), M.H.; project administration, M.H.; funding acquisition, M.H. All authors have read and agreed to the published version of the manuscript.

Funding: This research was funded by the IDEAS+ and QA4EO contracts funded by ESA-ESRIN. Jochem Verrelst was funded by the European Research Council (ERC) under the ERC-2017-STG SENTIFLEX project (grant agreement 755617).

Acknowledgments: Data analysis was supported by the Research and User Support (RUS) Service. The RUS Service is funded by the European Commission, managed by the European Space Agency, and operated by CSSI and its partners. The authors thank the Dutch Forestry Service (Staatsbosbeheer) for granting access to the site, and Devis Tuia and Benjamin Kellenberger for helpful discussions on machine learning. We thank two anonymous reviewers for their valuable comments.

Conflicts of Interest: The authors declare no conflict of interest. The funders had no role in the design of the study; in the collection, analyses, or interpretation of data; in the writing of the manuscript, or in the decision to publish the results.

Appendix A

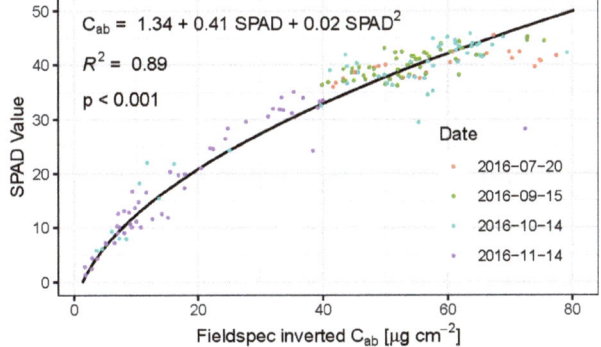

Figure A1. Comparison of Fieldspec inverted C_{ab} with SPAD meter value taken from the same leaf samples. Solid line, R^2, p-value and formula correspond to the quadratic fit. Dates correspond to the leaf sampling dates.

Table A1. List of TLS and satellite sampling dates.

TLS	Landsat 7	Landsat 8	Sentinel-2A
2015-11-18	2016-01-18	2016-02-11	2016-02-11
2016-04-01	2016-02-28	2016-03-07	2016-03-12
2016-04-08	2016-04-07	2016-05-01	2016-04-01
2016-04-11	2016-05-02	2016-07-20	2016-04-08
2016-04-14	2016-05-09	2016-08-30	2016-04-11
2016-04-18	2016-07-21	2016-09-06	2016-04-21
2016-04-21	2016-09-14	2016-09-15	2016-05-01
2016-05-01	2016-10-16	2016-11-25	2016-05-08
2016-05-04		2016-12-04	2016-05-11
2016-05-12			2016-07-10
2016-05-18			2016-07-17
2016-05-26			2016-07-20
2016-06-29			2016-08-16
2016-07-20			2016-08-26
2016-08-02			2016-09-05
2016-08-26			2016-09-08
2016-09-06			2016-09-15
2016-09-15			2016-09-25
2016-09-22			2016-10-05
2016-10-03			2016-12-04
2016-10-07			2016-12-27
2016-10-14			
2016-10-27			
2016-11-08			
2016-11-17			
2016-11-24			
2016-12-05			
2017-03-07			

References

1. Beer, C.; Reichstein, M.; Tomelleri, E.; Ciais, P.; Jung, M.; Carvalhais, N.; Rödenbeck, C.; Arain, M.A.; Baldocchi, D.; Bonan, G.B.; et al. Terrestrial gross carbon dioxide uptake: Global distribution and covariation with climate. *Science* **2010**, *329*, 834–838. [CrossRef] [PubMed]
2. Chen, J.; Black, T.A. Defining leaf area index for non-flat leaves. *Plant Cell Environ.* **1992**, *15*, 421–429. [CrossRef]
3. Malenovský, Z.; Rott, H.; Cihlar, J.; Schaepman, M.E.; García-Santos, G.; Fernandes, R.; Berger, M. Sentinels for science: Potential of Sentinel-1, -2, and -3 missions for scientific observations of ocean, cryosphere, and land. *Remote Sens. Environ.* **2012**, *120*, 91–101. [CrossRef]
4. Morisette, J.T.; Baret, F.; Privette, J.L.; Myneni, R.B.; Nickeson, J.E.; Garrigues, S.; Shabanov, N.V.; Weiss, M.; Fernandes, R.A.; Leblanc, S.G.; et al. Validation of global moderate-resolution LAI products: A framework proposed within the CEOS land product validation subgroup. *IEEE Trans. Geosci. Remote Sens.* **2006**, *44*, 1804–1814. [CrossRef]
5. Jacquemoud, S.; Verhoef, W.; Baret, F.; Bacour, C.; Zarco-Tejada, P.J.; Asner, G.P.; François, C.; Ustin, S.L. PROSPECT + SAIL models: A review of use for vegetation characterization. *Remote Sens. Environ.* **2009**, *113*, S56–S66. [CrossRef]
6. Roy, D.P.; Wulder, M.A.; Loveland, T.R.; Woodcock, C.E.; Allen, R.G.; Anderson, M.C.; Helder, D.; Irons, J.R.; Johnson, D.M.; Kennedy, R.; et al. Landsat-8: Science and product vision for terrestrial global change research. *Remote Sens. Environ.* **2014**, *145*, 154–172. [CrossRef]
7. Houborg, R.; McCabe, M.; Cescatti, A.; Gao, F.; Schull, M.; Gitelson, A. Joint leaf chlorophyll content and leaf area index retrieval from Landsat data using a regularized model inversion system (REGFLEC). *Remote Sens. Environ.* **2015**, *159*, 203–221. [CrossRef]
8. Li, W.; Weiss, M.; Waldner, F.; Defourny, P.; Demarez, V.; Morin, D.; Hagolle, O.; Baret, F. A Generic Algorithm to Estimate LAI, FAPAR and FCOVER Variables from SPOT4_HRVIR and Landsat Sensors: Evaluation of the Consistency and Comparison with Ground Measurements. *Remote Sens.* **2015**, *7*, 15494–15516. [CrossRef]

9. Yin, G.; Verger, A.; Qu, Y.; Zhao, W.; Xu, B.; Zeng, Y.; Liu, K.; Li, J.; Liu, Q. Retrieval of High Spatiotemporal Resolution Leaf Area Index with Gaussian Processes, Wireless Sensor Network, and Satellite Data Fusion. *Remote Sens.* **2019**, *11*, 244. [CrossRef]
10. Soudani, K.; François, C.; le Maire, G.; Le Dantec, V.; Dufrêne, E. Comparative analysis of IKONOS, SPOT, and ETM+ data for leaf area index estimation in temperate coniferous and deciduous forest stands. *Remote Sens. Environ.* **2006**, *102*, 161–175. [CrossRef]
11. Ganguly, S.; Nemani, R.R.; Zhang, G.; Hashimoto, H.; Milesi, C.; Michaelis, A.; Wang, W.; Votava, P.; Samanta, A.; Melton, F.; et al. Generating global Leaf Area Index from Landsat: Algorithm formulation and demonstration. *Remote Sens. Environ.* **2012**, *122*, 185–202. [CrossRef]
12. Delegido, J.; Verrelst, J.; Alonso, L.; Moreno, J. Evaluation of Sentinel-2 red-edge bands for empirical estimation of green LAI and chlorophyll content. *Sensors* **2011**, *11*, 7063–7081. [CrossRef] [PubMed]
13. Richter, K.; Hank, T.B.; Vuolo, F.; Mauser, W.; D'Urso, G. Optimal exploitation of the Sentinel-2 spectral capabilities for crop leaf area index mapping. *Remote Sens.* **2012**, *4*, 561–582. [CrossRef]
14. Verrelst, J.; Muñoz, J.; Alonso, L.; Delegido, J.; Rivera, J.P.; Camps-Valls, G.; Moreno, J. Machine learning regression algorithms for biophysical parameter retrieval: Opportunities for Sentinel-2 and -3. *Remote Sens. Environ.* **2012**, *118*, 127–139. [CrossRef]
15. Frampton, W.J.; Dash, J.; Watmough, G.; Milton, E.J. Evaluating the capabilities of Sentinel-2 for quantitative estimation of biophysical variables in vegetation. *ISPRS J. Photogramm. Remote Sens.* **2013**, *82*, 83–92. [CrossRef]
16. Campos-Taberner, M.; García-Haro, F.J.; Camps-Valls, G.; Grau-Muedra, G.; Nutini, F.; Crema, A.; Boschetti, M. Multitemporal and multiresolution leaf area index retrieval for operational local rice crop monitoring. *Remote Sens. Environ.* **2016**, *187*, 102–118. [CrossRef]
17. Korhonen, L.; Hadi; Packalen, P.; Rautiainen, M. Comparison of Sentinel-2 and Landsat 8 in the estimation of boreal forest canopy cover and leaf area index. *Remote Sens. Environ.* **2017**, *195*, 259–274. [CrossRef]
18. Brown, L.A.; Ogutu, B.O.; Dash, J. Estimating forest leaf area index and canopy chlorophyll content with Sentinel-2: An evaluation of two hybrid retrieval algorithms. *Remote Sens.* **2019**, *11*, 1752. [CrossRef]
19. Baret, F.; Hagolle, O.; Geiger, B.; Bicheron, P.; Miras, B.; Huc, M.; Berthelot, B.; Niño, F.; Weiss, M.; Samain, O.; et al. LAI, fAPAR and fCover CYCLOPES global products derived from VEGETATION. Part 1: Principles of the algorithm. *Remote Sens. Environ.* **2007**, *110*, 275–286. [CrossRef]
20. Durbha, S.S.; King, R.L.; Younan, N.H. Support vector machines regression for retrieval of leaf area index from multiangle imaging spectroradiometer. *Remote Sens. Environ.* **2007**, *107*, 348–361. [CrossRef]
21. Vuolo, F.; Neugebauer, N.; Bolognesi, S.F.; Atzberger, C.; D'Urso, G. Estimation of leaf area index using DEIMOS-1 data: Application and transferability of a semi-empirical relationship between two agricultural areas. *Remote Sens.* **2013**, *5*, 1274–1291. [CrossRef]
22. Lazaro-Gredilla, M.; Titsias, M.K.; Verrelst, J.; Camps-Valls, G. Retrieval of Biophysical Parameters With Heteroscedastic Gaussian Processes. *IEEE Geosci. Remote Sens. Lett.* **2014**, *11*, 838–842. [CrossRef]
23. Verrelst, J.; Camps-Valls, G.; Muñoz-Marí, J.; Rivera, J.P.; Veroustraete, F.; Clevers, J.G.; Moreno, J. Optical remote sensing and the retrieval of terrestrial vegetation bio-geophysical properties—A review. *ISPRS J. Photogramm. Remote Sens.* **2015**, *108*, 273–290. [CrossRef]
24. Haboudane, D.; Miller, J.R.; Pattey, E.; Zarco-Tejada, P.J.; Strachan, I.B. Hyperspectral vegetation indices and novel algorithms for predicting green LAI of crop canopies: Modeling and validation in the context of precision agriculture. *Remote Sens. Environ.* **2004**, *90*, 337–352. [CrossRef]
25. Widlowski, J.L.; Taberner, M.; Pinty, B.; Bruniquel-Pinel, V.; Disney, M.; Fernandes, R.; Gastellu-Etchegorry, J.P.; Gobron, N.; Kuusk, A.; Lavergne, T.; et al. Third Radiation Transfer Model Intercomparison (RAMI) exercise: Documenting progress in canopy reflectance models. *J. Geophys. Res. Atmos.* **2007**, *112*, 1–28. [CrossRef]
26. Widlowski, J.L.; Mio, C.; Disney, M.; Adams, J.; Andredakis, I.; Atzberger, C.; Brennan, J.; Busetto, L.; Chelle, M.; Ceccherini, G.; et al. The fourth phase of the radiative transfer model intercomparison (RAMI) exercise: Actual canopy scenarios and conformity testing. *Remote Sens. Environ.* **2015**, *169*, 418–437. [CrossRef]
27. Verrelst, J.; Sabater, N.; Rivera, J.; Muñoz-Marí, J.; Vicent, J.; Camps-Valls, G.; Moreno, J. Emulation of Leaf, Canopy and Atmosphere Radiative Transfer Models for Fast Global Sensitivity Analysis. *Remote Sens.* **2016**, *8*, 673. [CrossRef]
28. Koetz, B.; Baret, F.; Poilvé, H.; Hill, J. Use of coupled canopy structure dynamic and radiative transfer models to estimate biophysical canopy characteristics. *Remote Sens. Environ.* **2005**, *95*, 115–124. [CrossRef]

29. Rivera, J.P.; Verrelst, J.; Leonenko, G.; Moreno, J. Multiple cost functions and regularization options for improved retrieval of leaf chlorophyll content and LAI through inversion of the PROSAIL model. *Remote Sens.* **2013**, *5*, 3280–3304. [CrossRef]
30. Verrelst, J.; Rivera, J.P.; Leonenko, G.; Alonso, L.; Moreno, J. Optimizing LUT-based RTM inversion for semiautomatic mapping of crop biophysical parameters from Sentinel-2 and -3 data: Role of cost functions. *IEEE Trans. Geosci. Remote Sens.* **2014**, *52*, 257–269. [CrossRef]
31. Myneni, R.; Knyazikhin, Y.; Shabanov, N. Leaf Area Index and Fraction of Absorbed PAR Products from Terra and Aqua MODIS Sensors: Analysis, Validation, and Refinement. In *Land Remote Sensing and Global Environmental Change—NASA's Earth Observing System and the Science of ASTER and MODIS*; Ramachandran, B., Justice, C.O., Abrams, M.J., Eds.; Springer: New York, NY, USA; Dordrecht, The Netherlans; Heidelberg, Germany; London, UK, 2011; Chapter 27, pp. 603–633.
32. Brede, B.; Bartholomeus, H.; Suomalainen, J.; Clevers, J.; Verbesselt, J.; Herold, M.; Culvenor, D.; Gascon, F. The Speulderbos Fiducial Reference Site for Continuous Monitoring of Forest Biophysical Variables. In Proceedings of the Living Planet Symposium 2016, Prague, Czech Republic, 9–13 May 2016; p. 5.
33. Brede, B.; Gastellu-Etchegorry, J.P.; Lauret, N.; Baret, F.; Clevers, J.; Verbesselt, J.; Herold, M. Monitoring Forest Phenology and Leaf Area Index with the Autonomous, Low-Cost Transmittance Sensor PASTiS-57. *Remote Sens.* **2018**, *10*, 1032. [CrossRef]
34. Drusch, M.; Del Bello, U.; Carlier, S.; Colin, O.; Fernandez, V.; Gascon, F.; Hoersch, B.; Isola, C.; Laberinti, P.; Martimort, P.; et al. Sentinel-2: ESA's Optical High-Resolution Mission for GMES Operational Services. *Remote Sens. Environ.* **2012**, *120*, 25–36. [CrossRef]
35. Calders, K.; Schenkels, T.; Bartholomeus, H.; Armston, J.; Verbesselt, J.; Herold, M. Monitoring spring phenology with high temporal resolution terrestrial LiDAR measurements. *Agric. For. Meteorol.* **2015**, *203*, 158–168. [CrossRef]
36. Wilson, J.W. Estimation of foliage denseness and foliage angle by inclined point quadrats. *Aust. J. Bot.* **1963**, *11*, 95–105. [CrossRef]
37. Weiss, M.; Baret, F.; Smith, G.J.; Jonckheere, I.; Coppin, P. Review of methods for in situ leaf area index (LAI) determination Part II. Estimation of LAI, errors and sampling. *Agric. For. Meteorol.* **2004**, *121*, 37–53. [CrossRef]
38. Fernandes, R.; Plummer, S.; Nightingale, J.; Baret, F.; Camacho, F.; Fang, H.; Garrigues, S.; Gobron, N.; Lang, M.; Lacaze, R.; et al. Global Leaf Area Index Product Validation Good Practices. In *Best Practice for Satellite-Derived Land Product Validation*, 2.0.1 ed.; Schaepman-Strub, G., Román, M., Nickeson, J., Eds.; Land Product Validation Subgroup (WGCV/CEOS). Available online: https://lpvs.gsfc.nasa.gov/LAI/LAI_home.html (accessed on 13 January 2020).
39. Calders, K.; Armston, J.; Newnham, G.; Herold, M.; Goodwin, N. Implications of sensor configuration and topography on vertical plant profiles derived from terrestrial LiDAR. *Agric. For. Meteorol.* **2014**, *194*, 104–117. [CrossRef]
40. Bouriaud, O.; Soudani, K.; Bréda, N. Leaf area index from litter collection: Impact of specific leaf area variability within a beech stand. *Can. J. Remote Sens.* **2003**, *29*, 371–380. [CrossRef]
41. Irons, J.R.; Dwyer, J.L.; Barsi, J.A. The next Landsat satellite: The Landsat Data Continuity Mission. *Remote Sens. Environ.* **2012**, *122*, 11–21. [CrossRef]
42. Doxani, G.; Vermote, E.; Roger, J.C.; Gascon, F.; Adriaensen, S.; Frantz, D.; Hagolle, O.; Hollstein, A.; Kirches, G.; Li, F.; et al. Atmospheric Correction Inter-Comparison Exercise. *Remote Sens.* **2018**, *10*, 352. [CrossRef]
43. Claverie, M.; Ju, J.; Masek, J.G.; Dungan, J.L.; Vermote, E.F.; Roger, J.C.; Skakun, S.V.; Justice, C. The Harmonized Landsat and Sentinel-2 surface reflectance data set. *Remote Sens. Environ.* **2018**, *219*, 145–161. [CrossRef]
44. Verrelst, J.; Rivera, J.P.; Veroustraete, F.; Muñoz-Marí, J.; Clevers, J.G.; Camps-Valls, G.; Moreno, J. Experimental Sentinel-2 LAI estimation using parametric, non-parametric and physical retrieval methods—A comparison. *ISPRS J. Photogramm. Remote Sens.* **2015**, *108*, 260–272. [CrossRef]
45. Lehnert, L.W.; Meyer, H.; Bendix, J. hsdar: Manage, Analyse and Simulate Hyperspectral Data in R, R Package Version 0.5.1. Available online: https://rdrr.io/cran/hsdar/ (accessed on 13 January 2020).
46. Feret, J.B.; François, C.; Asner, G.P.; Gitelson, A.A.; Martin, R.E.; Bidel, L.P.R.; Ustin, S.L.; le Maire, G.; Jacquemoud, S. PROSPECT-4 and 5: Advances in the leaf optical properties model separating photosynthetic pigments. *Remote Sens. Environ.* **2008**, *112*, 3030–3043. [CrossRef]

47. Byrd, R.H.; Lu, P.; Nocedal, J.; Zhu, C. A Limited Memory Algorithm for Bound Constrained Optimization. *SIAM J. Sci. Comput.* **1995**, *16*, 1190–1208. [CrossRef]
48. Lauvernet, C.; Baret, F.; Hascoët, L.; Buis, S.; Le Dimet, F.X. Multitemporal-patch ensemble inversion of coupled surface-atmosphere radiative transfer models for land surface characterization. *Remote Sens. Environ.* **2008**, *112*, 851–861. [CrossRef]
49. Atzberger, C.; Richter, K. Spatially constrained inversion of radiative transfer models for improved LAI mapping from future Sentinel-2 imagery. *Remote Sens. Environ.* **2012**, *120*, 208–218. [CrossRef]
50. Gastellu-Etchegorry, J.P.; Demarez, V.; Pinel, V.; Zagolski, F. Modeling radiative transfer in heterogeneous 3D vegetation canopies. *Remote Sens. Environ.* **1996**, *58*, 131–156. [CrossRef]
51. Gastellu-Etchegorry, J.P.; Lauret, N.; Yin, T.; Landier, L.; Kallel, A.; Malenovsky, Z.; Bitar, A.A.; Aval, J.; Benhmida, S.; Qi, J.; et al. DART: Recent Advances in Remote Sensing Data Modeling With Atmosphere, Polarization, and Chlorophyll Fluorescence. *IEEE J. Sel. Top. Appl. Earth Obs. Remote Sens.* **2017**, *10*, 2640–2649. [CrossRef]
52. Gastellu-Etchegorry, J.P.; Martin, E.; Gascon, F. DART: A 3D model for simulating satellite images and studying surface radiation budget. *Int. J. Remote Sens.* **2004**, *25*, 73–96. [CrossRef]
53. Demarez, V.; Gastellu-Etchegorry, J. A Modeling Approach for Studying Forest Chlorophyll Content. *Remote Sens. Environ.* **2000**, *71*, 226–238. [CrossRef]
54. Malenovský, Z.; Homolová, L.; Zurita-Milla, R.; Lukeš, P.; Kaplan, V.; Hanuš, J.; Gastellu-Etchegorry, J.P.; Schaepman, M.E. Retrieval of spruce leaf chlorophyll content from airborne image data using continuum removal and radiative transfer. *Remote Sens. Environ.* **2013**, *131*, 85–102. [CrossRef]
55. Banskota, A.; Serbin, S.P.; Wynne, R.H.; Thomas, V.A.; Falkowski, M.J.; Kayastha, N.; Gastellu-Etchegorry, J.P.; Townsend, P.A. An LUT-Based Inversion of DART Model to Estimate Forest LAI from Hyperspectral Data. *IEEE J. Sel. Top. Appl. Earth Obs. Remote Sens.* **2015**, *8*, 3147–3160. [CrossRef]
56. Nagol, J.R.; Sexton, J.O.; Kim, D.H.; Anand, A.; Morton, D.; Vermote, E.; Townshend, J.R. Bidirectional effects in Landsat reflectance estimates: Is there a problem to solve? *ISPRS J. Photogramm. Remote Sens.* **2015**, *103*, 129–135. [CrossRef]
57. McKay, M.D.; Beckman, R.J.; Conover, W.J. Comparison of Three Methods for Selecting Values of Input Variables in the Analysis of Output from a Computer Code. *Technometrics* **1979**, *21*, 239–245. [CrossRef]
58. Sobol', I. On sensitivity estimation for nonlinear mathematical models. *Math. Model.* **1990**, *2*, 112–118.
59. Jansen, M.J. Analysis of variance designs for model output. *Comput. Phys. Commun.* **1999**, *117*, 35–43. [CrossRef]
60. Saltelli, A.; Annoni, P.; Azzini, I.; Campolongo, F.; Ratto, M.; Tarantola, S. Variance based sensitivity analysis of model output. Design and estimator for the total sensitivity index. *Comput. Phys. Commun.* **2010**, *181*, 259–270. [CrossRef]
61. Weiss, M.; Baret, F. S2ToolBox Level 2 Products: LAI, FAPAR, FCOVER, Version 1.1. Available online: step.esa.int/docs/extra/ATBD_S2ToolBox_L2B_V1.1.pdf (accessed on 13 January 2020).
62. Morton, D.C.; Nagol, J.; Carabajal, C.C.; Rosette, J.; Palace, M.; Cook, B.D.; Vermote, E.F.; Harding, D.J.; North, P.R.J. Amazon forests maintain consistent canopy structure and greenness during the dry season. *Nature* **2014**, *506*, 221–224. [CrossRef]
63. Brede, B.; Suomalainen, J.; Bartholomeus, H.; Herold, M. Influence of solar zenith angle on the enhanced vegetation index of a Guyanese rainforest. *Remote Sens. Lett.* **2015**, *6*, 972–981. [CrossRef]
64. Teets, D. Predicting sunrise and sunset times. *Coll. Math. J.* **2003**, *34*, 317–321. [CrossRef]
65. Breiman, L. Random forests. *Mach. Learn.* **2001**, *45*, 5–32. [CrossRef]
66. Belgiu, M.; Drăgu, L. Random forest in remote sensing: A review of applications and future directions. *ISPRS J. Photogramm. Remote Sens.* **2016**, *114*, 24–31. [CrossRef]
67. Smola, A.J.; Schölkopf, B. A tutorial on support vector regression. *Stat. Comput.* **2004**, *14*, 199–222. [CrossRef]
68. Verrelst, J.; Rivera, J.P.; Moreno, J.; Camps-Valls, G. Gaussian processes uncertainty estimates in experimental Sentinel-2 LAI and leaf chlorophyll content retrieval. *ISPRS J. Photogramm. Remote Sens.* **2013**, *86*, 157–167. [CrossRef]
69. Jupp, D.L.B.; Culvenor, D.S.; Lovell, J.L.; Newnham, G.J.; Strahler, A.H.; Woodcock, C.E. Estimating forest LAI profiles and structural parameters using a ground-based laser called 'Echidna'. *Tree Physiol.* **2009**, *29*, 171–181. [CrossRef]

70. Calders, K.; Origo, N.; Disney, M.; Nightingale, J.; Woodgate, W.; Armston, J.; Lewis, P. Variability and bias in active and passive ground-based measurements of effective plant, wood and leaf area index. *Agric. For. Meteorol.* **2018**, *252*, 231–240. [CrossRef]
71. Jonckheere, I.; Fleck, S.; Nackaerts, K.; Muys, B.; Coppin, P.; Weiss, M.; Baret, F. Review of methods for in situ leaf area index determination Part I. Theories, sensors and hemispherical photography. *Agric. For. Meteorol.* **2004**, *121*, 19–35. [CrossRef]
72. Woodgate, W.; Jones, S.D.; Suarez, L.; Hill, M.J.; Armston, J.D.; Wilkes, P.; Soto-Berelov, M.; Haywood, A.; Mellor, A. Understanding the variability in ground-based methods for retrieving canopy openness, gap fraction, and leaf area index in diverse forest systems. *Agric. For. Meteorol.* **2015**, *205*, 83–95. [CrossRef]
73. Leuschner, C.; Voß, S.; Foetzki, A.; Clases, Y. Variation in leaf area index and stand leaf mass of European beech across gradients of soil acidity and precipitation. *Plant Ecol.* **2006**, *186*, 247–258. [CrossRef]
74. Percival, G.C.; Keary, I.P.; Noviss, K. The potential of a chlorophyll content SPAD meter to quantify nutrient stress in foliar tissue of sycamore (Acer pseudoplatanus), English oak (Quercus robur), and European beech (Fagus sylvatica). *Arboric. Urban For.* **2008**, *34*, 89–100.
75. Buddenbaum, H.; Stern, O.; Paschmionka, B.; Hass, E.; Gattung, T.; Stoffels, J.; Hill, J.; Werner, W. Using VNIR and SWIR field imaging spectroscopy for drought stress monitoring of beech seedlings. *Int. J. Remote Sens.* **2015**, *36*, 4590–4605. [CrossRef]
76. Hintze, J.L.; Nelson, R.D. Violin Plots: A Box Plot-Density Trace Synergism. *Am. Stat.* **1998**, *52*, 181. [CrossRef]
77. Schlerf, M.; Atzberger, C. Inversion of a forest reflectance model to estimate structural canopy variables from hyperspectral remote sensing data. *Remote Sens. Environ.* **2006**, *100*, 281–294. [CrossRef]
78. Schlerf, M.; Atzberger, C. Vegetation Structure Retrieval in Beech and Spruce Forests Using Spectrodirectional Satellite Data. *IEEE J. Sel. Top. Appl. Earth Obs. Remote Sens.* **2012**, *5*, 8–17. [CrossRef]
79. Atzberger, C. Development of an invertible forest reflectance model The INFORM-Model. In Proceedings of the 20th EARSeL Symposium, A Decade of Trans-European Remote Sensing Cooperation, Dresden, Germany, 14–16 June 2000; pp. 39–44.

© 2020 by the authors. Licensee MDPI, Basel, Switzerland. This article is an open access article distributed under the terms and conditions of the Creative Commons Attribution (CC BY) license (http://creativecommons.org/licenses/by/4.0/).

Article

Assessment of UAV-Onboard Multispectral Sensor for Non-Destructive Site-Specific Rapeseed Crop Phenotype Variable at Different Phenological Stages and Resolutions

Sadeed Hussain [1], Kaixiu Gao [1], Mairaj Din [2], Yongkang Gao [1], Zhihua Shi [1,3] and Shanqin Wang [1,3,*]

1. College of Resource and Environment, Huazhong Agricultural University, Wuhan 430070, China; shyousafzai@webmail.hzau.edu.cn (S.H.); gaokaixiu@webmail.hzau.edu.cn (K.G.); xingyingao@webmail.hzau.edu.cn (Y.G.); pengshi@mail.hzau.edu.cn (Z.S.)
2. Department of Agronomy, University of Agriculture Faisalabad, Burewala 61010, Pakistan; dmairaj@uaf.edu.pk
3. Key Laboratory of Arable Land Conservation (Middle and Lower Reaches of Yangtze River), Ministry of Agriculture, Wuhan 430070, China
* Correspondence: sqwang@mail.hzau.edu.cn; Tel.: +86-13628680648

Received: 7 January 2020; Accepted: 22 January 2020; Published: 26 January 2020

Abstract: Unmanned aerial vehicles (UAVs) equipped with spectral sensors have become useful in the fast and non-destructive assessment of crop growth, endurance and resource dynamics. This study is intended to inspect the capabilities of UAV-onboard multispectral sensors for non-destructive phenotype variables, including leaf area index (LAI), leaf mass per area (LMA) and specific leaf area (SLA) of rapeseed oil at different growth stages. In addition, the raw image data with high ground resolution (20 cm) were resampled to 30, 50 and 100 cm to determine the influence of resolution on the estimation of phenotype variables by using vegetation indices (VIs). Quadratic polynomial regression was applied to the quantitative analysis at different resolutions and growth stages. The coefficient of determination (R^2) and root mean square error results indicated the significant accuracy of the LAI estimation, wherein the highest R^2 values were attained by RVI = 0.93 and MTVI2 = 0.89 at the elongation stage. The noise equivalent of sensitivity and uncertainty analyses at the different growth stages accounted for the sensitivity of VIs, which revealed the optimal VIs of RVI, MTVI2 and MSAVI in the LAI estimation. LMA and SLA, which showed significant accuracies at (R^2 = 0.85, 0.81) and (R^2 = 0.85, 0.71), were estimated on the basis of the predicted leaf dry weight and LAI at the elongation and flowering stages, respectively. No significant variations were observed in the measured regression coefficients using different resolution images. Results demonstrated the significant potential of UAV-onboard multispectral sensor and empirical method for the non-destructive retrieval of crop canopy variables.

Keywords: unmanned aircraft vehicle; multispectral sensor; vegetation indices; rapeseed crop; site-specific farming

1. Introduction

Applications of the unmanned aerial vehicle (UAV) platform in precision agriculture (PA) offered a precise and reliable solution for the optimisation of crop monitoring and management [1]. UAV in PA is a new but reliable remote sensing tool that can acquire high spectral, spatial and temporal resolution data and has the advantages of low cost, flexible platform and bird's eye view for rapid data collection [2,3]. UAV in PA can customise the imaging sensor to meet the spectral and spatial requirements and to achieve fast utilisation [4]. Compared with other remote sensing platforms, UAVs can fly at low altitudes and can capture high-resolution imagery, which offers very detailed spectral and spatial descriptions of the field [1,4]. For highly dynamic vegetation monitoring, the UAV-onboard sensor allows the precise determination of plant location during the growing season to act with corrective measures [2,4,5]. The timely monitoring of crop dynamics is critical in addressing various forms of environmental issues, agricultural practices, degasses and pest control that can eventually lead to enhanced production [6–8]. UAV, sensors and other geospatial techniques (e.g., GIS, remote sensing and GPS) help in determining field variations and optimum fertilisation, disease diagnosis and pest control for sustainable production; this type of determination process is called PA [2,9]. The application of remote sensing in agriculture has become an important research direction because it provides valuable information in terms of agronomic parameters, which facilitate sustainable crop monitoring. Remote sensing data also produce repeated and useful non-destructive crop biophysical attributes for PA [7,10]. Gathering these information required for the enhancement of agricultural production in PA is the key issue that requires the use of a proper design and decision-making system [2,11].

The data collected through the aerial and satellite approaches of remote sensing are useful; however, these platforms are limited by temporal and spatial resolutions, cloud covers and operational costs [12,13]. During the last decade, UAV has been typically used in crop monitoring with high-spatial resolution imagery, low operational cost and near-real-time data acquisition based on powerful sensor-bearing ideal platforms for mapping and monitoring [2,14,15]. In the practical applications of remote sensing techniques in PA, various agricultural equipment units have been developed but require major improvements to meet the variation requirements of seasonal patterns in numerous indices that are used in agronomic evaluations [7].

Leaf area index (LAI), leaf mass per area (LMA) and specific leaf area (SLA) are leaf functional traits that provide information about vegetation canopies in the functional diversity assessment and quantification of physiological processes to prescribe optimum management strategies [16]. These traits correspond to the analytical variables of the plant physiological process, such as growth rates, photosynthetic capacity and plant life strategies [17]. LAI refers to the one-sided green leaf area per unit of ground surface area and is a dimensionless parameter that characterises photosynthesis and respiration and defines the functional link to the canopy spectral reflectance [8,10,18]. LMA is calculated as the ratio of leaf dry mass to leaf area, which is a key biophysical variable involved in plant light capture and carbon gain [19]. Numerous studies identified its significance to the photosynthesis–nitrogen relationship and other essential characteristics, such as the leaf mass-based nitrogen content, which can be inferred from the remote sensing data as the leaf dry mass composed of several organic elements that absorb radiation at a fixed wavelength [19,20]. SLA refers to the leaf area per unit of dry leaf mass or the leaf mass per unit area, which are usually expressed in m^2/kg. SLA is an important trait associated to the plant growth rate, which provides information about photosynthetic capacity and leaf nitrogen variations [17]. This trait is indicative of the physiological processes, such as growth rate, light capture and survival [21].

Crop growth and yield models require the estimates of crop biophysical parameters and the association of infrared reflectance to the biophysical parameter of crop canopies that offer the tool for linking multispectral remote sensing to crop growth behaviour [8]. For instance, the soil and water assessment tool model uses the maximum LAI as the key parameter that influence crop growth and watershed flow routing; this parameter can also be changed according to the agronomists' feedback [22,23]. Agricultural studies need to compute and monitor the biochemical and biophysical properties of plants for crop growth, chlorophyll content, nitrogen content and LAI that provide clues to crop health and productivity. Compared with direct field surveys, the remote sensing techniques for the acquisition of these parameters are performed in real time, are non-destructive and provide the spatial details for the measurement and monitoring of these parameters. The empirical relationship between the biophysical parameters and spectral vegetation indices (VIs) has been developed and used to address the relationship between crop traits and canopy reflectance. The empirical relationship of spectral data is effective in predicting the crop biophysical traits because of its simplicity and ability to capture various canopy variations, and this approach is widely used because of its ease in computation [24].

VIs have been developed as a mathematical combination of various spectral bands of the electromagnetic spectrum and are related to various canopy parameters, which can enhance the vegetation signals by reducing the atmospheric and soil effects. The spectral VIs measure vegetation activity and demonstrate the spatial variations of the different seasons through space [7,25]. Substantial developments have been made in the interpretation and analysis of the spectral VIs, which have been applied from the field to the global level [25]. However, these indices are species-specific and are therefore not robust in different species with various leaf and canopy architectures [2,7,25].

Several studies regarding the application of UAV in rice crop yield and biomass estimation have been published [13]. The UAV application in sunflower crop has shown a significant correlation between normalized difference vegetation index (NDVI) and grain yield biomass and applied nitrogen content [2], whereas the UAV platform ('VIPtero') for onboard multispectral camera in vineyard management demonstrated a clear crop heterogeneity, which was consistent with the ground observations [9]. The multispectral sensor mounted on a UAV over potato crops demonstrated a significant correlation between NDVI and green normalized vegetation index (GNDVI) with LAI, plant cover and chlorophyll content [12].

Although the application of UAV has progressed considerably, it still has many limitations, such as payload, cost and operation consistency in agriculture. Thus, a substantial amount of work is still required. UAVs capture numerous high-resolution images that require lengthy processing times, high storage devices and high labour intensity even for professionals because of its low-altitude image acquisition capability [26,27]. Fine-resolution VIs demonstrated a significant influence on the accuracy of prediction models [28]. Although studies have determined that high-resolution images might not be the optimal choice for environmental variable and concluded that fine-resolution VIs obtain a similar prediction model accuracy [2,29,30]. The measuring scale is dependent on the association between biophysical traits and VIs from images with different resolutions, which can help select the appropriate resolution for site-specific management.

Remotely sensed VIs can offer significant information derivatives to agronomic parameters at the field scale. Therefore, the sensor that can monitor crops during the growth season at high spectral, spatial and temporal resolutions will provide useful information in the efficient and sustainable crop management with site-specific basis. Amongst the phenotypical parameters derived using remotely sensed data, LMA and SLA have received minimal attention. However, the retrieval of the mentioned parameters using the radiative transfer model (RTM) inversion could be ill-advised and computationally demanding because it requires a number of leaf and canopy variables. In this study, we optimised and validated numerous VIs for LAI and leaf dry weight (DW) prediction and investigated the potential of empirically estimated LAI and DW for the estimation of the two leaf functional traits, namely, LMA and SLA. Therefore, our study mainly aims to (1) evaluate the potential of a multispectral sensor onboard a UAV in relation to crop biophysical parameters, such as LAI and DW, during the growing season over rapeseed crops at different phenological stages, (2) assess the spectrally predicted leaf DW and LAI for the calculation of LMA and SLA at different phenological stages and (3) assess the effects of resolution on the LAI prediction over the phenological stages to provide simple, rapid and useful information for PA.

2. Materials and Methods

2.1. Study Area

The field experiment was conducted at the Zishi experimental station of Huazhong Agricultural University in Jingzhou (30°11′28.62″ N, 112°23′31.61″ E) located in the south-central part of Hubei Province, China from 18 December 2017 to 10 March 2018. This area has a subtropical monsoon climate with an average rainfall of 1100–1300 mm, average annual temperature of 15.9 °C to 16.6 °C, hydrothermal synchronisation and consistent good climatic condition suitable for agricultural activities. Winter rapeseed was used as the experimental material in this study. To imitate the high variability of growing conditions and investigate the spectral VIs for the estimation of crop phenotype parameters, the data were collected from 16 different fields and 170 point locations. The experimental fields and sampling points are presented in Figure 1.

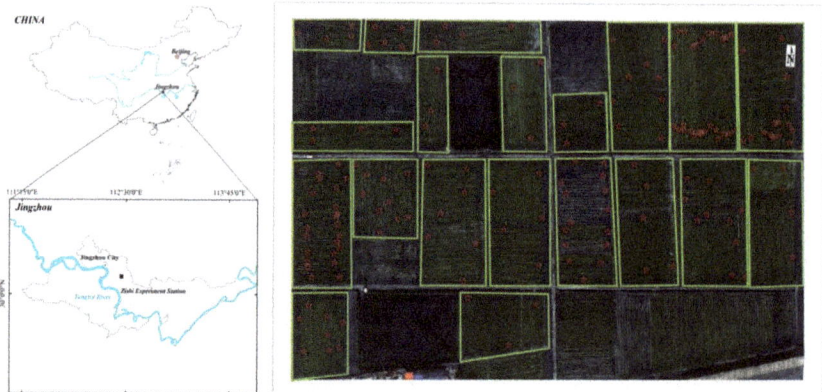

Figure 1. Zishi experimental station and field view.

2.2. Biophysical Parameter Measurements

The non-destructive LAI was measured using a plant canopy analyser (SunScan, Probe type SS1, Delta-T Devices, England) over all point locations (Figure 1) after 55 days of plantation on 18 December 2017, after 79 day of plantation on 11 January 2018, 106 days after plantation on 7 February 2018 and after 138 day of plantation on 10 March 2018 according to the crop growth stages from seedling to maturity during the growing season. Measurements were obtained on clear cloudless days. At each phenological stage, 30 subsampling points were selected randomly, and four plants were sampled destructively with their roots. The plant density was also recorded in each point, location, latitude and longitude. The samples were stored in bags and moved to the laboratory. The leaves, stems and roots of the samples were separated to obtain their fresh weights and then placed in an oven at 70 °C for 48 hours to obtain the dry mass. The fresh and dry weights of each leaf sample were determined using a high-precision digital scale. LMA and SLA were calculated as follows:

$$LMA = Dw/LA \qquad (1)$$

where Dw and LA are the leaf DW and leaf area, respectively.

$$SLA = LA/Dw \qquad (2)$$

where LA and Dw are the leaf area and the corresponding Dw, respectively [17,31].

2.3. UAV Image Acquisition

Multispectral images were also acquired on the days of LAI measurement using the UAV platform. Prior to the flight campaign, the camera was mounted on the UAV under suitable working condition. A lightweight RedEdge MicaSense multispectral camera (MicaSense RedEdge®) onboard a fixed-wing UAV (T-EZ; Golden Wing UAS Co., Ltd.; Chengdu, China) was used to capture the field images shown in Figure S1. A stable platform was used to adjust the camera pointing towards the nadir. The spectral bands covered the wavelength intervals 450 nm centre, 20 nm bandwidth (blue), 560 nm centre, 20 nm bandwidth (green), 668 nm centre, 10 nm bandwidth (red), 717 nm centre, 10 nm bandwidth (RedEdge), 840 nm centre and 40 nm bandwidth (near-infrared). After take-off, the UAV was programmed to follow the predefined route and complete the flight campaign. The flight altitude was 300 m and resulted in a ground resolution of 20 cm.

2.4. Image Processing

The raw multispectral images acquired by the RedEdge camera were processed in terms of image mosaicking, vignetting correction and raw digital number (DNs) for the reflectance conversion using the Pix4Dmapper (Pix4D SA, Switzerland) software. The raw digital numbers of the acquired images were firstly converted into radiance and then into surface reflectance based on the linear regression models by using the calibration target data, wherein the surface reflectance is a linear function of the digital numbers [32]. In each stage, 1024 images were captured and mosaicked from the study area, and the georeferenced images of the study area were acquired at the end of the fourth crop season. Moreover, the raw images with the ground resolution of 20 cm were resampled to 30, 50 and 100 cm by using the pixel aggregate interpolation method to assess the influence of image resolution on VIs. Lastly, the VIs were calculated.

2.5. Multispectral VIs

The canopy spectral data were used to develop the VIs, which are sensitive to canopy structure, pigments and chlorophyll, for the LAI, LMA and SLA estimation in rapeseed crops. Nine VIs were calculated, several of which have been proposed as surrogates for LAI estimation [10,25,32]. The calculated VIs, their formula and description are listed in Table 1. For all VIs, the images with the same resolution were stacked, and the mean VI values centred on the 1 m² location were extracted for 160 point locations in ENVI 5.1 (Exelis Visual Information Solution, Inc.; Boulder, CO, USA).

Table 1. Description and formulas of the investigated VIs.

Indices	Formulas	Description	References
RVI	$RVI = \rho_{nir}/\rho_{red}$	Sensitive to nitrogen	[10,33]
NDVI	$NDVI = (\rho_{nir} - \rho_{red})/(\rho_{nir} + \rho_{red})$	Structure (LAI, fraction) Chlorophyll content	[34]
GNDVI	$GNDVI = (\rho_{nir} - \rho_{green})/(\rho_{nir} + \rho_{green})$	Biomass, LAI, photosynthesis and plant stress	[7,35]
BNDVI	$BNDVI = (\rho_{nir} - \rho_{blue})/(\rho_{nir} + \rho_{blue})$	Chlorophyll content	[36]
SAVI	$SAVI = \frac{(\rho_{nir} - \rho_{red})}{(\rho_{nir} + \rho_{red} + 0.5)}(1 + 0.5)$	Structure (LAI, fraction)	[37,38]
OSAVI	$OSAVI = (1 + 0.16)\frac{(\rho_{nir} - \rho_{red})}{(\rho_{nir} + \rho_{red} + 0.16)}$	Structure (LAI, fraction)	[39,40]
MSAVI	$MSAVI = \rho_{nir} + 0.5 - (0.5\sqrt{(2*\rho_{nir}+1)^2 - 8(\rho_{nir}-(2*\rho_{red})})$	Structure (LAI, sensitive to canopy effects)	[41]
MSAVI2	$MSAVI2 = \frac{1}{2}\left[2\rho_{nir}+1-\sqrt{(2\rho_{nir}+1)^2-8(\rho_{nir}-\rho_{red})}\right]$	Structure (LAI, fraction)	[41]
MTVI2	$MTVI2 = \frac{1.5[1.2(\rho_{nir}-\rho_{green})-(2.5\rho_{red}-\rho_{green})]}{\sqrt{(2\rho_{nir}+1)^2-(6\rho_{nir}-5\sqrt{\rho_{red}})-0.5}}$	Structure (Sensitive to LAI, resistant to chlorophyll influence)	[38]

Note: RVI, ratio vegetation index; NDVI, normalized difference vegetation index; GNDVI, green normalized vegetation index; BNDVI, blue normalized difference vegetation index; SAVI, soil-adjusted vegetation index; OSAVI, optimised soil-adjusted vegetation index; MSAVI, modified soil-adjusted vegetation index; MSAVI2, modified soil-adjusted vegetation index 2; MTVI2, modified triangular vegetation index 2.

2.6. Statistical Analysis

Regression models were used for the quantitative analysis between spectral VIs as independent and LAI and leaf DW as dependent variables at the different phenological stages. A total of 140 and 30 sample points for LAI and LD were used for model calibration, respectively, whereas 30 sample points each were used for validation. To assess the model performance, the coefficient of determination (R^2), root mean square error (RMSE) and relative RMSE (RRMSE) were used and calculated using Equations. (3), (4) and (5), respectively [18,42,43]. The large R^2 value and low RMSE and RRMSE values indicated the high precision and accuracy of the model as follows:

$$R^2 = 1 - \frac{\sum_{i=1}^{n}(y_i - \hat{y}_i)^2}{\sum_{i=1}^{n}(y_i - \bar{y})^2} \qquad (3)$$

$$RMSE = \sqrt{\frac{\sum_{i=1}^{n}(y_i - \hat{y}_i)^2}{n}} \tag{4}$$

$$RRMSE = \frac{RMSE}{\bar{y}} \times 100 \tag{5}$$

where R^2 and RMSE measure the relationship of the fitting function. However, for the nonlinear functions amongst LAI and VIs, the R^2 values may be misleading because sensitivity changes continuously [10]. Thus, a different accuracy matric noise equivalent (NE) was needed to evaluate and determine the LAI estimation accuracy. NE also accounts for the slope and scattering of data points in the best-fit function as follows [10,25,44]:

$$NE\Delta LAI = \frac{RMSE\{VI\ VS.\ LAI\}}{d(VI)/d(LAI)} \tag{6}$$

where RMSE is the root mean square error of the best-fit function between the spectral Vis and LAI, and $d(VI)/d(LAI)$ is the first-order derivative of the observed relationship.

3. Results

3.1. Association of VIs to the Phenolgical Stages

A wide range of variations in LAI, DW, LMA and SLA due to different canopy architectures as well as variations due to leaf sizes and shapes at different phenological stages was recorded. Table 2 presents the descriptive statistics of canopy LAI, DW, LMA and SLA of rapeseed oil at the phenological stages in the Zishi experimental station.

All the VIs investigated at the different growth stages revealed close associations with the crop reflectance characteristics. The spectral VIs demonstrated significant variations in all the phenological stages from seedling to maturity, and the results are compiled in Table 3. The VIs attained its peak values at the elongation stage and started to decline afterwards. All the VIs demonstrated a significant relation with LAI at the seedling stage. Furthermore, the MSAVI2, MTVI2 and SAVI showed the most sensitive VIs with an R^2 value of 0.87, followed by MSAVI ($R^2 = 0.86$), whilst NDVI and OSAVI achieved $R^2 = 0.85$, GNDVI $R^2 = 0.82$ and BNDVI obtained an R^2 value of 0.81 at the seedling stage. RVI, NDVI, MSAVI and MTVI2 showed the most influenced VIs to LAI from the seedling to maturity stages and attained the maximum R^2 (0.93–0.89, 0.88–0.86, 0.87–0.86 and 0.89–0.86) from the elongation to flowering stages. The minimum R^2 values at the maturity stage were expected due to crop senescence.

Table 2. Descriptive statistics of canopy LAI, DW, LMA and SLA of rapeseed oil at the different phenological stages in the Zishi experimental station.

Phonological stage	Statistics	LAI (m^{2}m^{-2})	DW (g·cm^{-2})	LMA (g·cm^{-2})	SLA (cm^{2}g^{-1})
Seedling stage	Minimum value	0.2	0.005	0.004	0.001
	Maximum value	4.9	0.051	0.037	0.055
	Mean value	1.72	0.025	0.016	0.012
	Standard deviation	1.13	0.014	0.009	0.012
Elongation stage	Minimum value	0.2	0.008	0.003	0.001
	Maximum value	5.03	0.083	0.087	0.024
	Mean value	2.23	0.038	0.023	0.005
	Standard deviation	1.15	0.019	0.014	0.004
Flowering stage	Minimum value	0.1	0.010	0.002	0.001
	Maximum value	4.83	0.044	0.082	0.032
	Mean value	1.60	0.027	0.023	0.006
	Standard deviation	0.99	0.010	0.014	0.006
Maturity stage	Minimum value	0.1	0.001	0.002	0.001
	Maximum value	3.72	0.052	0.046	0.045
	Mean value	1.74	0.022	0.013	0.007
	Standard deviation	0.70	0.010	0.009	0.008

Table 3. Relationship of spectral VIs at the different phenological stages.

	Phenological stages							
	R^2				RMSE			
VIs	Seedling	Elongation	Flowering	Maturity	Seedling	Elongation	Flowering	Maturity
RVI	0.86	0.93	0.89	0.45	0.44	0.30	0.32	0.49
NDVI	0.86	0.88	0.87	0.41	0.47	0.40	0.35	0.54
GNDVI	0.82	0.87	0.82	0.41	0.50	0.41	0.42	0.51
BNDVI	0.82	0.86	0.79	0.44	0.50	0.42	0.44	0.50
SAVI	0.86	0.85	0.83	0.45	0.43	0.44	0.41	0.49
OSAVI	0.85	0.86	0.84	0.49	0.45	0.42	0.40	0.48
MSAVI	0.85	0.87	0.86	0.47	0.44	0.40	0.37	0.49
MSAVI2	0.86	0.85	0.82	0.48	0.43	0.44	0.42	0.48
MTVI2	0.88	0.89	0.86	0.52	0.40	0.38	0.36	0.46

3.2. Evaluation of Spectral VIs for Estimation of Rapeseed LAI

The optimal spectral VIs for the estimation of the LAI best-fit model relationship for VIs and LAI at all the phenological stages are shown in Table 4. The spectral pattern variations and close association with crop reflectance revealed the potential capability of the explored VIs in LAI estimation. All the VIs demonstrated a high significant relationship to the temporal distribution of LAI at the different growth stages, except at the maturity stage, which revealed the lowest coefficient of determination ($R^2 < 0.50$) and were omitted from the calculations. Amongst the investigated phenological stages, all the VIs attained their highest coefficient of determination at the elongation stage, whereas the highest R^2 value was achieved by RVI ($R^2 = 0.93$), followed by MTVI2 ($R^2 = 0.89$) at the different phenological stages, as provided in Table 3. At the next flowering phenological stage, R^2 started to decrease until maturity, and the best-fit model of LAI with spectral VIs provided RVI ($R^2 = 0.89$), followed by NDVI ($R^2 = 0.87$) and MTVI2, MSAVI ($R^2 = 0.86$) and OSAVI ($R^2 = 0.84$), whereas SAVI attained $R^2 = 0.83$ and GNDVI and MSAVI2 attained $R^2 = 0.82$. The least R^2 value was attained by BNDVI ($R^2 = 0.79$). MSAVI and MTVI2 have the modifying factor to deal with the saturation problem when the LAI surpasses the saturation level (LAI > 3), whilst the sensitivity of NDVI suffered and levelled out.

Table 4. Quadratic polynomial regression and summary statistics of relationship between LAI and studied VIs based on calibration data set at different phenological stages.

Phenological Stage	VIs	Model	R^2	RRMSE
Seedling Stage	RVI	$y = 0.0462x^2 - 0.1042x + 0.3819$	0.86	24%
	NDVI	$y = 43.705x^2 - 47.037x + 13.056$	0.84	26%
	GNDVI	$y = 65.417x^2 - 54.631x + 11.848$	0.82	28%
	BNDVI	$y = 104.91x^2 - 130.44x + 40.965$	0.82	28%
	SAVI	$y = 46.485x^2 - 30.462x + 5.4871$	0.86	24%
	OSAVI	$y = 44.048x^2 - 38.005x + 8.6542$	0.85	25%
	MSAVI	$y = 24.965x^2 - 6.9701x + 0.9874$	0.86	25%
	MSAVI2	$y = 28.964x^2 - 16.621x + 2.9121$	0.86	24%
	MTVI2	$y = 22.307x^2 - 4.5415x + 0.7493$	0.88	22%
Elongation Stage	RVI	$y = 0.011x^2 + 0.5195x - 1.6884$	0.93	13%
	NDVI	$y = 68.09x^2 - 79.03x + 23.531$	0.88	17%
	GNDVI	$y = 60.213x^2 - 48.337x + 10.064$	0.87	18%
	BNDVI	$y = 119.49x^2 - 142.14x + 42.779$	0.86	18%
	SAVI	$y = 31.046x^2 - 17.785x + 2.8609$	0.85	19%
	OSAVI	$y = 43.247x^2 - 37.807x + 8.7777$	0.86	18%
	MSAVI	$y = 18.643x^2 - 3.1952x + 0.433$	0.88	17%
	MSAVI2	$y = 19.239x^2 - 8.6972x + 1.2314$	0.85	19%
	MTVI2	$y = 17.139x^2 - 1.3713x + 0.3066$	0.89	16%
Flowering Stage	RVI	$y = 0.1402x^2 - 0.273x - 0.2612$	0.89	19%
	NDVI	$y = 49.059x^2 - 46.541x + 11.289$	0.87	21%
	GNDVI	$y = 42.303x^2 - 27.235x + 4.1857$	0.82	25%
	BNDVI	$y = 94.642x^2 - 102.03x + 27.677$	0.79	27%
	SAVI	$y = 27.79x^2 - 17.512x + 2.8845$	0.83	25%
	OSAVI	$y = 38.042x^2 - 31.109x + 6.6067$	0.84	24%
	MSAVI	$y = 12.155x^2 + 1.2127x - 0.3412$	0.86	22%
	MSAVI2	$y = 26.44x^2 - 17.359x + 3.1088$	0.82	25%
	MTVI2	$y = 15.947x^2 - 0.3135x + 0.0678$	0.86	22%

3.3. Sensitivity Analysis

The sensitivity analysis results demonstrated that BNDVI and GNDVI presented the maximum insensitivity, whereas OSAVI exhibited moderately higher sensitivity and was unreliable during the entire range of LAI. NDVI and SAVI exhibited higher sensitivity for LAI < 2.5 m^{2}m^{-2} and a decreasing sensitivity trend was observed for higher LAI > 3.0 m^{2}m^{-2}, whereas MSAVI2 performed better than NDVI and SAVI even at LAI > 3.0 m^{2}m^{-2} [25,45]. RVI showed higher sensitivity to LAI and is the best index for the detection of numerical variations in LAI because other indices are not robust at higher LAI values. MSAVI and MTVI2 exhibited higher and consistent sensitivity for LAI < 3.0 m^{2}m^{-2}, and a decreasing trend was observed for higher LAI > 3.0 m^{2}m^{-2} for MSAVI. Therefore, RVI and MTVI2 demonstrated the lowest NE values to ensure the highest sensitivities to LAI and the best multispectral indices for quantitatively detecting variations in LAI (Figure 2).

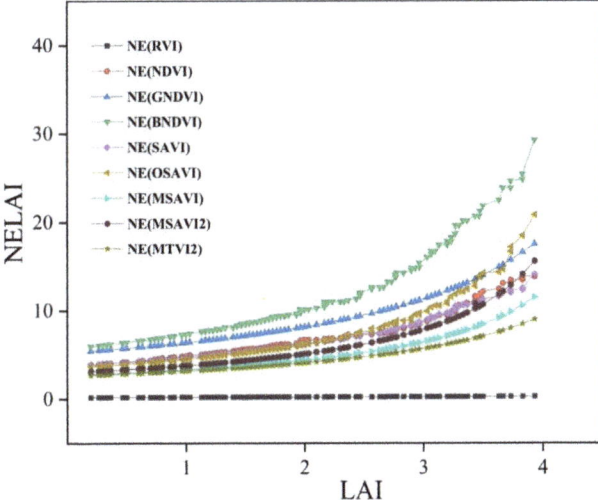

Figure 2. Sensitivity analysis for VIs evaluated for LAI estimation.

3.4. Relationship of VIs and Leaf DW

To evaluate the optimal VIs in the estimation of leaf DW (gm^{-2}), the best-fit model relationship was used between the leaf DW and VIs presented in Table 1 at the different phenological stages demonstrated in Table 5. The best-fit function demonstrated the nonlinear relationship between VIs and leaf DW with R^2 that range from 0.34 to 0.68 and maximum R^2 (0.68) attained by using SAVI at the seedling stage. At the early seedling stage, SAVI performed well with the maximum R^2 value, which was demonstrated at the early growth stage, wherein the canopy is not fully developed, and the soil contributed to the reflectance. The highest R^2 (0.75) value was obtained by RVI and GNDVI, followed by NDVI with R^2 = 0.70 at the elongation stage and then began to decline afterwards. At the flowering stage, the maximum R^2 (0.72) value was achieved by MSAVI, followed by NDVI, SAVI and MSAVI2 at 0.71. All the VIs showed similar asymptotic patterns with a certain degree of scattering from the nonlinear fits.

Table 5. Power function models and summary statistics of the relationship between LD and the explored VIs based on the calibration data set at different phenological stages.

Phenological Stage	VIs	Model	R^2	RRMSE
Seedling Stage	RVI	$y = 23.32562x^{1.31245}$	0.59	38%
	NDVI	$y = 1047.82027x^{3.91124}$	0.60	38%
	GNDVI	$y = 1803.29932x^{3.25556}$	0.53	41%
	BNDVI	$y = 1626.95299x^{5.62912}$	0.59	38%
	SAVI	$y = 1403.98804x^{2.28148}$	0.68	34%
	OSAVI	$y = 1135.78629x^{2.72497}$	0.62	37%
	MSAVI	$y = 947.33922x^{1.16294}$	0.59	38%
	MSAVI2	$y = 1014.56073x^{1.83346}$	0.63	36%
	MTVI2	$y = 725.25602x^{0.84641}$	0.45	44%
Elongation Stage	RVI	$y = 16.96029x^{1.69583}$	0.75	27%
	NDVI	$y = 2130.44779x^{5.10758}$	0.70	29%
	GNDVI	$y = 4586.364x^{4.33196}$	0.75	26%
	BNDVI	$y = 3529.73179x^{6.18505}$	0.66	31%
	SAVI	$y = 2522.0947x^{2.78668}$	0.67	31%
	OSAVI	$y = 2119.82125x^{3.48711}$	0.67	30%
	MSAVI	$y = 1879.90964x^{1.59398}$	0.68	30%
	MSAVI2	$y = 1548.72146x^{2.08531}$	0.65	31%
	MTVI2	$y = 1885.3379x^{1.47159}$	0.68	30%
Flowering Stage	RVI	$y = 18.43701x^{1.82708}$	0.70	25%
	NDVI	$y = 1884.4817x^{3.95822}$	0.71	24%
	GNDVI	$y = 2335.00995x^{3.18035}$	0.66	26%
	BNDVI	$y = 3603.56195x^{5.83675}$	0.61	28%
	SAVI	$y = 1742.20092x^{2.77482}$	0.71	24%
	OSAVI	$y = 1710.05981x^{3.23342}$	0.66	26%
	MSAVI	$y = 1496.06954x^{1.47261}$	0.72	24%
	MSAVI2	$y = 1466.35551x^{2.51978}$	0.71	24%
	MTVI2	$y = 1295.23183x^{1.1978}$	0.70	24%

3.5. Estimation LMA and SLA using Spectral VIs

The performance of all the indices were compared at the different phenological stages to select the optimum index in the prediction of LAI and DW. The best-fit regression model illustrated the good relationship in the prediction of LAI and DW (Tables 4 and 5), respectively, at nearly all growth stages, except the maturity stage, which was excluded from the data. The best-fit model equations of the optimal VIs were applied for the prediction of LAI and DW at each phenological stage and used in Equations (1) and (2) in the estimation of LMA and SLA, respectively. The spatial distributions of LAI, LMA and SLA are shown in Figures S2–S4. The optimal prediction equation was used in the LAI estimation, whereas LMA and SLA were utilised in Equations (1) and (2) for the generation of maps at all the explored phenological stages, respectively. Field measurement with the estimations from composited maps provided the validation results and illustrated the potential use of LAI, LMA and SLA derivation at all the growth stages.

3.6. Validation of LAI, LMA and SLA Estimates

To validate the calibration model for LAI and estimated LMA, SLA with the predicted LAI and DW, multispectral images and ground measured data were used whilst the determined optimal VIs were considered in the prediction of LAI and DW at each growth stage and used in the estimation of LMA and SLA. The observed and estimated LAI, LMA and SLA were compared in Figures 3–6. The predictions of optimal VIs were selected on the basis of the highest R^2 and lowest RMSE, which were plotted against the ground observed LAI, LMA and SLA. The optimal VIs, RVI and MTVI2 suggested consistency between the measured and estimated LAI ($R^2 > 0.80$) of rapeseed oil crops at all the growth stages. By contrast, the model validation demonstrated less adequate results for LMA and SLA at the seedling stage ($R^2 = 0.62, 0.25$) but realised consistent results at the elongation ($R^2 = 0.85$,

0.80) and flowering (R^2 = 0.85, 0.71) stages (Figure 6). Plotting the measured–estimated LMA values demonstrated less deviation from the 1:1 line, whereas SLA was strong under and over the prediction that occurs at higher SLA values, thereby suggesting the saturation of the spectral signal [46].

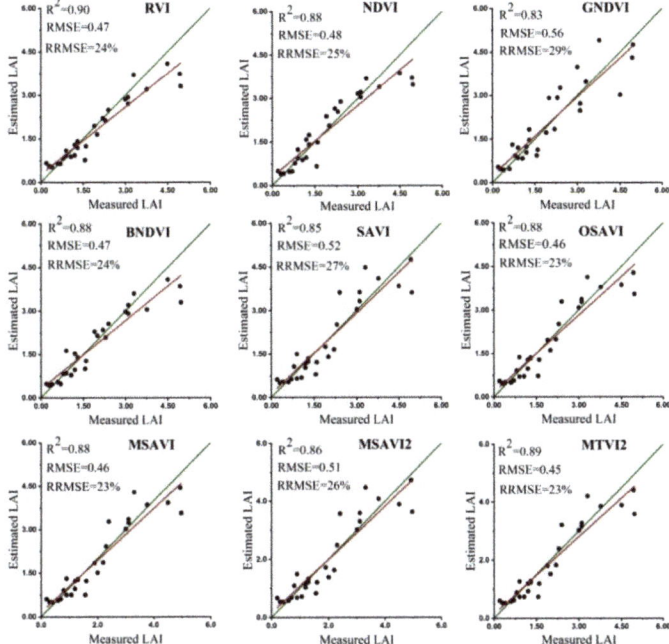

Figure 3. Estimated vs. measured LAI at the seedling stage. The green line shows the 1:1 correlation of the estimated and measured variables, whereas the red lines present the linear regression models.

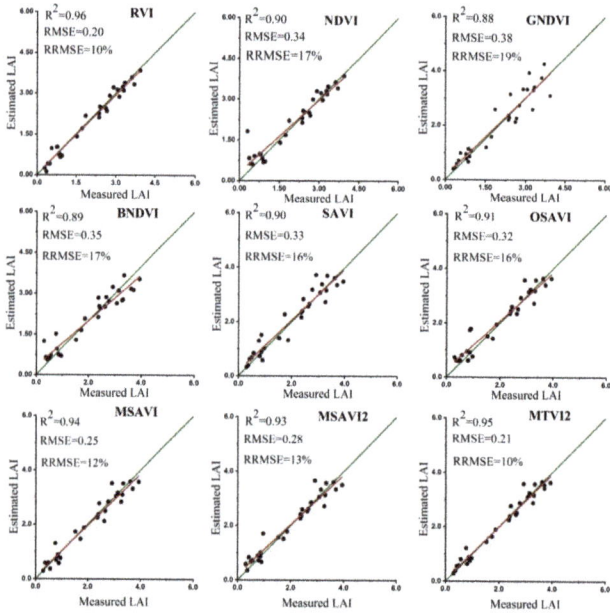

Figure 4. Estimated vs. measured LAI at the elongation stage. The green line shows the 1:1 correlation of the estimated and measured variables, whereas the red lines present the linear regression models.

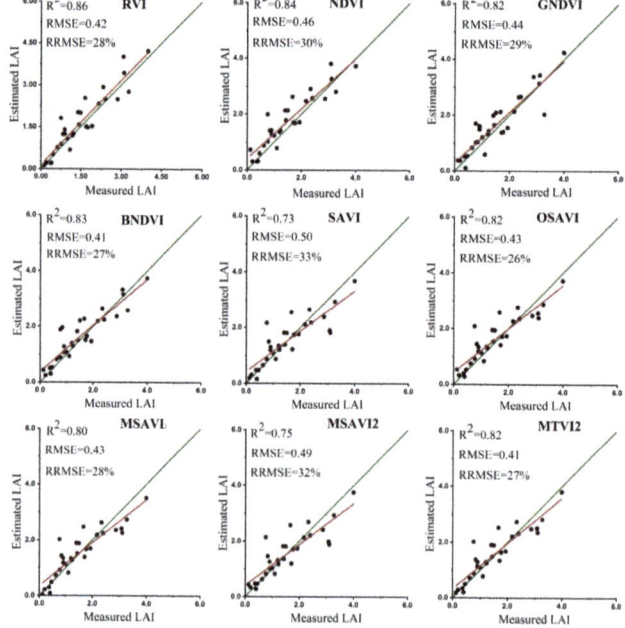

Figure 5. Estimated vs. measured LAI at the flowering stage. The green line shows the 1:1 correlation of the estimated and measured variables, whereas the red lines represent the linear regression models.

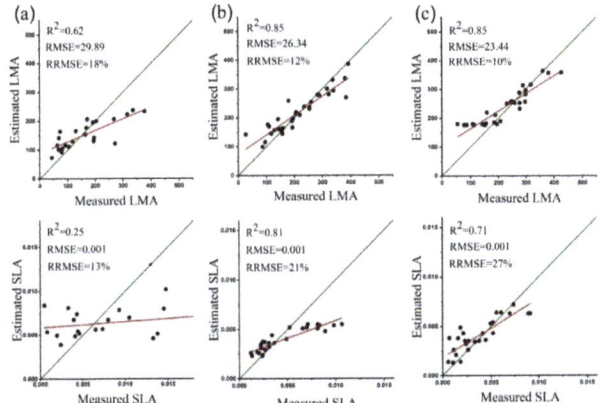

Figure 6. Estimated vs. measured LMA and SLA at the (**a**) seedling, (**b**) elongation and (**c**) flowering stages. The green line shows the 1:1 correlation of the estimated and measured variables, whereas the red lines are the linear regression models.

3.7. Evaluation of Image Resolution Effect

Four images with 20 cm pixel resolution were acquired over the rapeseed crops during the growing season and resampled to 30, 50 and 100 cm pixel resolutions. To evaluate the effects of image resolutions on the remote estimation of LAI over the rapeseed crops, the quantitative relationship between VIs and LAI were obtained at different image resolutions (20, 30, 50 and 100 cm). At all the mentioned image resolutions, the estimated LAI using VIs had no significant differences. The precision of the best-fit regression model between the LAI and VIs determined no significant variations and was similar at all the image resolutions. Thus, a significant high correlation between spectral VIs and LAI was acquired at the elongation and flowering stages, which are independent of image resolution. The best-fit model R^2 and RMSE are presented in Figure 7.

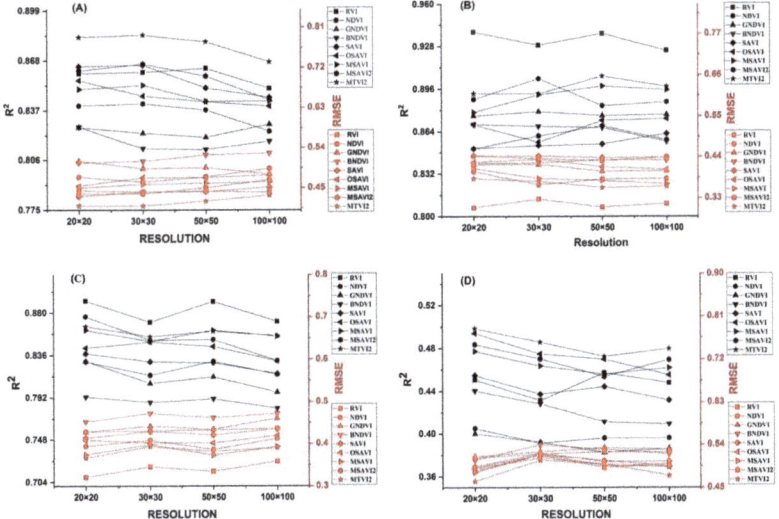

Figure 7. Best-fit model R^2 and RMSE that describe the relationship between spectral VIs and LAI with respect to the image resolution at the (**a**) seedling, (**b**) elongation, (**c**) flowering and (**d**) maturity stages.

4. Discussion

As a remote sensing platform, UAVs are ideal for mapping and monitoring in PA [9,15]. With the recent developments in UAV technologies, numerous studies have been conducted using remote sensors onboard UAVs to evaluate its application for PA and subsequently obtained considerable attention from researchers [2,9,12–15,25,31,47].

The key canopy biophysical and biochemical parameter retrievals can be attained adequately using certain spectral bands, that is, the green spectrum is ideal for leaf chlorophyll estimation [35]. Reflectance in the near-infrared (NIR) is recognised as sensitive to variant biomass and canopy structures, whereas the red and green reflectance values respond to the background variations and senescent vegetation [48].

This study aimed to determine the optimal VIs derived from UAV multispectral images that could be used for the remote and empirical non-destructive estimation of biophysical parameters, such as LAI, LMA and SLA over rapeseed crops at the different growth stages. The empirical method for the estimation of biophysical parameters is proficient and delivers precise estimations [24,32]. Moreover, this type of method has been widely applied for the improved quantitative accuracy of spectral VIs and used for the evaluation of crop attributes, such as LAI, N status and biomass for various crops [10,49]. In this study, the potential capability of multispectral VIs mentioned in Table 1 was evaluated over rapeseed crops at different growth stages. The results revealed that the crop variations from seedling to maturity are similar to the canopy reflectance characteristics. Variations in the crop canopy reflectance attains its peak value at the elongation stage and begins to decline afterwards until the maturity stage; this behaviour is due to crop senescence, as reported previously in [10,50]. Variations in canopy reflectance at the different spectral wave bands are synchronised with LAI because of the modification of leaf chlorophyll contents at different phenological stages [51]. Spectral reflectance from the canopy is affected not only by the biophysical characteristics of foliage but also by the direction of incidence radiation, canopy architecture and soil background [52]. The small variations in spectral reflectance at the early phenological stage in the visible region are likely due to the crop nitrogen content and soil water background. However, variations in NIR reflectance with variant leaf orientations at certain growth stages are due to the overlapping leaves that decrease the active photosynthetic size as the LAI reaches a plateau [53,54].

Table 2 demonstrates the highly significant relationship between LAI with all the aforementioned spectral VIs derived from the multispectral remote sensing data. However, a close relationship was obtained by RVI with LAI because LAI had a value of approximately 3 m^2m^{-2}, followed by NDVI and MTVI2 [55]. When LAI reached its saturation point, the reduced variability of red and NIR reflectance make RVI and NDVI insensitive. However, GNDVI apparently did not reach the saturation level even at LAI with moderate to high (4–5) values. The precision of the LAI estimate was better when green and blue bands were used instead of that that with the red band even at LAI values greater than 3 m^2m^{-2} [42]. The NIR band has a considerably strong impact on the relationship between LAI and spectral reflectance and must be considered under varying crop situations [7,10,56]. The nonlinear best-fit function between LAI and VIs demonstrated inconsistent sensitivity, and R^2 and RMSE values for estimating the accuracy of LAI can be misleading. Thus, the sensitivity analysis (NE) can be applied as the precision indicator to verify the performance of VIs in LAI estimation [57,58].

LMA and SLA are important in determining plant composition and eco-physiological characterisation [17,59–62]. Compared with the general empirical approach for the estimation of SLA, as reported in [32,48], its indirect estimation using the predicted LAI and DW illustrated highly significant retrieval accuracy. The SLA results showed remarkable variations between the different growth stages, as shown in Figure 6. The estimation results of SLA at the elongation and flowering stages are consistent with those of previously reported studies [63]. Soil background has less affected the correlation results as the data were captured at mature seedling (rosette) stage. The slight change in the performance of multispectral VIs from the UAV images could be awarded to the different viewing geometries and instrument spectral response functions [32].

Remotely sensed LMA ($R^2 = 0.85$, nRMSE= 0.10) and ($R^2 = 0.85$, nRMSE = 0.12) at the elongation and flowering stages was calculated with high estimation accuracy, as previously reported in [47]. The study estimated LMA using the LUT-based PROSAIL inversion method over the rapeseed crop. The LMA estimation results were also consistent with [19] who estimated LMA across a wide range of plant species through continuous wavelet analysis, and the study demonstrated remarkable implications in the prediction of dry matter content at the canopy level.

In terms of image resolution, our study concluded that the application of very fine-resolution remote sensing images does not reflect any significant difference. Thus, the fine-resolution images would not essentially increase the prediction accuracy of the regression model between LAI and spectral VIs. The precision of regression model between LAI and VIs did not change significantly by using images with different resolutions, as shown in Figure 7. The end users of UAV images need to consider the spatial, spectral and temporal resolutions and processing time for site-specific crop monitoring and management, as previously reported in [2,29,64–66].

5. Conclusions

In summary, herein we reported the application of multispectral sensors onboard UAV in acquiring images of rapeseed crops and retrieving their biophysical characteristics, such as LAI, LMA and SLA, throughout the phenological stages. The results revealed that the strongest relationship of spectral VIs was obtained in the elongation stage. The sensitivity analysis between LAI and VIs revealed that RVI, MSAVI and MTVI2 were the optimal VIs for LAI estimation. LMA and SLA results demonstrated the significant estimation accuracy by using the predicted leaf DW and LAI at the elongation and flowering stages, respectively. In terms of image resolutions, robust results can still be obtained when the maximum 100 cm resolution imagery is used for oil rapeseed crop characterisation. Therefore, high-altitude images will be preferred in obtaining a decreased number of images, which will significantly influence image acquisition and processing time.

Compared with earlier remote sensing platforms, the sensor onboard UAVs had the basic advantage of reaching the targeted site and acquiring very comprehensive information throughout the growing season. Unlike satellite- and aircraft-based remote sensing platforms, the UAV remote sensing platform demonstrated better operational advantages, such as the provision of high-spatial resolution imagery with low operational cost and near-real-time data acquisition. Although satellite data were limited because of temporal resolution, cloud cover, availability at the ideal time and the operational cost for high-resolution imagery, this technology has become an important tool for large area mapping and monitoring because of its distinctive capabilities in satellite remote sensing imagery that offers extensive information in a synoptic and frequent manner.

Supplementary Materials: The following are available online at http://www.mdpi.com/2072-4292/12/3/397/s1, Figure S1: The UAV used for image acquisition, Figure S2: Map of LAI, LMA and SLA at seedling stage, Figure S3: Map of LAI, LMA and SLA at elongation stage, Figure S4: Map of LAI, LMA and SLA at flowering stage.

Author Contributions: Conceptualization, S.S. and S.W.; Data curation, S.S.; Formal analysis, S.S., K.G., M.D. and Y.G.; Funding acquisition, S.W.; Investigation, K.G. and S.W.; Methodology, S.S.; Resources, Z.S. and S.W.; Software, S.H.; Supervision, Z.S. and S.W.; Validation, S.H.; Writing—original draft, S.H.; Writing—review and editing, Z.S. and S.W. All authors have read and agreed to the published version of the manuscript.

Funding: This work was supported by the National Key Research and Development Program of China (Grant No. 2018YFD0200900).

Acknowledgments: We would like to thanks and acknowledge management of Zishi experiment station, Jingzhou, China and special thanks to Muhammad Yaseen, Muhammad Shakir, Muhammad Asim, Nauman and Sajjad for their help at various stages of experiments.

Conflicts of Interest: The authors declare no conflict of interest.

References

1. Turner, D.; Lucieer, A.; Watson, C. An Automated Technique for Generating Georectified Mosaics from Ultra-High Resolution Unmanned Aerial Vehicle (UAV) Imagery, Based on Structure from Motion (SfM) Point Clouds. *Remote Sens.* **2012**, *4*, 1392–1410. [CrossRef]

2. Vega, F.A.; Ramirez, F.C.; Saiz, M.P.; Rosua, F.O. ScienceDirect Multi-temporal imaging using an unmanned aerial vehicle for monitoring a sunflower crop. *Biosyst. Eng.* **2015**, *132*, 19–27. [CrossRef]
3. Zhang, C.; Kovacs, J.M. The application of small unmanned aerial systems for precision agriculture: A review. *Precis. Agric.* **2015**, *13*, 693–712. [CrossRef]
4. Tremblay, N.; Vigneault, P.; Bélec, C.; Fallon, E.; Bouroubi, M.Y. A comparison of performance between UAV and satellite imagery for N status assessment in corn. In Proceedings of the 12th International Conference on Precision Agriculture, Sacramento, CA, USA, 20–23 July 2014.
5. Hunt, E.R.; Daughtry, C.S.T.; Mirsky, S.B.; Hively, W.D. Remote sensing with unmanned aircraft systems for precision agriculture applications. In Proceedings of the 2013 Second International Conference on Agro-Geoinformatics (Agro-Geoinformatics), Fairfax, VA, USA, 12–16 August 2013.
6. Zhao, G.; Miao, Y.; Wang, H.; Su, M.; Fan, M.; Zhang, F.; Jiang, R.; Zhang, Z.; Liu, C.; Liu, P.; et al. A preliminary precision rice management system for increasing both grain yield and nitrogen use efficiency. *Field Crops Res.* **2013**, *154*, 23–30. [CrossRef]
7. Hatfield, J.L.; Prueger, J.H. Value of Using Different Vegetative Indices to Quantify Agricultural Crop Characteristics at Different Growth Stages under Varying Management Practices. *Remote Sens.* **2010**, *2*, 562–578. [CrossRef]
8. Warren, G.; Metternicht, G. Agricultural Applications of High-Resolution Digital Multispectral Imagery: Evaluating Within-Field Spatial Variability of Canola (Brassica napus) in Western Australia. *Photogramm. Eng. Remote Sens.* **2005**, *71*, 595–602. [CrossRef]
9. Primicerio, J.; Di Gennaro, S.F.; Fiorillo, E.; Genesio, L.; Lugato, E.; Matese, A.; Vaccari, F.P. A flexible unmanned aerial vehicle for precision agriculture. *Precis. Agric.* **2012**, *13*, 517–523. [CrossRef]
10. Din, M.; Zheng, W.; Rashid, M.; Wang, S.; Shi, Z. Evaluating Hyperspectral Vegetation Indices for Leaf Area Index Estimation of Oryza sativa L. at Diverse Phenological Stages. *Front. Plant Sci.* **2017**, *8*, 1–17. [CrossRef]
11. Lan, Y.; Thomson, S.J.; Huang, Y.; Hoffmann, W.C.; Zhang, H. Current status and future directions of precision aerial application for site-specific crop management in the USA. *Comput. Electron. Agric.* **2010**, *74*, 34–38. [CrossRef]
12. Hunt, E.-R.; Horneck, D.; Gadler, D.; Bruce, A.; Turner, R.; Spinelli, C.; Brungardt, J.; Hamm, P. Detection of nitrogen deficiency in potatoes using small unmanned aircraft systems. In Proceedings of the 12th International Conference on Precision Agriculture, Sacramento, CA, USA, 20–23 July 2014.
13. Swain, K.C.; Zaman, Q.U. Rice Crop Monitoring with Unmanned Helicopter Remote Sensing Images. In *Remote Sensing of Biomass—Principles and Applications*; Fatoyinbo, T., Ed.; InTech: Rijeca, Croatia, 2012; pp. 253–272.
14. Laliberte, A. Unmanned aerial vehicle-based remote sensing for rangeland assessment, monitoring, and management. *J. Appl. Remote Sens.* **2009**, *3*, 033542. [CrossRef]
15. Lelong, C.; Burger, P.; Jubelin, G.; Roux, B.; Labbé, S.; Baret, F. Assessment of Unmanned Aerial Vehicles Imagery for Quantitative Monitoring of Wheat Crop in Small Plots. *Sensors* **2008**, *8*, 3557–3585. [CrossRef] [PubMed]
16. Tilman, D. Functional Diversity. In *Encyclopedia of Biodiversity*; Elsevier: Amsterdam, The Netherlands, 2001; Volume 3, pp. 109–120. [CrossRef]
17. Ali, A.M.; Darvishzadeh, R.; Skidmore, A.K.; van Duren, I.; Heiden, U.; Heurich, M. Prospect inversion for indirect estimation of leaf dry matter content and specific leaf area. *Int. Arch. Photogramm. Remote Sens. Spat. Inf. Sci.* **2015**, *40*, 277–284. [CrossRef]
18. Jin, X.; Diao, W.; Xiao, C.; Wang, F.; Chen, B.; Wang, K.; Li, S. Estimation of Wheat Agronomic Parameters using New Spectral Indices. *PLoS ONE* **2013**, *8*, e72736. [CrossRef] [PubMed]
19. Cheng, T.; Rivard, B.; Sánchez-Azofeifa, A.G.; Féret, J.-B.; Jacquemoud, S.; Ustin, S.L. Deriving leaf mass per area (LMA) from foliar reflectance across a variety of plant species using continuous wavelet analysis. *ISPRS J. Photogramm. Remote Sens.* **2014**, *87*, 28–38. [CrossRef]
20. De Riva, E.G.; Olmo, M.; Poorter, H.; Ubera, J.L.; Villar, R. Leaf Mass per Area (LMA) and Its Relationship with Leaf Structure and Anatomy in 34 Mediterranean Woody Species along a Water Availability Gradient. *PloS ONE* **2016**, *11*, 1–18. [CrossRef]
21. Homolova, L.; Malenovsky, Z.; Clevers, J.G.P.W.; Garcıá-Santos, G. Review of optical-based remote sensing for plant trait mapping. *Ecol. Complex.* **2013**, *15*, 1–16. [CrossRef]

22. Wei, Z.; Zhang, B.; Liu, Y.; Xu, D. The Application of a Modified Version of the SWAT Model at the Daily Temporal Scale and the Hydrological Response unit Spatial Scale: A Case Study Covering an Irrigation District in the Hei River Basin. *Water* **2018**, *10*, 1064. [CrossRef]
23. Singh, G.; Saraswat, D. Development and evaluation of targeted marginal land mapping approach in SWAT model for simulating water quality impacts of selected second generation biofeedstock. *Environ. Model. Softw.* **2016**, *81*, 26–39. [CrossRef]
24. Xie, Q.; Huang, W.; Liang, D.; Chen, P.; Wu, C.; Yang, G.; Zhang, J.; Huang, L.; Zhang, D. Leaf area index estimation using vegetation indices derived from airborne hyperspectral images in winter wheat. *IEEE J. Sel. Top. Appl. Earth Obs. Remote Sens.* **2014**, *7*, 3586–3594. [CrossRef]
25. Viña, A.; Gitelson, A.A.; Nguy-Robertson, A.L.; Peng, Y. Comparison of different vegetation indices for the remote assessment of green leaf area index of crops. *Remote Sens. Environ.* **2011**, *115*, 3468–3478.
26. Marques, P.; Martins, M.; Baptista, A.; Torres, J.P.N. Communication Antenas for UAVs. *J. Eng. Sci. Technol. Rev.* **2018**, *11*, 90–102. [CrossRef]
27. Hardin, P.J.; Hardin, T.J. Small-Scale Remotely Piloted Vehicles in Environmental Research. *Geogr. Compass* **2010**, *4*, 1297–1311. [CrossRef]
28. Taylor, J.A.; Jacob, F.; Galleguillos, M.; Prévot, L.; Guix, N.; Lagacherie, P. The utility of remotely-sensed vegetative and terrain covariates at different spatial resolutions in modelling soil and watertable depth (for digital soil mapping). *Geoderma* **2013**, *193–194*, 83–93. [CrossRef]
29. Xu, Y.; Smith, S.E.; Grunwald, S.; Abd-Elrahman, A.; Wani, S.P. Evaluating the effect of remote sensing image spatial resolution on soil exchangeable potassium prediction models in smallholder farm settings. *J. Environ. Manag.* **2017**, *200*, 423–433. [CrossRef]
30. Kim, J.; Grunwald, S.; Rivero, R.G. Soil Phosphorus and Nitrogen Predictions Across Spatial Escalating Scales in an Aquatic Ecosystem Using Remote Sensing Images. *IEEE Trans. Geosci. Remote Sens.* **2014**, *52*, 6724–6737.
31. Vergara-Díaz, O.; Zaman-Allah, M.A.; Masuka, B.; Hornero, A.; Zarco-Tejada, P.; Prasanna, B.M.; Cairns, J.E.; Araus, J.L. A Novel Remote Sensing Approach for Prediction of Maize Yield Under Different Conditions of Nitrogen Fertilization. *Front. Plant Sci.* **2016**, *7*, 1–13.
32. Liu, S.; Li, L.; Gao, W.; Zhang, Y.; Liu, Y.; Wang, S.; Lu, J. Diagnosis of Nitrogen Status In Winter Oilseed Rape (Brassica napus L.) Using In-situ Hyperspectral Data and Unmanned Aerial Vehicle (UAV) Multispectral Images. *Comput. Electron. Agric.* **2018**, *151*, 185–195. [CrossRef]
33. Pearson, R.L.; Miller, L.D. *Remote Mapping of Standing Crop Biomass for Estimation of the Productivity of the Shortgrass Prairie*; Colorado State University: Fort Collins, CO, USA, 1972.
34. J Rouse, J.W.; Haas, R.W.; Schell, J.A.; Deering, D.H.; Harlan, J.C. *Monitoring the Vernal Advancement and Retrogradation (Greenwave effect) of Natural Vegetation*; NASA/GSFC: Greenbelt, MD, USA, 1974.
35. Gitelson, A.A.; Kaufman, Y.J.; Merzlyak, M.N. Use of a green channel in remote sensing of global vegetation from EOS-MODIS. *Remote Sens. Environ.* **1996**, *58*, 289–298. [CrossRef]
36. Wang, F.M.; Huang, J.F.; Tang, Y.L.; Wang, X.Z. New Vegetation Index and Its Application in Estimating Leaf Area Index of Rice. *Rice Sci.* **2007**, *14*, 195–203. [CrossRef]
37. Huete, A. A soil-adjusted vegetation index (SAVI). *Remote Sens. Environ.* **1988**, *25*, 295–309. [CrossRef]
38. Haboudane, D. Hyperspectral vegetation indices and novel algorithms for predicting green LAI of crop canopies: Modeling and validation in the context of precision agriculture. *Remote Sens. Environ.* **2004**, *90*, 337–352. [CrossRef]
39. Rondeaux, G.; Steven, M.; Baret, F. Optimization of soil-adjusted vegetation indices. *Remote Sens. Environ.* **1996**, *55*, 95–107. [CrossRef]
40. Whiting, M.L.; Ustin, S.L.; Zarco-Tejada, P.; Palacios-Orueta, A.; Vanderbilt, V.C. Hyperspectral mapping of crop and soils for precision agriculture. In *Remote Sensing and Modeling of Ecosystems for Sustainability III*; Gao, W., Ustin, S.L., Eds.; SPIE: Bellingham, WA, USA, 2006; Volume 6298, p. 62980B.
41. Qi, J.; Chehbouni, A.; Huete, A.R.; Keer, Y.H.; Sorooshian, S. A modified soil adusted vegetation index. *Remote Sens. Environ.* **1994**, *48*, 119–126. [CrossRef]
42. Li, F.; Mao, L.; Hennig, S.D.; Gnyp, M.L.; Chen, X.P.; Jia, L.L.; Bareth, G. Evaluating hyperspectral vegetation indices for estimating nitrogen concentration of winter wheat at different growth stages. *Precis. Agric.* **2010**, *11*, 335–357. [CrossRef]

43. Din, M.; Ming, J.; Hussain, S.; Ata-Ul-Karim, S.T.; Rashid, M.; Tahir, M.N.; Hua, S.; Wang, S. Estimation of dynamic canopy variables using hyperspectral derived vegetation indices under varying N rates at diverse phenological stages of rice. *Front. Plant Sci.* **2019**, *9*, 1–16. [CrossRef]
44. Taylor, P.; Gitelson, A.A. Remote estimation of crop fractional vegetation cover: the use of noise equivalent as an indicator of performance of vegetation indices. *Int. J. Remote Sens.* **2013**, *34*, 37–41.
45. Nguy-Robertson, A.L.; Peng, Y.; Gitelson, A.A.; Arkebauer, T.J.; Pimstein, A.; Herrmann, I.; Karnieli, A.; Rundquist, D.C.; Bonfil, D.J. Estimating green LAI in four crops: Potential of determining optimal spectral bands for a universal algorithm. *Agric. For. Meteorol.* **2014**, *192–193*, 140–148. [CrossRef]
46. Roelofsen, H.D.; Van Bodegom, P.M.; Kooistra, L.; Witte, J.P.M. Predicting leaf traits of herbaceous species from their spectral characteristics. *Ecol. Evol.* **2014**, *4*, 706–719. [CrossRef]
47. Wang, S.Q.; Gao, W.H.; Ming, J.; Li, L.T.; Xu, D.H.; Liu, S.S.; Lu, J.W. A TPE based inversion of PROSAIL for estimating canopy biophysical and biochemical variables of oilseed rape. *Comput. Electron. Agric.* **2018**, *152*, 350–362. [CrossRef]
48. Houborg, R.; Anderson, M.; Daughtry, C. Utility of an image-based canopy reflectance modeling tool for remote estimation of LAI and leaf chlorophyll content at the field scale. *Remote Sens. Environ.* **2009**, *113*, 259–274. [CrossRef]
49. Li, F.; Mistele, B.; Hu, Y.; Chen, X.; Schmidhalter, U. Comparing hyperspectral index optimization algorithms to estimate aerial N uptake using multi-temporal winter wheat datasets from contrasting climatic and geographic zones in China and Germany. *Agric. For. Meteorol.* **2013**, *180*, 44–57. [CrossRef]
50. Xiong, D.; Wang, D.; Liu, X.; Peng, S.; Huang, J.; Li, Y. Leaf density explains variation in leaf mass per area in rice between cultivars and nitrogen treatments. *Ann. Bot.* **2016**, *117*, 963–971. [CrossRef] [PubMed]
51. Jégo, G.; Pattey, E.; Liu, J. Using Leaf Area Index, retrieved from optical imagery, in the STICS crop model for predicting yield and biomass of field crops. *Field Crops Res.* **2012**, *131*, 63–74.
52. Carvalho, S.; Van der Putten, W.H.; Hol, W.H.G. The Potential of Hyperspectral Patterns of Winter Wheat to Detect Changes in Soil Microbial Community Composition. *Front. Plant Sci.* **2016**, *7*, 1–11. [CrossRef]
53. Huang, M.; Yang, C.; Ji, Q.; Jiang, L.; Tan, J.; Li, Y. Tillering responses of rice to plant density and nitrogen rate in a subtropical environment of southern China. *Field Crops Res.* **2013**, *149*, 187–192. [CrossRef]
54. Tian, Y.-C.; Yang, J.; Yao, X.; Zhu, Y.; Cao, W.-X. Quantitative relationships between hyper-spectral vegetation indices and leaf area index of rice. *J. Appl. Ecol.* **2009**, *20*, 1685–1690.
55. Zhang, F.; Zhou, G.; Nilsson, C. Remote estimation of the fraction of absorbed photosynthetically active radiation for a maize canopy in Northeast China. *J. Plant Ecol.* **2015**, *8*, 429–435. [CrossRef]
56. Darvishzadeh, R.; Atzberger, C.; Skidmore, A.K.; Abkar, A.A. Leaf Area Index derivation from hyperspectral vegetation indicesand the red edge position. *Int. J. Remote Sens.* **2009**, *30*, 6199–6218. [CrossRef]
57. Xiao, Y.; Zhao, W.; Zhou, D.; Gong, H. Sensitivity Analysis of Vegetation Reflectance to Biochemical and Biophysical Variables at Leaf, Canopy, and Regional Scales. *IEEE Trans. Geosci. Remote Sens.* **2014**, *52*, 4014–4024. [CrossRef]
58. Marshall, M.; Thenkabail, P.; Biggs, T.; Post, K. Hyperspectral narrowband and multispectral broadband indices for remote sensing of crop evapotranspiration and its components (transpiration and soil evaporation). *Agric. For. Meteorol.* **2016**, *218–219*, 122–134. [CrossRef]
59. Reich, P.B.; Ellsworth, D.S.; Walters, M.B. Leaf structure (specific leaf area) modulates photosynthesis–nitrogen relations: Evidence from within and across species and functional groups. *Funct. Ecol.* **1998**, *12*, 948–958. [CrossRef]
60. Nautiyal, P.C.; Rao, N.; Joshi, Y.C. Moisture-deficit-induced changes in leaf-water content, leaf carbon exchange rate and biomass production in groundnut cultivars differing in specific leaf area. *Field Crops Res.* **2002**, *74*, 67–79. [CrossRef]
61. Vergara-díaz, O.; Zaman-allah, M.A.; Masuka, B.; Hornero, A. A Novel Remote Sensing Approach for Prediction of Maize Yield Under Different Conditions of Nitrogen Fertilization. *Front. Plant Sci.* **2016**, *7*, 1–13.
62. Li, C.; Wulf, H.; Schmid, B.; He, J.; Schaepman, M.E.; Member, S. Estimating Plant Traits of Alpine Grasslands on the Qinghai-Tibetan Plateau Using Remote Sensing. *IEEE J. Sel. Top. Appl. Earth Obs. Remote Sens.* **2018**, *11*, 1–13. [CrossRef]
63. Anser, G.P.; Martin, R.E. Airborne spectranomics: mapping canopy chemical and taxonomic diversity in tropical forests. *Environ. Front. Ecol.* **2009**, *7*, 269–276.
64. Myneni, R.B.; Williams, D.L. On the relationship between FAPAR and NDVI. *Remote Sens. Environ.* **1994**, *49*, 200–211. [CrossRef]

65. Steinberg, A.; Chabrillat, S.; Stevens, A.; Segl, K.; Foerster, S. Prediction of common surface soil properties based on Vis-NIR airborne and simulated EnMAP imaging spectroscopy data: Prediction accuracy and influence of spatial resolution. *Remote Sens.* **2016**, *8*, 613. [CrossRef]
66. Maynard, J.J.; Johnson, M.G. Scale-dependency of LiDAR derived terrain attributes in quantitative soil-landscape modeling: Effects of grid resolution vs. neighborhood extent. *Geoderma* **2014**, *230–231*, 29–40. [CrossRef]

© 2020 by the authors. Licensee MDPI, Basel, Switzerland. This article is an open access article distributed under the terms and conditions of the Creative Commons Attribution (CC BY) license (http://creativecommons.org/licenses/by/4.0/).

Article

Retrieval of the Fraction of Radiation Absorbed by Photosynthetic Components ($FAPAR_{green}$) for Forest Using a Triple-Source Leaf-Wood-Soil Layer Approach

Siyuan Chen [1,2], Liangyun Liu [1,*], Xiao Zhang [1,3], Xinjie Liu [1], Xidong Chen [1,3], Xiaojin Qian [1,3], Yue Xu [4] and Donghui Xie [4]

1. State Key Laboratory of Remote Sensing Science, Institute of Remote Sensing and Digital Earth, Chinese Academy of Sciences, Beijing 100094, China; rs_chensiyuan@163.com (S.C.); zhangxiao@radi.ac.cn (X.Z.); liuxj@radi.ac.cn (X.L.); chenxd@radi.ac.cn (X.C.); qianxj@radi.ac.cn (X.Q.)
2. College of Geomatics, Xi'an University of Science and Technology, Xi'an 710054, China
3. College of Resources and Environment, University of Chinese Academy of Sciences, Beijing 100049, China
4. State Key Laboratory of Remote Sensing Science, Beijing Key Laboratory of Environmental Remote Sensing and Digital City, School of Geography, Beijing Normal University, Beijing 100875, China; 201821051067@mail.bnu.edu.cn (Y.X.); xiedonghui@bnu.edu.cn (D.X.)
* Correspondence: liuly@radi.ac.cn; Tel.: +86-10-8217-8163

Received: 12 September 2019; Accepted: 18 October 2019; Published: 23 October 2019

Abstract: The fraction of absorbed photosynthetically active radiation (FAPAR) is generally divided into the fraction of radiation absorbed by the photosynthetic components ($FAPAR_{green}$) and the fraction of radiation absorbed by the non-photosynthetic components ($FAPAR_{woody}$) of the vegetation. However, most global FAPAR datasets do not take account of the woody components when considering the canopy radiation transfer. The objective of this study was to develop a generic algorithm for partitioning $FAPAR_{canopy}$ into $FAPAR_{green}$ and $FAPAR_{woody}$ based on a triple-source leaf-wood-soil layer (TriLay) approach. The LargE-Scale remote sensing data and image simulation framework (LESS) model was used to validate the TriLay approach. The results showed that the TriLay $FAPAR_{green}$ had higher retrieval accuracy, as well as a significantly lower bias (R^2 = 0.937, Root Mean Square Error (RMSE) = 0.064, and bias = −6.02% for black-sky conditions; R^2 = 0.997, RMSE = 0.025 and bias = −4.04% for white-sky conditions) compared to the traditional linear method (R^2 = 0.979, RMSE = 0.114, and bias = −18.04% for black-sky conditions; R^2 = 0.996, RMSE = 0.106 and bias = −16.93% for white-sky conditions). For FAPAR that did not take account of woody components ($FAPAR_{noWAI}$), the corresponding results were R^2 = 0.920, RMSE = 0.071, and bias = −7.14% for black-sky conditions, and R^2 = 0.999, RMSE = 0.043, and bias = −6.41% for white-sky conditions. Finally, the dynamic $FAPAR_{green}$, $FAPAR_{woody}$, $FAPAR_{canopy}$ and $FAPAR_{noWAI}$ products for a North America region were generated at a resolution of 500 m for every eight days in 2017. A comparison of the results for $FAPAR_{green}$ against those for $FAPAR_{noWAI}$ and $FAPAR_{canopy}$ showed that the discrepancy between $FAPAR_{green}$ and other FAPAR products for forest vegetation types could not be ignored. For deciduous needleleaf forest, in particular, the black-sky $FAPAR_{green}$ was found to contribute only about 23.86% and 35.75% of $FAPAR_{canopy}$ at the beginning and end of the year (from January to March and October to December, JFM and OND), and 75.02% at the peak growth stage (from July to September, JAS); the black-sky $FAPAR_{noWAI}$ was found to be overestimated by 38.30% and 28.46% during the early (JFM) and late (OND) part of the year, respectively. Therefore, the TriLay approach performed well in separating $FAPAR_{green}$ from $FAPAR_{canopy}$, which is of great importance for a better understanding of the energy exchange within the canopy.

Keywords: the fraction of radiation absorbed by photosynthetic components ($FAPAR_{green}$); triple-source; leaf area index (LAI); woody area index (WAI); clumping index (CI); Moderate Resolution Imaging Spectroradiometer (MODIS); soil albedo

1. Introduction

The fraction of absorbed photosynthetically active radiation (FAPAR) is a significant biochemical and physiological variable used in tracing the exchanges of energy, mass, and momentum, and is also widely used in many climate, ecological, biogeochemical, agricultural, and hydrology models [1,2]. FAPAR is, therefore, an important input parameter and widely used in satellite-based Production Efficiency Models (PEMs) [3–6] to estimate gross primary productivity (GPP) or net primary production (NPP).

In general, the FAPAR inversion algorithms could be divided into two types: empirical statistical models based on vegetation indexes and physical methods based on the canopy radiation transfer model. Although the empirical statistical model based on vegetation indexes is relatively simple, involves only a few parameters and has high computational efficiency, it is subject to many uncertainties due to factors such as the atmospheric environment, vegetation type, and quality of remote sensing data. The physically based methods could be further divided into two categories. The first type is the direct inversion method, which uses the canopy radiation transfer model to link FAPAR with the canopy spectra [7–9]. For example, the Moderate Resolution Imaging Spectroradiometer (MODIS) algorithm uses the three-dimensional radiation transmission model to invert FAPAR from the bi-directional reflectance [10–12]. The Joint Research Centre (JRC) FAPAR algorithm is also based on a physical model that uses a continuous vegetation canopy model [13] to link land surface reflectance with FAPAR. However, these methods are mostly based on the radiative transfer model; thus, the inversion process is complicated for retrieval of FAPAR. The main problem with such methods is that it is difficult to overcome the uncertainty caused by model coupling and spatial heterogeneity. The second type of physically based method is the forward modeling method [14–19]. Most models of this type are based on the gap fraction model, which determines FAPAR according to canopy structure parameters such as LAI and the clumping index. The disadvantage of this approach is that it relies too much on the accuracy of the canopy structure parameters. Furthermore, it is difficult to accurately determine the soil albedo and extinction coefficient, which are also important parameters needed to determine the contribution of multiple scattering between the soil and canopy to FAPAR [17].

Recently, several global FAPAR products have become available, including the Moderate Resolution Imaging Spectroradiometer (MODIS) [20,21], Energy Balance Residual (EBR) [15], Multi-angle Imaging SpectroRadiometer (MISR) [11], CYCLOPES [22], GLOBCARBON [23], Global Land Surface Satellite (GLASS) [14], the Medium Resolution Imaging Spectrometer (MERIS) [24], Joint Research Center Two Stream Inversion Package (JRC-TIP) [25], and European Space Agency (ESA) products [26]. These global FAPAR products have been widely validated, with reported errors varying from 0.08 to 0.23 [14,27–32]. However, most of the global FAPAR products do not consider the effect of non-photosynthetic components in the radiative transfer process, which introduces errors, especially for forest types. Moreover, many researchers have used the fraction of radiation absorbed by photosynthetic components ($FAPAR_{green}$) instead of the fraction of radiation absorbed by the canopy ($FAPAR_{canopy}$) to monitor and estimate the light use efficiency (LUE), radiation use efficiency (RUE), and productivity at different temporal scales [33–36].

The forest vegetation ecosystem plays an important role in the global ecosystem. However, quantifying the temporal variation in $FAPAR_{green}$ for a forest ecosystem represents an important challenge for remote sensing and ecology researchers as it is extremely difficult to measure $FAPAR_{green}$ at large scales over plant growing seasons directly. Also, previous studies have shown that the contribution of woody components is relatively large: for instance, Asner et al. [37] found that stems increased $FAPAR_{canopy}$ by 10–40%. Therefore, the partitioning of absorbed radiation into photosynthetic and non-photosynthetic parts is very important for better modeling of vegetation photosynthesis and energy exchange within the canopy.

Already, some studies have looked at the estimation of $FAPAR_{green}$ from remote sensing data. Hall et al. [38] estimated $FAPAR_{green}$ using a simple linear relationship between $FAPAR_{canopy}$ and LAI_{green}/LAI_{total} (LAI_{total} denotes the total leaf area index including green and senescent leaves, while LAI_{green} represents the green leaf area index). However, this simple partitioning is problematic because the green and woody components within the canopy do not constitute a simple linear mix in terms of radiation transfer. Zhang et al. [39] first retrieved the biophysical and biochemical variables using the modified PROSPECT model coupled with the SAIL-2 model (hereafter called PROSAIL-2 model), and then calculated $FAPAR_{green}$ and $FAPAR_{canopy}$ using the forward simulation approach. However, $FAPAR_{green}$ retrieval using the PROSAIL-2 model is relatively complex and needs several physiological and biochemical parameters as model inputs. Gitelson et al. [40] also separated $FAPAR_{canopy}$ into photosynthetically active green components ($FAPAR_{green}$) and non-photosynthetic active components using the ratio LAI_{green}/LAI_{total} for maize and soybeans. The relationship between vegetation indices and $FAPAR_{green}$ was also used to retrieve $FAPAR_{green}$ [41,42]. Nevertheless, to date, the current $FAPAR_{green}$ products do not take into account the effect of non-photosynthetic components on canopy radiative transfer.

In this study, we aim to develop an operational algorithm for partitioning $FAPAR_{canopy}$ into $FAPAR_{green}$ and $FAPAR_{woody}$ for forest types. A simple triple-source leaf–wood–soil layer model (TriLay) that describes the radiation transfer within the canopy-soil system is presented. $FAPAR_{canopy}$ is first separated into the fraction of PAR absorbed by the canopy for downwelling radiation ($FAPAR_{canopy\downarrow}$) and the fraction of PAR absorbed by the canopy for the upwelling radiation reflected by the soil background ($FAPAR_{canopy\uparrow}$). Then, $FAPAR_{canopy\downarrow}$ and $FAPAR_{canopy\uparrow}$ are further split into the fraction of radiation absorbed by photosynthetic components ($FAPAR_{green}$) and that absorbed by non-photosynthetic components ($FAPAR_{woody}$) using the TriLay model. Finally, the $FAPAR_{green}$, $FAPAR_{woody}$, and $FAPAR_{canopy}$ products are generated using the MODIS albedo (MCD43A3), LAI (MCD15A2H), land cover (MCD12Q1), clumping index (CI), and soil albedo products based on the TriLay approach, and the discrepancies between different FAPAR products are used to investigate the contributions of woody components to the canopy-absorbed radiation. The partitioning of absorbed radiation into green and woody parts using the TriLay model is done not just to provide $FAPAR_{green}$ and $FAPAR_{woody}$—which is of great importance for better understanding the energy exchange within the canopy. The consideration of woody components should also improve the accuracy of $FAPAR_{green}$ estimates, which is important for better modeling of vegetation photosynthesis.

2. Materials and Methods

2.1. Satellite Datasets

In order to produce $FAPAR_{green}$ products, several satellite datasets were utilized in this study, including MODIS LAI products, land cover products, CI products, and soil albedo products from 2017. Data simulated by the LESS model [43] was used to validate the retrieved $FAPAR_{green}$ products. A mid-latitude region (tile h10v05, covering 30.00° N–40.00° N and 80.00° W–104.43° W) was selected to investigate the discrepancy between $FAPAR_{green}$ and other FAPAR products because there were abundant forest vegetation types within this MODIS tile.

2.1.1. MODIS LAI/FAPAR Products (MCD15A2H)

MCD15A2H V006 is a MODIS eight-day composite LAI/FAPAR product that includes FAPAR, LAI, and quality control (QC) data with a resolution of 500 m [44]. The main retrieval algorithm for LAI and FAPAR contains a Look-up-Table (LUT) based on a 3D radiation transfer model [21] that uses the atmospherically corrected Red and near-infrared (NIR) Bidirectional Reflectance Function (BRF) [45]. A back-up algorithm based on the empirical relationships between the Normalized Difference Vegetation Index (NDVI) and LAI and FAPAR at the canopy scale is used at the same time. Also, for the biome types and typical conditions that are considered, observed, and modeled spectral BRFs and soil patterns are compared for each pixel. The LAI and FAPAR values that lie within a fixed level of uncertainty are then taken to be acceptable. Finally, averaged values of LAI and FAPAR are used as the eventually retrieved values [20].

2.1.2. MODIS Land Cover Product (MCD12Q1)

The MODIS Land Cover Type Product (MCD12Q1) provides land cover maps with a temporal resolution of one year and a spatial resolution of 500m at a global scale from 2001 until the present; it includes several classification schemes (International Geosphere-Biosphere Programme (IGBP), University of Maryland (UMD), LAI, BIOME-Biogeochemical Cycles (DBC), Plant Functional Types (PFT), FAO-Land Cover Classification System land cover (LCCS1), etc.). The main algorithm used in MCD12Q1 is a supervised classification method (decision tree) combined with a boosting technique [20] based on MODIS reflectance data [46,47]. The International Geosphere-Biosphere Program (IGBP) classification scheme, which defines 17 land cover types, was used in this study.

2.1.3. Global Clumping Index (CI) Product

The clumping index (CI) signifies the characteristics of groups of foliage in the canopy, and thus, it is a significant parameter describing the structure of the vegetation canopy [48]. According to Jiao et al. [49], a method based on the MODIS Bidirectional Reflectance Distribution Function (BRDF) and a linear relationship between the CI and the normalized difference between the angular indexes of the hotspot and dark spot (NDHD) could be utilized to generate CIs within a valid range (0.33 to 1.00); a back-up algorithm is also used to substitute values for the invalid CIs [49]. A global CI dataset with an eight-day temporal resolution and 500 m spatial resolution covering the period from 2002 to the present has been produced by Jiao et al. [49].

2.1.4. Global Soil Albedo Product

A non-linear spectral mixture model (NSM) model proposed by Liu & Zhang (2018) [15,50] was used to retrieve the global visible (VIS) soil albedo. The main idea in the NSM model is the dual-source vegetation–soil layer approach. In this approach, it is assumed that the leaves are located in the upper canopy, while the soil components are found in the lower canopy. Based on this assumption, the canopy albedo can be approximated as a non-linear mixture of the "pure" vegetation and soil parts [15]. For pixels with abnormal values (smaller than 0.02 or greater than 0.3), prior values acquired by a global database of land surface parameters at 1 km resolution (ECOCLIMAP) [51,52] and the yearly composite value were used instead. Finally, the estimated global VIS soil albedo based on the NSM model is obtained, having a spatial resolution of 500 m and a temporal resolution of eight days.

2.2. Data Simulated by the LESS Model

To quantitatively evaluate the performance of the proposed TriLay approach for estimating $FAPAR_{green}$, the LargE-Scale remote sensing data and image simulation (LESS) framework model [43] was employed to generate a simulated dataset that covered most of the conditions found in forests. LESS is a ray-tracing based 3D radiative transfer model which can simulate remote sensing data and images over large-scale and realistic 3D scenes (http://lessrt.org/). LESS employs a weighted forward photon tracing (FPT) method to simulate multispectral bidirectional reflectance factor (BRF) or flux-related data (e.g., downwelling radiation) and a backward path tracing (BPT) method to generate sensor images (e.g., fisheye images) or large-scale (e.g., 1 km^2) spectral images. The accuracy of LESS is evaluated with other models as well as field measurements in terms of directional BRFs and pixel-wise simulated image comparisons, which shows very good agreement [43,53].

The modeling area is located in the Genhe Forestry Reserve (Genhe) (120°12′ to 122°55′ E, 50°20′ to 52°30′ N), Greater Khingan of Inner Mongolia, Northeastern China. It has a hilly terrain with 75% forest cover, which is mainly composed of Dahurian Larch (Larix gmelinii) and White Birch (Betula platyphylla Suk.). A pure plot of Dahurian Larch (L9) is established and is selected for modeling, with its location in Figure 1. The position, crown width, breast diameter, tree height, and transmittance for trees were entered into the LESS model. A total of eight scenes were constructed, with each individual tree in a scene consisting of branches and leaves. The branches and size of the woody area were kept the same in all the scenes, but the leaf area was changed to produce scenes with different values of the LAI. For each scene, different FAPAR values (including $FAPAR_{canopy}$, $FAPAR_{green}$ and $FAPAR_{woody}$)

were calculated under black-sky (the ratio of diffuse light is zero, and nine different solar zenith angles of 0°–80° at 10° intervals were set) and white-sky conditions (the ratio of diffuse light is 1). The main input parameters used in the LargE-Scale remote sensing data and image simulation (LESS) model are listed as Table 1. Finally, a total of 72 black-sky simulations and eight white-sky simulations were achieved for the different conditions giving.

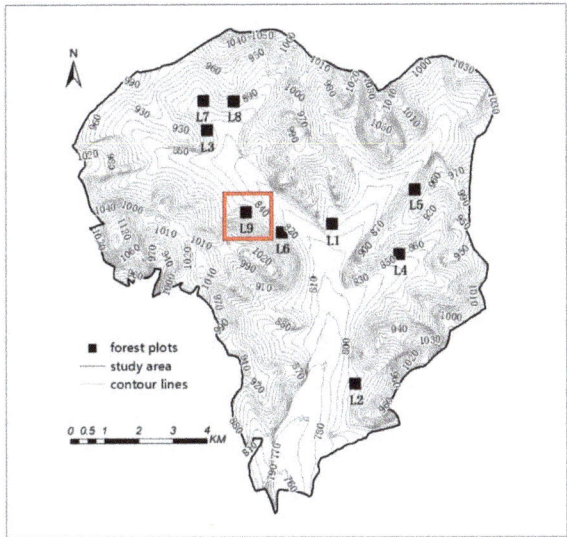

Figure 1. Locations of the selected Larch plots (L9) in Genhe Forestry Reserve.

Table 1. The main input parameters used in the LargE-Scale remote sensing data and image simulation (LESS) model simulations.

Parameter	Definition	Units	Range or Values
	Canopy		
LAI	leaf area index	m²/m²	1.31–8.69
WAI	woody area index	m²/m²	1.65
	Leaf layer		
Reflectance		—	0.041–0.205
Transmittance		—	0.001–0.286
	Soil layer		
Reflectance		—	0.001–0.134
	Woody layer		
Reflectance		—	0.069–0.237
	Imaging Geometry		
SZA	sun zenith angle	degrees	0, 10, 20, 30, 40, 50, 60, 70, 80
$Ratio_{Sky}$	ratio of diffuse light	—	0, 1

2.3. Algorithms for Estimating Global $FAPAR_{green}$ and $FAPAR_{woody}$ Datasets

To split the fraction of radiation absorbed by photosynthetic components ($FAPAR_{green}$) from $FAPAR_{canopy}$, a novel triple-source leaf–wood–soil layer model was proposed for generating global $FAPAR_{green}$ products for forest vegetation types. Figure 2 is a flowchart of the process used to retrieve global $FAPAR_{green}$ products.

First, $FAPAR_{canopy}$ is split into two parts: the fraction of PAR absorbed by the canopy for the downwelling radiation ($FAPAR_{canopy\downarrow}$) and that absorbed by the canopy for the upwelling radiation reflected by the soil background ($FAPAR_{canopy\uparrow}$). Then, $FAPAR_{canopy\downarrow}$ and $FAPAR_{canopy\uparrow}$ are further split into the fraction of radiation absorbed by photosynthetic components ($FAPAR_{green}$) and that absorbed by non-photosynthetic components (e.g., branches and stems, hereafter called $FAPAR_{woody}$). Finally, $FAPAR_{green}$ and $FAPAR_{woody}$ can be calculated separately using the TriLay approach.

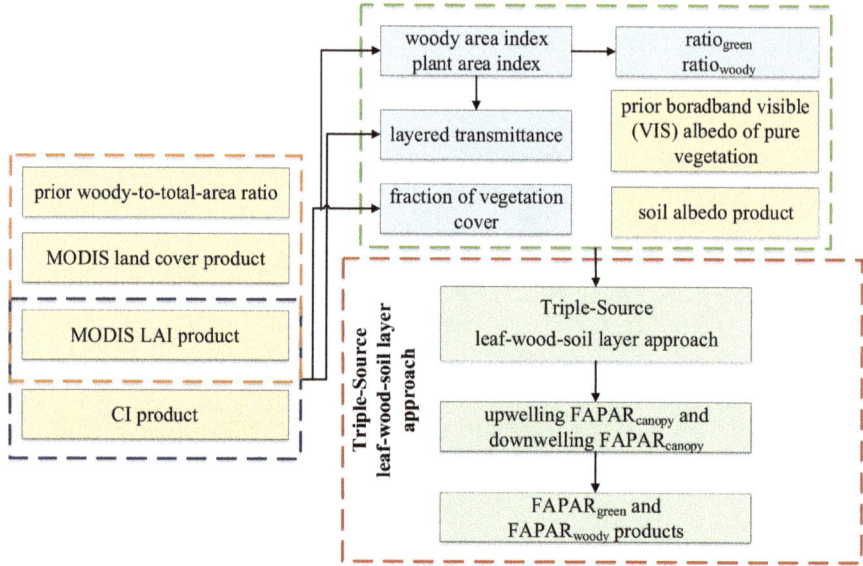

Figure 2. Flowchart illustrating the Triple-source leaf–wood–soil layer (TriLay) method for estimating the fraction of radiation absorbed by photosynthetic components ($FAPAR_{green}$) and the fraction of radiation absorbed by woody components ($FAPAR_{woody}$).

2.3.1. The Triple-Source Leaf–Wood–Soil Layer Model

A triple-source leaf–wood–soil layer model (TriLay) was developed to model the radiation transfer within the vegetation–soil system—this model is illustrated as Figure 3.

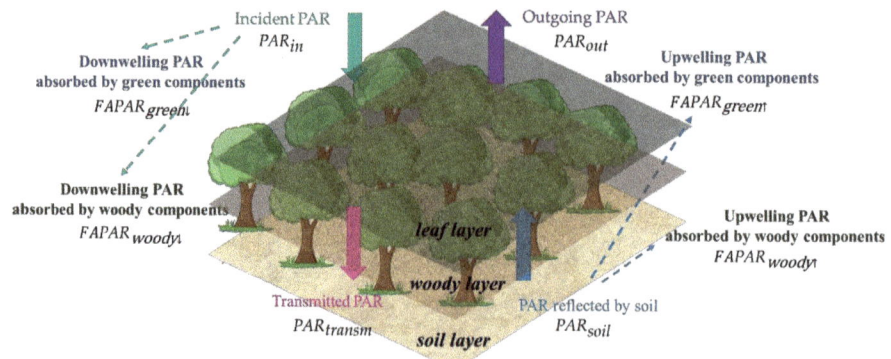

Figure 3. Illustration of the triple-source leaf–wood–soil layer model.

In this study, we used the layer approach to illustrate the distribution of leaves and soil in the whole canopy, which is consistent with the approach used in our previous study [15]. The layer approach assumes that the canopy consists only of green components and woody components and that all leaves are found above the canopy. The NSM model [15] was then used to simulate the canopy albedo.

The fraction of PAR absorbed by the canopy, $FAPAR_{canopy}$, can be separated into the fraction of PAR absorbed by the canopy for the downwelling radiation ($FAPAR_{canopy\downarrow}$) and that for the upwelling radiation reflected from the soil background ($FAPAR_{canopy\uparrow}$):

$$FAPAR_{canopy} = FAPAR_{canopy\downarrow} + FAPAR_{canopy\uparrow} \quad (1)$$

where $FAPAR_{canopy\downarrow}$ describes the fraction of downwelling radiation absorbed within the canopy assuming the soil background is dark, and $FAPAR_{canopy\uparrow}$ describes the fraction of upwelling radiation absorbed within the canopy due to the interaction between the ground (soil and understory) and the canopy. A point worth emphasizing is that, assuming a black soil background, the $FAPAR_{canopy\downarrow}$ is equal to the $FAPAR_{canopy}$.

Therefore, $FAPAR_{canopy\downarrow}$ can be given by:

$$FAPAR_{canopy} = (1 - \tau_{PAI}) * (1 - Albedo_{pure} * FVC) \quad (2)$$

where τ_{PAI} is the transmittance of the whole canopy, which contains green and woody parts, FVC is the fraction of vegetation cover, and $Albedo_{pure}$ is the visible (VIS) albedo for pure vegetation, which is represented by the VIS albedo for vegetation with a "saturated" LAI value (e.g., LAI ≥ 6) [14]. According to the statistical results by Liu et al. [15], $Albedo_{pure}$ was set to 0.025 (for white-sky conditions) and 0.020 (for black-sky conditions) for all woody vegetation types (including evergreen needleleaf forest, evergreen broadleaf forest, deciduous needleleaf forest, and deciduous broadleaf forest).

τ_{PAI} can be calculated as the product of τ_{LAI} and τ_{WAI}, and the directional transmittance can be determined using the gap fraction model [16,53,54]:

$$\tau_{PAI} = \tau_{LAI} \times \tau_{WAI} \quad (3)$$

$$\tau_{LAI} = e^{-k_1 \times G(\theta) \times CI \times LAI / \cos(\theta)} \quad (4)$$

$$\tau_{WAI} = e^{-k_2 \times G(\theta) \times CI \times WAI / \cos(\theta)} \quad (5)$$

where LAI and WAI are the leaf area index and woody area index, respectively, k_1 and k_2 are extinction coefficients for the green components and woody components, respectively, and θ is the solar zenith angle. k_1 can be determined using the leaf absorptance in the VIS band and was set to 0.88 based on simulations made using the PROSPECT-5 model and also the Leaf optical properties experiment 93 (LOPEX'93) and Leaf Optical Properties Database (an experiment conducted at the National Institute for Agricultural Research in Angers, France in June 2003) [55–57]. The woody components were assumed to be opaque with a constant extinction coefficient (k_2) of 0.91 based on the simulated LESS data. $G(\theta)$ is the projection of the unit foliage area on the plane perpendicular to the solar incident direction, θ. For green leaves, $G(\theta)$ is normally given a value of 0.5 for canopies with a spherical leaf angle distribution. For woody components, we assuming the woody components have the same angular distribution as green leaves, which also assumed by Chen et al. [58], Kucharik et al. [59], and Sea et al. [60]. CI is the clumping index; we also assume the same CI values for both green leaves and woody components based on the findings of Chen et al. [61] and Zou et al. [62].

The gap fraction model is also used to calculate FVC for a fixed solar zenith angle of 0° and a fixed value of 1 for the leaf extinction coefficient:

$$FVC = 1 - e^{-G(\theta) \times LAI \times CI} \quad (6)$$

Similarly, $FAPAR_{canopy\uparrow}$ can be calculated as follows:

$$FAPAR_{canopy\uparrow} = (1 - \tau_{PAI}) * (1 - Albedo_{pure} * FVC) * \tau_{PAI}^{ws} * Albedo_{soil} \quad (7)$$

where $Albedo_{soil}$ is the soil visible albedo, which can be generated by the NSM model proposed by Liu et al. [15]. τ_{PAI}^{ws} is the canopy transmittance under white-sky conditions, which can be calculated as the product of the white-sky transmittance of leaves (τ_{LAI}^{ws}) and the white-sky transmittance of woody components (τ_{WAI}^{ws}):

$$\tau_{LAI}^{ws} = 2 \times \int_0^{\frac{\pi}{2}} \left(e^{-k_1 \times G(\theta) \times LAI \times CI / \cos(\theta)}\right) \times \sin(\theta) \times \cos(\theta) d\theta \qquad (8)$$

$$\tau_{WAI}^{ws} = 2 \times \int_0^{\frac{\pi}{2}} \left(e^{-k_2 \times G(\theta) \times WAI \times CI / \cos(\theta)}\right) \times \sin(\theta) \times \cos(\theta) d\theta \qquad (9)$$

$$\tau_{PAI}^{ws} = \tau_{LAI}^{ws} \times \tau_{WAI}^{ws} \qquad (10)$$

2.3.2. Determination of Woody Area Index

In general, it is expensive and time-consuming to make accurate estimates of the woody area index (WAI), and destructive sampling is often the only option available for the quantification of the WAI in tropical evergreen forests [63]. Therefore, for generating global $FAPAR_{green}$ datasets, the use of accurate estimates of WAI is unrealistic. Hence, in this study, we aimed to determine the global WAI using MCD15A2H LAI data and assumed that the WAI was constant within a given year.

First, we assumed that the woody-to-total area ratio is measured during the peak growth stage (July–August–September, JAS) when the LAI has its maximum value for the year. The woody-to-total area ratio was determined according to the forest types listed in the MODIS land cover product (MCD12Q1), including evergreen needleleaf forest (ENF), evergreen broadleaf forest (EBF), deciduous broadleaf forest (DBF), deciduous needleleaf forest (DNF) and mixed forest (MF). Therefore, WAI values for different forest vegetation types were calculated using a simple linear relationship between the plant area index (PAI) and the WAI:

$$WAI = PAI \times ratio_{woody} = \frac{LAI_{max}}{1 - ratio_{woody}} \times ratio_{woody} \qquad (11)$$

where $ratio_{woody}$ is the mean value of the woody-to-total area ratio for various forest types, as given in the literature [58,64,65]. LAI_{max} is the maximum value of LAI within a given year; this was acquired from MCD15A2H LAI products.

2.3.3. Separating $FAPAR_{green}$ and $FAPAR_{woody}$ from $FAPAR_{canopy}$

In a similar way to Equation (1), the fraction of PAR absorbed by the green and woody components can also be separated into the fraction of PAR absorbed by the canopy for the downwelling radiation and that for the upwelling radiation reflected from the soil background. Firstly $FAPAR_{canopy\downarrow}$ is split into $FAPAR_{green\downarrow}$ and $FAPAR_{woody\downarrow}$:

$$FAPAR_{canopy\downarrow} = FAPAR_{green\downarrow} + FAPAR_{woody\downarrow} \qquad (12)$$

where $FAPAR_{green\downarrow}$ and $FAPAR_{woody\downarrow}$ can be obtained as

$$FAPAR_{green\downarrow} = FAPAR_{canopy\downarrow} \times ratio_{green} \times w_{1\downarrow} \qquad (13)$$

$$FAPAR_{woody\downarrow} = FAPAR_{canopy\downarrow} \times ratio_{woody} \times w_{2\downarrow} \qquad (14)$$

where $ratio_{green}$ and $ratio_{woody}$ are the ratio of leaf area index to plant area index and woody area index to plant area index, respectively. $w_{1\downarrow}$ and $w_{2\downarrow}$ are the weighting coefficients for the green (i.e., photosynthetic) and woody components in terms of the radiation transfer within the canopy. The terms involved in equations (10) and (11) can be given as:

$$ratio_{green} = \frac{LAI}{PAI} \qquad (15)$$

$$ratio_{woody} = \frac{WAI}{PAI} \qquad (16)$$

$$w_{2\downarrow} = w_{1\downarrow} \times \tau_{LAI} \qquad (17)$$

$$w_{1\downarrow} \times ratio_{green} + w_{2\downarrow} \times ratio_{woody} = 1 \tag{18}$$

$FAPAR_{green\downarrow}$ and $FAPAR_{woody\downarrow}$ can then be obtained by solving equations (13)–(18):

$$FAPAR_{green\downarrow} = \frac{ratio_{green} \times FAPAR_{canopy\downarrow}}{ratio_{green} + \tau_{LAI} \times ratio_{woody}} \tag{19}$$

$$FAPAR_{woody\downarrow} = \frac{ratio_{woody} \times FAPAR_{canopy\downarrow} \times \tau_{LAI}}{ratio_{green} + \tau_{LAI} \times ratio_{woody}} \tag{20}$$

Similarly, $FAPAR_{green\uparrow}$ and $FAPAR_{woody\uparrow}$ can also be acquired:

$$FAPAR_{green\uparrow} = \frac{ratio_{green} \times FAPAR_{canopy\uparrow} \times \tau_{WAI}}{ratio_{woody} + \tau_{WAI} \times ratio_{green}} \tag{21}$$

$$FAPAR_{woody\uparrow} = \frac{ratio_{woody} \times FAPAR_{canopy\uparrow}}{ratio_{woody} + \tau_{WAI} \times ratio_{green}} \tag{22}$$

Finally, $FAPAR_{green}$ and $FAPAR_{woody}$ can be calculated as:

$$FAPAR_{green} = FAPAR_{green\uparrow} + FAPAR_{green\downarrow} \tag{23}$$

$$FAPAR_{woody} = FAPAR_{woody\uparrow} + FAPAR_{woody\downarrow} \tag{24}$$

3. Results

3.1. Validation of the TriLay Method using Simulations made by the LESS Model

It is too challenging to obtain in-situ measurements of $FAPAR_{green}$ for forests, and so the simulated dataset (Table 1) derived using the LESS model was used for the validation of the TriLay model. Seventy-two black-sky simulations of $FAPAR_{canopy}$, $FAPAR_{green}$, and $FAPAR_{woody}$ together with eight white-sky simulations were available for validation.

Figure 4 illustrates the validation results for $FAPAR_{canopy}$, $FAPAR_{green}$, and $FAPAR_{woody}$. The estimated and simulated FAPAR values are distributed close to the 1:1 line. Also, it can be seen that the TriLay approach can produce accurate estimates of $FAPAR_{canopy}$, giving Root Mean Square Error (RMSEs) of 0.048 and 0.024 for black-sky and white-sky conditions, respectively, as against the LESS-simulated values. The corresponding R^2 values are 0.945 and 0.999. For $FAPAR_{green}$, the validation results give RMSEs of 0.064 and 0.025, respectively, for black-sky and white-sky FAPAR. Finally, it can be seen that $FAPAR_{woody}$ can also be accurately estimated: the RMSE and R^2 values are 0.042 and 0.709, respectively, for black-sky conditions, and 0.014 and 0.992 for white-sky conditions. These results show that the TriLay approach can be used to accurately estimate $FAPAR_{canopy}$, $FAPAR_{green}$, and $FAPAR_{woody}$ for forest land cover types.

Furthermore, as illustrated in Figure 4b, there is a slight underestimation for $FAPAR_{green}$ at smaller SZAs (0°–60°) and a slight overestimation for larger SZAs (70°–80°). From Figure 4c, it can be seen that $FAPAR_{woody}$ is also slightly underestimated at smaller SZAs (0°–40°) and overestimated at larger angles (50°–80°).

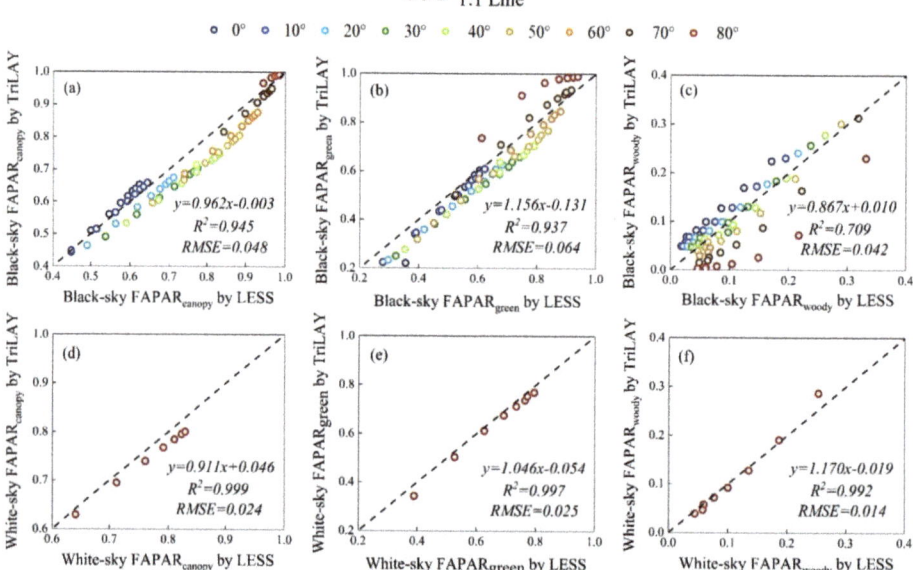

Figure 4. Validation of $FAPAR_{canopy}$, $FAPAR_{green}$ and $FAPAR_{woody}$ estimates made by the Trilay model against the LargE-Scale remote sensing data and image simulation (LESS)—simulated FAPARs: (a–c) black-sky $FAPAR_{canopy}$, $FAPAR_{green}$, and $FAPAR_{woody}$; (d–f) white-sky $FAPAR_{canopy}$, $FAPAR_{green}$, and $FAPAR_{woody}$.

3.2. Comparison of Different Methods using the LESS Simulations

Hall et al. [38] estimated $FAPAR_{green}$ based on a simple linear relationship between $FAPAR_{canopy}$ and $FAPAR_{green}$. They used the ratio LAI_{green}/LAI_{total} to determine $FAPAR_{green}$:

$$FAPAR_{green} = FAPAR_{canopy} \times \frac{LAI_{green}}{total\ LAI} \tag{25}$$

where total LAI is the PAI mentioned above. In order to test the accuracy of the linear mixture method, we also validated the $FAPAR_{green}$ estimated by the linear mixture method [38] using LESS-simulated data.

Figure 5 shows the accuracy assessment results for the linear mixture method. The results show a noticeable underestimation for $FAPAR_{green}$ and a huge overestimation for $FAPAR_{woody}$. Although the FAPARs retrieved using the linear mixture method are highly correlated with the LESS simulation, with R^2 values of 0.979, 0.996 for $FAPAR_{green}$, and 0.934, 0.985 for $FAPAR_{woody}$ under black-sky and white-sky conditions, the corresponding RMSE values are much higher than those found using our TriLay approach—0.114, 0.106 for $FAPAR_{green}$ as against 0.064, 0.025, and 0.113, 0.106 for $FAPAR_{woody}$ as against 0.042, 0.014 under the black-sky and white-sky conditions, respectively.

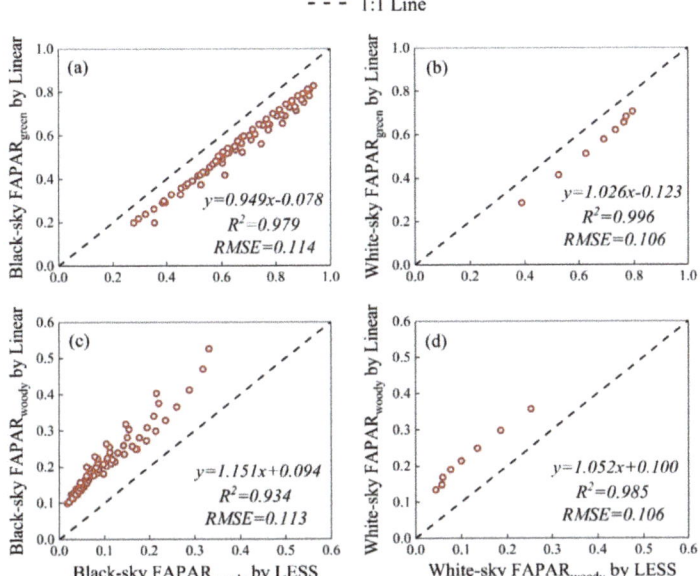

Figure 5. Validation of $FAPAR_{green}$ and $FAPAR_{woody}$ estimated by the linear method using the LESS-simulated values of FAPAR: (a–b) black-sky $FAPAR_{green}$ and $FAPAR_{woody}$ against the LESS-simulated $FAPAR_{green}$; (c–d) white-sky $FAPAR_{green}$ and $FAPAR_{woody}$ against the LESS-simulated $FAPAR_{woody}$.

The FAPAR values derived without considering the woody components ($FAPAR_{noWAI}$) were also validated using the LESS simulations; the results are shown in Figure 6. These results give an RMSE of 0.071 and R^2 of 0.920 for the black-sky conditions and the corresponding values of 0.043 and 0.999 for the white-sky conditions. Although the R^2 values are higher, the RMSEs are still greater than those found using the TriLay approach.

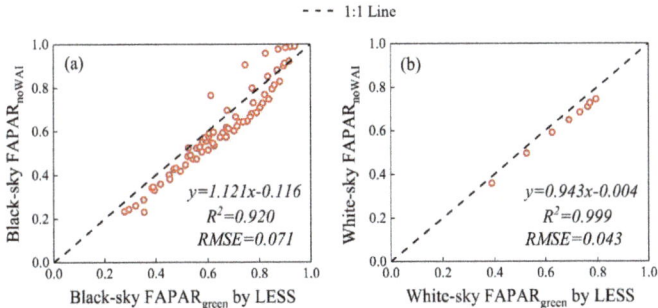

Figure 6. Validation of $FAPAR_{noWAI}$ using the LESS-simulated FAPAR values: (a) black-sky and (b) white-sky $FAPAR_{noWAI}$ against the LESS-simulated $FAPAR_{green}$.

In order to further compare the performances of different methods, we also calculated the bias for $FAPAR_{green}$, $FAPAR_{woody}$, and $FAPAR_{noWAI}$ using the LESS-simulated FAPAR values. Table 2 summarizes the retrieval accuracy for these three methods. The results show that the TriLay method gave the best FAPAR retrieval results, having the smallest bias and RMSE values (RMSE = 0.064 and 0.025, and bias = −6.02%, −4.04% for $FAPAR_{green}$ under black-sky and white-sky conditions, respectively).

Table 2. Retrieval accuracy of $FAPAR_{green}$, $FAPAR_{woody}$, and $FAPAR_{noWAI}$ validated using the LESS simulations.

(a) For $FAPAR_{green}$ Products

$FAPAR_{green}$	TriLay		Linear		noWAI	
	Black-Sky	White-Sky	Black-Sky	White-Sky	Black-Sky	White-Sky
R^2	0.937	0.997	0.979	0.996	0.920	0.999
RMSE	0.064	0.025	0.114	0.106	0.071	0.043
Bias	−6.02%	−4.04%	−18.04%	−16.93%	−7.14%	−6.41%

(b) For $FAPAR_{woody}$ products

$FAPAR_{woody}$	TriLay		Linear	
	Black-Sky	White-Sky	Black-Sky	White-Sky
R^2	0.709	0.992	0.934	0.985
RMSE	0.042	0.014	0.113	0.106
Bias	6.87%	−4.64%	153.84%	123.47%

3.3. Temporal Variations in Different FAPAR Products

Using the TriLay approach, black-sky and white-sky products, including $FAPAR_{green}$, $FAPAR_{woody}$, and $FAPAR_{canopy}$, and also FAPAR without woody components ($FAPAR_{noWAI}$), were generated for tile h10v05 in 2017. The black-sky FAPAR products were determined according to the SZA values at 10:30 am local time (the overpass time of the Terra satellite).

To investigate the seasonal variations in $FAPAR_{green}$ and the other FAPAR products, only the mean black-sky FAPAR values were calculated for the different forest vegetation types, as shown in Figure 7. The proportion of black-sky $FAPAR_{green}$ in $FAPAR_{canopy}$ and $FAPAR_{noWAI}$, and also the bias between the black-sky $FAPAR_{green}$, $FAPAR_{canopy}$, and $FAPAR_{noWAI}$ were calculated for different periods during 2017 in order to analyze the differences between different black-sky FAPAR products, as well as to quantify the contribution of the woody components for several forest vegetation types (deciduous broadleaf forest, deciduous needleleaf forest, evergreen broadleaf forest, and evergreen needleleaf forest, referred to as DBF, DNF, EBF, and ENF, respectively). The results are illustrated in Figure 8 and Table 3.

In general, the black-sky $FAPAR_{green}$ and $FAPAR_{noWAI}$ exhibit typical seasonal variations for the selected forest types, with low values during the early and late period (January–February–March (JFM) and October–November–December (OND)) and high values during the peak growth stage (July–August–September (JAS)). The black-sky $FAPAR_{canopy}$ is obviously higher than the black-sky $FAPAR_{green}$ during the whole year and has a much smaller seasonal variation. The black-sky $FAPAR_{woody}$ behaves in the opposite way; for deciduous seasons (JFM and OND), $FAPAR_{woody}$ is close to or higher than its value during the peak growth stage (JAS) because the black-sky FAPAR increases with increasing SZA value. (The mean SZA at 10:30 am varies from 22.98° (22 December) to 62.25° (22 June) within tile h10v05.)

The black-sky $FAPAR_{green}$ is about 52.59% and 60.60% of the black-sky $FAPAR_{canopy}$ for deciduous broadleaf forest during JFM and OND, respectively; the corresponding figures for deciduous needleleaf forest are only about 23.86% and 35.75%. During the peak growth stage (JAS), the black-sky $FAPAR_{green}$ is about 93.36%, 75.02%, 90.93% and 87.14% of $FAPAR_{canopy}$ for DBF, DNF, EBF, and ENF, respectively.

There are also small discrepancies between $FAPAR_{noWAI}$ and $FAPAR_{green}$ (Figure 8). In particular, for deciduous needleleaf forests, the black-sky $FAPAR_{noWAI}$ is overestimated by 38.30% and 28.46% during the early and late stages of the year (JFM and OND). For evergreen forests, the difference can be neglected as there is only a very slight underestimation of 0.68% to 2.39% during the whole year.

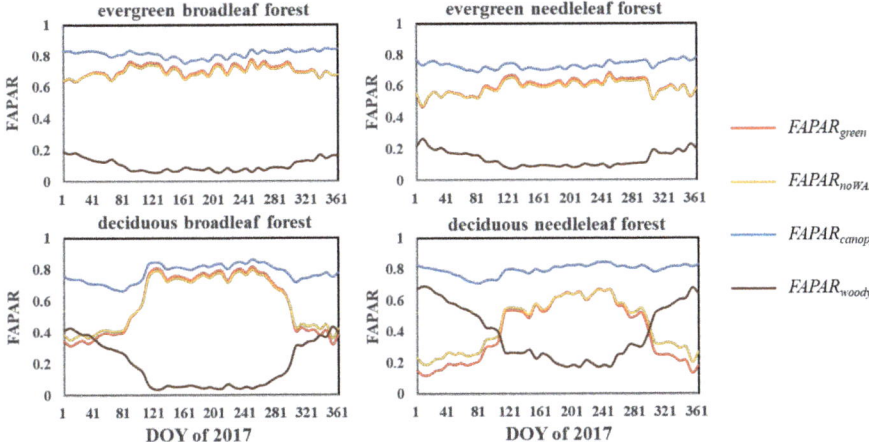

Figure 7. Temporal variations in the mean black-sky $FAPAR_{green}$, $FAPAR_{noWAI}$, $FAPAR_{canopy}$, and $FAPAR_{woody}$ products for forest vegetation types within tile h10v05 during 2017 (tile h10v05 is located in North America, covering 30.00° N–40.00° N and 80.00° W–104.4° W).

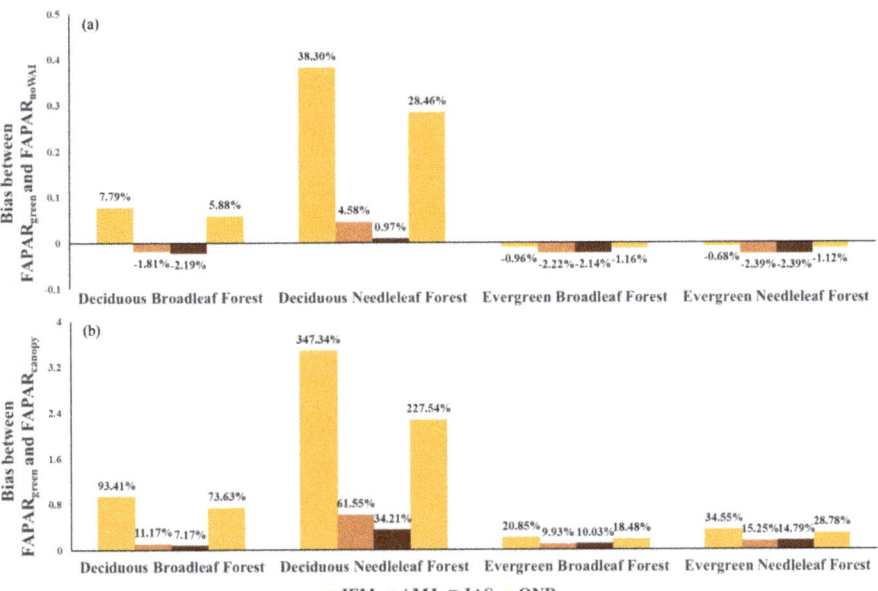

Figure 8. Bias between the black-sky $FAPAR_{green}$ and other black-sky FAPAR products within tile h10v05 during different periods in 2017: (**a**) bias between $FAPAR_{green}$ and $FAPAR_{noWAI}$; (**b**) bias between $FAPAR_{green}$ and $FAPAR_{canopy}$. JFM, AMJ, JAS, and OND represent the four seasons January to March, April to June, July to September and October to December, respectively.

Table 3. The ratios of black-sky $FAPAR_{green}$ to black-sky $FAPAR_{canopy}$ (R_{canopy}) and to $FAPAR_{noWAI}$ (R_{noWAI}) for selected forest types for different periods of 2017.

Period of Year	DBF		DNF		EBF		ENF	
	R_{canopy} (%)	R_{noWAI}	R_{canopy} (%)	R_{noWAI}	R_{canopy} (%)	R_{noWAI}	R_{canopy} (%)	R_{noWAI}
JFM	52.59	93.14	23.86	73.90	82.94	101.01	74.55	100.72
AMJ	90.74	101.93	64.13	96.32	91.03	102.27	86.85	102.46
JAS	93.36	102.24	75.02	99.19	90.93	102.19	87.14	102.46
OND	60.60	95.50	35.75	81.65	84.54	101.20	78.00	101.19

JFM, AMJ, JAS, and OND represent the four seasons January to March, April to June, July to September and October to December, respectively.

4. Discussion

4.1. Uncertainty in Determining WAI

Currently, methods for measuring the woody area index (WAI) and woody-to-total area ratio can be classified into direct methods (e.g., destructive sampling) and indirect methods [66]. However, both the direct and indirect methods can only be applied at small scales and to a limited range of vegetation types. In this study, we used a constant value of the woody-to-total area ratio for each forest type and derived the WAI from the PAI value for the peak growth stage. We obtained woody-to-total area ratios for different forest types (ENF, EBF, DBF, and DNF) from an extensive literature review [58,64–67]—the statistical metrics, including the mean, standard deviation, and coefficient of variation, are shown in Table 4. These results show that the woody-to-total area ratios for different forest types vary from 0.158 to 0.3. The variation in this ratio within each of the forest types is also large—10.00% to 82.22%. Therefore, the woody-to-total area ratio not only varies with the forest type but also changes a lot for each individual forest type. This means that there is definitely some uncertainty due to these factors. Even so, Zou et al. [66] showed that the woody-to-total area ratio is relatively stable for the same forest stand and thus the assumption of a fixed woody-to-total area ratio for each forest type is reasonable. Differences in in situ measurement methods can also contribute to the variation in the ratio for a given forest type. The retrieval of $FAPAR_{green}$ can thus be improved if an accurate woody-to-total area ratio dataset is available.

Table 4. Statistical details of prior woody-to-total area ratios.

Forest Vegetation Type	ENF	EBF	DNF	DBF
number of samples	35	8	3	4
mean value	0.185	0.18	0.3	0.158
standard deviation	0.062	0.148	0.03	0.101
coefficient of variation	33.51%	82.22%	10.00%	63.92%

4.2. Uncertainty Caused by the Use of Fixed Values of the Extinction Coefficients and $Albedo_{pure}$

The canopy directional transmittance can be determined using the gap fraction model [16,53,54]. Splitting the canopy into green components and woody components requires that the transmittances are also calculated separately. For the green components, the leaf absorptance at the VIS band varies from 0.79 to 0.94 as the chlorophyll content varies (from 20 to 100 µg/cm^2 under natural conditions), according to figures obtained using the PROSPECT-5 model [56,57]. Also, because no remote sensing leaf chlorophyll content product is available, the extinction coefficient for leaves (k_1) was assumed to have a fixed value of 0.88 (corresponding to a chlorophyll content of 35 µg/cm^2 [15]). For the extinction coefficient of woody components (k_2), we also used a fixed value of 0.91, based on the data simulated by LESS (a needleleaf forest scene). However, Suwa et al. [68] reported a smaller extinction coefficient of 0.77 for brighter woody stems. Therefore, the extinction coefficient of woody components may vary with forest type, and thus the use of a fixed value for k_2 is also a source of error. At present, it is still very challenging to determine extinction coefficients for different forest canopy types.

In addition, a fixed value of $Albedo_{pure}$ was used in the TriLay model for woody vegetation types (i.e., ENF, EBF, DBF, and DNF), and was approximated based on the dense vegetation [15]. According to the statistical results obtained from the MCD43A3 albedo product by Liu et al. [15], the visible albedo of dense vegetation with a "saturated" LAI value (e.g., LAI = 6) is very low and stable, with a mean value of 0.025 and a variance of 0.007 for white-sky condition, and a mean value of 0.020 and a variance of 0.006 for black-sky conditions. Therefore, the use of the prior VIS albedo values for "pure" vegetation may introduce a very small error, but it should be negligible [14].

4.3. Setting the Clumping Index for Photosynthetic and Woody Components

The clumping index characterizes the grouping of foliage within distinct canopy structures (such as tree crowns, shrubs, and row crops) relative to a random spatial distribution of leaves and is an important structural parameter for plant canopies that can influence canopy radiation regimes [49]. In our TriLay approach, the clumping index for woody components was assumed to be the same as for green leaves, and so is another definite source of error [69]. However, it is currently challenging to obtain the clumping effects of woody components within forest canopies [62]. Furthermore, Chen et al. [61] indicated that the clumping of shoots in branches has a similar effect to the clumping of leaves within shoots. What's more, Zou et al. [62] found that the differences between the estimated CI for canopy and woody components was below 6% at the zenithal ranges of 0°–75°, and the difference was only 2% in the range of 30°–60°, which is quite small at most medium zenithal ranges thus can represent most actual conditions. Based on these results and the unavailability of CI datasets for forest canopies at large scales, we directly used the value of CI from Jiao et al. [49] to describe the clumping effect for both leaves and woody components. Therefore, the use of the same clumping index for both leaves and woody components is reasonable and credibe enough.

5. Conclusions

In this paper, a triple-source leaf–wood–soil layer (TriLay) method for separating $FAPAR_{green}$ and $FAPAR_{woody}$ from $FAPAR_{canopy}$ using the MODIS LAI, land cover, and non-linear spectral mixture model (NSM)-retrieved soil albedo [15] together with global CI products [49] was proposed.

According to the validation carried out using LESS-simulated FAPAR values, the TriLay $FAPAR_{green}$ was more accurate (R^2 = 0.937, RMSE = 0.064 and bias = −6.02% for black-sky conditions; R^2 = 0.997, RMSE = 0.025 and bias = −4.04% for white-sky conditions) than the traditional linear method (R^2 = 0.979, RMSE = 0.114 and bias = −18.04% for black-sky conditions; R^2 = 0.996, RMSE = 0.106 and bias = −16.93% for white-sky conditions), and also more accurate than FAPAR obtained without the consideration of woody components ($FAPAR_{noWAI}$) (R^2 = 0.920, RMSE = 0.071 and bias = −7.14% for black-sky conditions; R^2 = 0.999, RMSE = 0.043 and bias = −6.41% for white-sky conditions). A comparison of the results for black-sky $FAPAR_{green}$ against $FAPAR_{noWAI}$ and $FAPAR_{canopy}$ showed that the discrepancies between the black-sky $FAPAR_{green}$ and other FAPAR products could not be ignored for forest types. In particular, for deciduous needleleaf forest, the black-sky $FAPAR_{green}$ contributed only about 23.86% and 35.75% of $FAPAR_{canopy}$ during the early and late stages (JFM and OND) of the year, respectively, and 75.02% during the peak growth stage (JAS). There were also smaller discrepancies between the black-sky $FAPAR_{noWAI}$ and $FAPAR_{green}$. For deciduous needleleaf forests, in particular, the black-sky $FAPAR_{noWAI}$ was overestimated by 38.30% and 28.46%, respectively, during the early and late stages of the year (JFM and OND).

Overall, this study provides a new method for partitioning $FAPAR_{canopy}$ into $FAPAR_{green}$ and $FAPAR_{woody}$ for forest types and will improve the understanding of energy exchange within the canopy. In addition, the exclusion of the contribution of woody components may certainly improve the accuracy of the $FAPAR_{green}$ estimates for forest types, which is significant in terms of the better modeling of vegetation photosynthesis.

Author Contributions: Conceptualization—S.C. and L.L.; methodology—S.C. and L.L.; software—S.C., X.Z. and X.C.; validation—S.C., Y.X. and D.X.; formal analysis—S.C. and X.Q.; investigation—S.C. and L.L.; resources—S.C.; writing—original draft preparation—S.C.; writing—review and editing—L.L. and X.L.

Funding: This research was funded by the National Key Research and Development Program of China, grant number 2017YFA0603001, the Key Research Program of the Chinese Academy of Sciences (ZDRW-ZS-2019-1), the National Natural Science Foundation of China (41825002).

Acknowledgments: The authors gratefully acknowledge the CI products provided by Ziti Jiao at Beijing Normal University, and the MODIS products obtained from the Land Products Land Processes Distributed Active Archive Center (LP DAAC).

Conflicts of Interest: The authors declare no conflict of interest.

References

1. Monteith, J.L. Vegetation and the atmosphere. Volume 1. Principles. *J. Appl. Ecol.* **1977**, *14*, 655.
2. Sellers, P.; Dickinson, R.; Randall, D.; Betts, A.; Hall, F.; Berry, J.; Collatz, G.; Denning, A.; Mooney, H.; Nobre, C. Modeling the exchanges of energy, water, and carbon between continents and the atmosphere. *Science* **1997**, *275*, 502–509. [CrossRef] [PubMed]
3. Prince, S.D.; Goward, S.N. Global primary production: A remote sensing approach. *J. Biogeogr.* **1995**, *22*, 815–835. [CrossRef]
4. Ruimy, A.; Dedieu, G.; Saugier, B. Turc: A diagnostic model of continental gross primary productivity and net primary productivity. *Glob. Biogeochem. Cycles* **1996**, *10*, 269–285. [CrossRef]
5. Running, S.W.; Nemani, R.R.; Heinsch FAZhao, M.S.; Reeves, M.; Hashimoto, H. A continuous satellite-derived measure of global terrestrial primary production. *Bioscience* **2004**, *54*, 547–560. [CrossRef]
6. Potter, C.S.; Randerson, J.T.; Field, C.B.; Matson, P.A.; Vitousek, P.M.; Mooney, H.A.; Klooster, S.A. Terrestrial ecosystem production: A process model based on global satellite and surface data. *Glob. Biogeochem. Cycles* **1993**, *7*, 811–841. [CrossRef]
7. Verhoef, W.; Bach, H. Coupled soil-leaf-canopy and atmosphere radiative transfer modeling to simulate hyperspectral multi-angular surface reflectance and toa radiance data. *Remote Sens. Environ.* **2007**, *109*, 166–182. [CrossRef]
8. Liu, R.; Huang, W.; Ren, H.; Yang, G.; Xie, D.; Wang, J. Photosynthetically active radiation vertical distribution model in maize canopy. *Trans. Chin. Soc. Agric. Eng.* **2011**, *27*, 115–121.
9. Li, W.; Fang, H. Estimation of direct, diffuse, and total fpars from landsat surface reflectance data and ground-based estimates over six fluxnet sites. *J. Geophys. Res. Biogeosci.* **2015**, *120*, 96–112. [CrossRef]
10. Myneni, R.B.; Ramakrishna, R.; Nemani, R.; Running, S.W. Estimation of global leaf area index and absorbed par using radiative transfer models. *IEEE Trans. Geosci. Remote Sens.* **2002**, *35*, 1380–1393. [CrossRef]
11. Knyazikhin, Y.; Martonchik, J.V.; Diner, D.J.; Myneni, R.B.; Verstraete, M.; Pinty, B.; Gobron, N. Estimation of vegetation canopy leaf area index and fraction of absorbed photosynthetically active radiation from atmosphere-corrected misr data. *J. Geophys. Res. Space Phys.* **1998**, *103*, 32257–32275. [CrossRef]
12. Tian, Y.; Zhang, Y.; Knyazikhin, Y.; Myneni, R.B.; Glassy, J.M.; Dedieu, G.; Running, S.W. Prototyping of modis lai and fpar algorithm with lasur and landsat data. *IEEE Trans. Geosci. Remote Sens.* **2000**, *38*, 2387–2401. [CrossRef]
13. Gobron, N.; Pinty, B.; Verstraete, M.M.; Govaerts, Y. A semidiscrete model for the scattering of light by vegetation. *J. Geophys. Res. Space Phys.* **1997**, *102*, 9431–9446. [CrossRef]
14. Xiao, Z.; Liang, S.; Rui, S.; Wang, J.; Bo, J. Estimating the fraction of absorbed photosynthetically active radiation from the modis data based glass leaf area index product. *Remote Sens. Environ.* **2015**, *171*, 105–117. [CrossRef]
15. Liu, L.; Zhang, X.; Xie, S.; Liu, X.; Song, B.; Chen, S.; Peng, D. Global white-sky and black-sky fapar retrieval using the energy balance residual method: Algorithm and validation. *Remote Sens.* **2019**, *11*, 1004. [CrossRef]
16. Chen, J.M. Canopy architecture and remote sensing of the fraction of photosynthetically active radiation absorbed by boreal conifer forests. *IEEE IEEE Trans. Geosci. Remote Sens.* **1996**, *34*, 1353–1368. [CrossRef]
17. Chen, L.; Liu, Q.; Fan, W.; Li, X.; Xiao, Q.; Yan, G.; Tian, G. A bi-directional gap model for simulating the directional thermal radiance of row crops. *Sci. China* **2002**, *45*, 1087–1098. [CrossRef]
18. Fan, W.; Yuan, L.; Xu, X.; Chen, G.; Zhang, B. A new fapar analytical model based on the law of energy conservation: A case study in china. *IEEE J. Sel. Top. Appl. Earth Obs. Remote Sens.* **2014**, *7*, 3945–3955. [CrossRef]
19. Fang, H.; Liang, S.; Mcclaran, M.P.; Leeuwen, W.J.D.V.; Drake, S.; Marsh, S.E.; Thomson, A.M.; Izaurralde, R.C.; Rosenberg, N.J. Biophysical characterization and management effects on semiarid rangeland observed from landsat etm+ data. *IEEE Trans. Geosci. Remote Sens.* **2005**, *43*, 125–134. [CrossRef]

20. Myneni, R.B.; Hoffman, S.; Knyazikhin, Y.; Privette, J.L.; Glassy, J.; Tian, Y.; Wang, Y.; Song, X.; Zhang, Y.; Smith, G.R.; et al. Global products of vegetation leaf area and fraction absorbed par from year one of modis data. *Remote Sens. Environ.* **2002**, *83*, 214–231. [CrossRef]
21. Knyazikhin, Y.; Martonchik, J.; Myneni, R.B.; Diner, D.; Running, S.W. Synergistic algorithm for estimating vegetation canopy leaf area index and fraction of absorbed photosynthetically active radiation from modis and misr data. *J. Geophys. Res. Space Phys.* **1998**, *103*, 32257–32275. [CrossRef]
22. Baret, F.; Hagolle, O.; Geiger, B.; Bicheron, P.; Miras, B.; Huc, M.; Berthelot, B.; Niño, F.; Weiss, M.; Samain, O. Lai, fapar and fcover cyclopes global products derived from vegetation: Part 1: Principles of the algorithm. *Remote Sens. Environ.* **2009**, *110*, 275–286. [CrossRef]
23. Plummer, S.; Arino, O.; Simon, M.; Steffen, W. Establishing a earth observation product service for the terrestrial carbon community: The globcarbon initiative. *Mitig. Adapt. Strat. Glob. Chang.* **2006**, *11*, 97–111. [CrossRef]
24. Gobron, N.; Pinty, B.; Verstraete, M.; Govaerts, Y. The MERIS Global Vegetation Index (MGVI): Description and preliminary application. *Int. J. Remote Sens.* **1999**, *20*, 1917–1927. [CrossRef]
25. Pinty, B.; Clerici, M.; Andredakis, I.; Kaminski, T.; Taberner, M.; Verstraete, M.M.; Gobron, N.; Plummer, S.; Widlowski, J.-L. Exploiting the MODIS albedos with the Two-stream Inversion Package (JRC-TIP): 2. Fractions of transmitted and absorbed fluxes in the vegetation and soil layers. *J. Geophys. Res. Space Phys.* **2011**, *116*. [CrossRef]
26. Disney, M.; Muller, J.-P.; Kharbouche, S.; Kaminski, T.; Voßbeck, M.; Lewis, P.; Pinty, B. A New Global fAPAR and LAI Dataset Derived from Optimal Albedo Estimates: Comparison with MODIS Products. *Remote Sens.* **2016**, *8*, 275. [CrossRef]
27. Wang, Y.; Tian, Y.; Zhang, Y.; Elsaleous, N.; Knyazikhin, Y.; Vermote, E.; Myneni, R.B. Investigation of product accuracy as a function of input and model uncertainties—Case study with seawifs and modis lai/fpar algorithm. *Remote Sens. Environ.* **2000**, *78*, 299–313. [CrossRef]
28. Camacho, F.; Cernicharo, J.; Lacaze, R.; Baret, F.; Weiss, M. GEOV1: LAI, FAPAR essential climate variables and FCOVER global time series capitalizing over existing products. Part 2: Validation and intercomparison with reference products. *Remote Sens. Environ.* **2013**, *137*, 310–329. [CrossRef]
29. Gobron, N.; Pinty, B.; Aussedat, O.; Taberner, M.; Faber, O.; Melin, F.; Lavergne, T.; Robustelli, M.; Snoeij, P. Uncertainty estimates for the FAPAR operational products derived from MERIS—Impact of top-of-atmosphere radiance uncertainties and validation with field data. *Remote Sens. Environ.* **2008**, *112*, 1871–1883. [CrossRef]
30. Fritsch, S.; Machwitz, M.; Ehammer, A.; Conrad, C.; Dech, S. Validation of the collection 5 MODIS FPAR product in a heterogeneous agricultural landscape in arid Uzbekistan using multitemporal RapidEye imagery. *Int. J. Remote Sens.* **2012**, *33*, 6818–6837. [CrossRef]
31. Pickett-Heaps, C.A.; Canadell, J.; Briggs, P.R.; Gobron, N.; Haverd, V.; Paget, M.J.; Pinty, B.; Raupach, M.R. Evaluation of six satellite-derived Fraction of Absorbed Photosynthetic Active Radiation (FAPAR) products across the Australian continent. *Remote Sens. Environ.* **2014**, *140*, 241–256. [CrossRef]
32. Tao, X.; Liang, S.; Wang, D. Assessment of five global satellite products of fraction of absorbed photosynthetically active radiation: Intercomparison and direct validation against ground-based data. *Remote Sens. Environ.* **2015**, *163*, 270–285. [CrossRef]
33. Gitelson, A.A.; Peng, Y.; Arkebauer, T.J.; Suyker, A.E. Productivity, absorbed photosynthetically active radiation, and light use efficiency in crops: Implications for remote sensing of crop primary production. *J. Plant Physiol.* **2015**, *177*, 100–109. [CrossRef] [PubMed]
34. Tewes, A.; Schellberg, J. Towards remote estimation of radiation use efficiency in maize using uav-based low-cost camera imagery. *Agronomy* **2018**, *8*, 16. [CrossRef]
35. Miao, G.; Guan, K.; Xi, Y.; Bernacchi, C.J.; Masters, M.D. Sun-induced chlorophyll fluorescence, photosynthesis, and light use efficiency of a soybean field. *J. Geophys. Res. Biogeosci.* **2018**, *123*, 610–623. [CrossRef]
36. Gitelson, A.A.; Arkebauer, T.J.; Suyker, A.E. Convergence of daily light use efficiency in irrigated and rainfed C3 and C4 crops. *Remote Sens. Environ.* **2018**, *217*, 30–37. [CrossRef]
37. Asner, G.P.; Wessman, C.A.; Archer, S. Scale dependence of absorption of photosynthetically active radiation in terrestrial ecosystems. *Ecol. Appl.* **1998**, *8*, 1003–1021. [CrossRef]
38. Hall, F.G.; Huemmrich, K.F.; Goetz, S.J.; Sellers, P.J.; Nickeson, J.E. Satellite remote sensing of surface energy balance: Success, failures, and unresolved issues in FIFE. *J. Geophys. Res. Space Phys.* **1992**, *97*, 19061–19089. [CrossRef]

39. Zhang, Q.; Xiao, X.; Braswell, B.; Linder, E.; Baret, F.; Moore, B. Estimating light absorption by chlorophyll, leaf and canopy in a deciduous broadleaf forest using MODIS data and a radiative transfer model. *Remote Sens. Environ.* **2005**, *99*, 357–371. [CrossRef]
40. Gitelson, A.A. Remote estimation of fraction of radiation absorbed by photosynthetically active vegetation: Generic algorithm for maize and soybean. *Remote Sens. Lett.* **2019**, *10*, 283–291. [CrossRef]
41. Gitelson, A.A.; Peng, Y.; Arkebauer, T.J.; Schepers, J. Relationships between gross primary production, green LAI, and canopy chlorophyll content in maize: Implications for remote sensing of primary production. *Remote Sens. Environ.* **2014**, *144*, 65–72. [CrossRef]
42. Gitelson, A.A.; Peng, Y.; Huemmrich, K.F. Relationship between fraction of radiation absorbed by photosynthesizing maize and soybean canopies and NDVI from remotely sensed data taken at close range and from MODIS 250m resolution data. *Remote Sens. Environ.* **2014**, *147*, 108–120. [CrossRef]
43. Qi, J.; Xie, D.; Guo, D.; Yan, G. A large-scale emulation system for realistic three-dimensional (3-d) forest simulation. *IEEE J. Sel. Top. Appl. Earth Obs. Remote Sens.* **2017**, *10*, 4834–4843. [CrossRef]
44. Myneni, R.; Knyazikhin, Y.; Park, T. *MCD15A2H MODIS/Terra + Aqua leaf area index/FPAR 8-day L4 Global 500 m SIN Grid V006, NASA EOSDIS Land Processes DAAC.* 2015. Available online: http://doi.org/10.5067/MODIS/MCD15A2H.006 (accessed on 12 September 2019).
45. Vermote, E.; Vermeulen, A. Atmospheric correction algorithm: Spectral reflectances (mod09). *ATBD Version* **1999**, *4*, 1–107.
46. Friedl, M.; McIver, D.; Hodges, J.; Zhang, X.; Muchoney, D.; Strahler, A.; Woodcock, C.; Gopal, S.; Schneider, A.; Cooper, A.; et al. Global land cover mapping from MODIS: Algorithms and early results. *Remote Sens. Environ.* **2002**, *83*, 287–302. [CrossRef]
47. Friedl, M.A.; Sulla-Menashe, D.; Tan, B.; Schneider, A.; Ramankutty, N.; Sibley, A.; Huang, X. MODIS Collection 5 global land cover: Algorithm refinements and characterization of new datasets. *Remote Sens. Environ.* **2010**, *114*, 168–182. [CrossRef]
48. Chen, J.; Menges, C.; Leblanc, S. Global mapping of foliage clumping index using multi-angular satellite data. *Remote Sens. Environ.* **2005**, *97*, 447–457. [CrossRef]
49. Jiao, Z.; Dong, Y.; Schaaf, C.B.; Chen, J.M.; Roman, M.; Wang, Z.; Zhang, H.; Ding, A.; Erb, A.; Hill, M.J.; et al. An algorithm for the retrieval of the clumping index (CI) from the MODIS BRDF product using an adjusted version of the kernel-driven BRDF model. *Remote Sens. Environ.* **2018**, *209*, 594–611. [CrossRef]
50. Liu, L.; Zhang, X. Dynamic Mapping of Broadband Visible Albedo of Soil Background at Global 500-m Scale from MODIS Satellite Products. In *Land Surface and Cryosphere Remote Sensing IV*; International Society for Optics and Photonics: Washington, DC, USA, 2018; p. 107770L.
51. Irons, J.R.; Ranson, K.J.; Daughtry, C.S.T. Estimating big bluestem albedo from directional reflectance measurements. *Remote Sens. Environ.* **1988**, *25*, 185–199. [CrossRef]
52. Carrer, D.; Meurey, C.; Ceamanos, X.; Roujean, J.L.; Calvet, J.C. Dynamic mapping of snow-free vegetation and bare soil albedos at global 1 km scale from 10-year analysis of modis satellite products. *Remote Sens. Environ.* **2014**, *140*, 420–432. [CrossRef]
53. Widlowski, J.-L. On the bias of instantaneous FAPAR estimates in open-canopy forests. *Agric. For. Meteorol.* **2010**, *150*, 1501–1522. [CrossRef]
54. Lhomme, J.-P.; Chehbouni, A. Comments on dual-source vegetation—Atmosphere transfer models. *Agric. For. Meteorol.* **1999**, *94*, 269–273. [CrossRef]
55. Hosgood, B.; Jacquemoud, S.; Andreoli, G.; Verdebout, J.; Pedrini, G.; Schmuck, G. *Leaf Optical Properties Experiment 93 (LOPEX93)*; Report EUR—16095-EN; Joint Research Centre, Institute for Remote Sensing Applications: Ispra, Italy; European Commission: Luxembourg, 1995.
56. Jacquemoud, S.; Baret, F. PROSPECT: A model of leaf optical properties spectra. *Remote Sens. Environ.* **1990**, *34*, 75–91. [CrossRef]
57. Féret, J.-B.; François, C.; Asner, G.P.; Gitelson, A.A.; Martin, R.E.; Bidel, L.P.; Ustin, S.L.; Le Maire, G.; Jacquemoud, S. PROSPECT-4 and 5: Advances in the leaf optical properties model separating photosynthetic pigments. *Remote Sens. Environ.* **2008**, *112*, 3030–3043. [CrossRef]
58. Chen, J.M. Optically-based methods for measuring seasonal variation of leaf area index in boreal conifer stands. *Agric. For. Meteorol.* **1996**, *80*, 135–163. [CrossRef]
59. Kucharik, C.J.; Norman, J.M.; Gower, S.T. Measurements of branch area and adjusting leaf area index indirect measurements. *Agric. For. Meteorol.* **1998**, *91*, 69–88. [CrossRef]

60. Sea, W.B.; Choler, P.; Beringer, J.; Weinmann, R.A.; Hutley, L.B.; Leuning, R. Documenting improvement in leaf area index estimates from MODIS using hemispherical photos for Australian savannas. *Agric. For. Meteorol.* **2011**, *151*, 1453–1461. [CrossRef]
61. Chen, J.M.; Leblanc, S.G. A four-scale bidirectional reflectance model based on canopy architecture. *IEEE Trans. Geosci. Remote Sens.* **1997**, *35*, 1316–1337. [CrossRef]
62. Jie, Z.; Yan, G.; Ling, C. Estimation of canopy and woody components clumping indices at three mature. *IEEE J. Sel. Top. Appl. Earth Obs. Remote Sens.* **2015**, *8*, 1–10.
63. Clark, D.B.; Olivas, P.C.; Oberbauer, S.F.; Clark, D.A.; Ryan, M.G. First direct landscape-scale measurement of tropical rain forest leaf area index, a key driver of global primary productivity. *Ecol. Lett.* **2008**, *11*, 163–172. [CrossRef]
64. Zheng, G.; Ma, L.; Wei, H.; Eitel, J.U.H.; Moskal, L.M.; Zhang, Z. Assessing the contribution of woody materials to forest angular gap fraction and effective leaf area index using terrestrial laser scanning data. *IEEE Trans. Geosci. Remote Sens.* **2016**, *54*, 1475–1487. [CrossRef]
65. Zou, J.; Zhuang, Y.; Chianucci, F.; Mai, C.; Lin, W.; Leng, P.; Luo, S.; Yan, B. Comparison of Seven Inversion Models for Estimating Plant and Woody Area Indices of Leaf-on and Leaf-off Forest Canopy Using Explicit 3D Forest Scenes. *Remote Sens.* **2018**, *10*, 1297. [CrossRef]
66. Zou, J.; Yan, G.; Zhu, L.; Zhang, W. Woody-to-total area ratio determination with a multispectral canopy imager. *Tree Physiol.* **2009**, *29*, 1069–1080. [CrossRef] [PubMed]
67. Ma, L.; Zheng, G.; Eitel, J.U.; Magney, T.S.; Moskal, L.M. Determining woody-to-total area ratio using terrestrial laser scanning (TLS). *Agric. For. Meteorol.* **2016**, *228*, 217–228. [CrossRef]
68. Suwa, R. Canopy photosynthesis in a mangrove considering vertical changes in light-extinction coefficients for leaves and woody organs. *J. For. Res.* **2011**, *16*, 26–34. [CrossRef]
69. Chen, J.M.; Cihlar, J. Retrieving leaf area index of boreal conifer forests using landsat tm images. *Remote Sens. Environ.* **1996**, *55*, 153–162. [CrossRef]

© 2019 by the authors. Licensee MDPI, Basel, Switzerland. This article is an open access article distributed under the terms and conditions of the Creative Commons Attribution (CC BY) license (http://creativecommons.org/licenses/by/4.0/).

Article

Climate Data Records of Vegetation Variables from Geostationary SEVIRI/MSG Data: Products, Algorithms and Applications

Francisco Javier García-Haro [1,*], Fernando Camacho [2], Beatriz Martínez [1], Manuel Campos-Taberner [1], Beatriz Fuster [2], Jorge Sánchez-Zapero [2] and María Amparo Gilabert [1]

[1] Earth Physics and Thermodynamics Departmnet, Faculty of Physics, Universitat de València, Dr. Moliner, 46100 Burjassot, València, Spain
[2] Earth Observation Laboratory (EOLAB), Parc Científic de la Universitat de València, Catedrático Agustín Escardino, 9, 46980 Paterna, València, Spain
* Correspondence: j.garcia.haro@uv.es

Received: 26 July 2019; Accepted: 6 September 2019; Published: 9 September 2019

Abstract: The scientific community requires long-term data records with well-characterized uncertainty and suitable for modeling terrestrial ecosystems and energy cycles at regional and global scales. This paper presents the methodology currently developed in EUMETSAT within its Satellite Application Facility for Land Surface Analysis (LSA SAF) to generate biophysical variables from the Spinning Enhanced Visible and InfraRed Imager (SEVIRI) on board MSG 1-4 (Meteosat 8-11) geostationary satellites. Using this methodology, the LSA SAF generates and disseminates at a time a suite of vegetation products, such as the leaf area index (LAI), the fraction of the photosynthetically active radiation absorbed by vegetation (FAPAR) and the fractional vegetation cover (FVC), for the whole Meteosat disk at two temporal frequencies, daily and 10-days. The FVC algorithm relies on a novel stochastic spectral mixture model which addresses the variability of soils and vegetation types using statistical distributions whereas the LAI and FAPAR algorithms use statistical relationships general enough for global applications. An overview of the LSA SAF SEVIRI/MSG vegetation products, including expert knowledge and quality assessment of its internal consistency is provided. The climate data record (CDR) is freely available in the LSA SAF, offering more than fifteen years (2004-present) of homogeneous time series required for climate and environmental applications. The high frequency and good temporal continuity of SEVIRI products addresses the needs of near-real-time users and are also suitable for long-term monitoring of land surface variables. The study also evaluates the potential of the SEVIRI/MSG vegetation products for environmental applications, spanning from accurate monitoring of vegetation cycles to resolving long-term changes of vegetation.

Keywords: meteosat second generation (MSG); biophysical parameters (LAI; FVC; FAPAR); SEVIRI; climate data records (CDR); stochastic spectral mixture model (SSMM); Satellite Application Facility for Land Surface Analysis (LSA SAF)

1. Introduction

Observations and monitoring are essential components of Global Framework for Climate Services initiative [1,2] that contribute significantly to meet the needs of climate research services. The role that satellites have played in observing the variability and change of the Earth system has increased significantly in the last decades. Thanks to the Earth observation (EO) satellite systems, a large number of variables related to the atmosphere, oceanic and terrestrial domains are accessible [3].

The Satellite Application Facility (SAF) for Land Surface Analysis (LSA) is a leading center for development and operational retrieval of land surface variables from European Organization for the Exploitation of Meteorological Satellites (EUMETSAT), specifically aimed to understand and quantify terrestrial processes and land-atmosphere interactions. Land surface models' assessment and improvement of high quality and robust EO products are among the target applications of LSA SAF products. Particularly, the geostationary Spinning Enhanced Visible and Infrared Imager (SEVIRI) on board the Meteosat Second Generation (MSG) platform provides a very high temporal resolution (15 minutes) required for resolving the diurnal cycles of radiation, surface albedo and surface temperature [4]. The MSG program covers a series of four identical satellites MSG 1-4 (Meteosat 8-11) launched in 2002, 2005, 2012, and 2015 respectively. SEVIRI operates in 12 spectral channels [5], although only channels 1 (VIS0.6), 2 (VIS0.8) and 3 (NIR1.6), centered at about 0.635 µm (red), 0.81 µm (NIR) and 1.64 µm (SWIR), respectively situate in the optical domain.

The LSA SAF provides also variables for the characterization of terrestrial ecosystems and their role in the energy balance of Earth, such as land surface fluxes and three important biophysical variables related with the amount, structure and state of vegetation: Leaf Area Index (LAI), Fraction of Absorbed Photosynthetically Active Radiation (FAPAR) and Fractional Vegetation Cover (FVC) [6]. These variables are widely used in many land-surface vegetation, climate, and crop production models [7–13]. FVC represents the fraction of green vegetation covering a unit area of horizontal soil, corresponding to the gap fraction in the nadir direction. LAI is a measure of the amount of live foliage present in the canopy per unit ground surface. The FAPAR is the fraction of photosynthetically active radiation (400-700 nm) absorbed by the green parts of the canopy, and therefore constitutes an indicator of the presence, health and productivity of live vegetation.

Nowadays, consistent long-term data records of these variables with well-characterized uncertainty are currently required by the scientific community to model terrestrial ecosystems and energy cycles at regional and global scales [14]. Models related with land surface and climate usually require Near-Real-Time (NRT) datasets of these variables at high temporal frequencies, and with regional to global coverage [15]. Both, LAI and FAPAR are recognized as essential climate variables in the terrestrial domain by Global Climate Observing System (GCOS) [16].

Several methods have been proposed for estimating biophysical parameters from large scale optical sensors, some of which have been implemented in operational processing lines from POLarization and Directionality of Earth Reflectances (POLDER) [17], Moderate Resolution Imaging Spectroradiometer (MODIS) [18], MISR (Multi-angle Imaging SpectroRadiometer) [19], MEdium Resolution Imaging Spectrometer (MERIS) [20,21], Sea-viewing Wide Field-of-view *Sensor* (SEAWIFS) [22], VEGETATION [23,24] and Advanced Very High Resolution Radiometer (AVHRR) [25]. These methods can be roughly classified in two main categories: (i) methods based on optimized vegetation indices [26], and (ii) methods based on the inversion of physical models. Both approaches present advantages and drawbacks. It is referred to [27] for a detail intercomparison on the spread of currently available physical models. Although methods that invert physical models can potentially be applied to varying surface types, they require specifying a large number of model parameters, which may lead to unstable solutions [28]. Most inversion methods use one-dimensional (turbid) models (e.g., PROSAIL [29]), thereby assuming surface homogeneity within an image pixel. They are thus limited to address the sub-pixel mixing of land cover types, which is common at coarse resolutions satellites. The process of identifying the sub-pixel proportions of the constituent components is called spectral mixture model (SMM). SMM approaches are methods especially adequate for global studies, since the spatial variability within pixel is high [30,31].

In this paper, we describe the methodology currently used in LSA SAF to operationally generate the suite of SEVIRI/MSG (FVC, LAI and FAPAR) vegetation products. The high frequency of acquisition provided by the SEVIRI instrument assures the availability of cloud-free data for appropriately monitoring the vegetation dynamics in large regions. When one is choosing a retrieval method, one needs to consider the limited spatial and spectral resolution of SEVIRI instrument (i.e., only 3 optical channels and a spatial resolution of 3 km at the sub-satellite point that declines rapidly at higher latitudes). Furthermore, the challenges that the MSG geostationary satellite poses as compared to polar-orbiting satellite sensors due to diurnal variations in the solar zenith angle need to be considered. The algorithm to retrieve FVC relies on a novel stochastic SMM method, which addresses the large variability of soils and vegetation types using statistical (sum of Gaussian) distributions. The LSA SAF operational scheme to retrieve LAI and FAPAR uses statistical relationships general enough for global applications, proposed by Roujean and Lacaze [16] and Roujean and Bréon [32], respectively.

LSA SAF has been providing in NRT vegetation products (FVC, LAI and FAPAR) at two different time resolutions (daily and 10-days) since 2008 over the geostationary Meteosat disk, covering Europe, Africa, the Middle East and parts of South America. The suite of SEVIRI/MSG vegetation products is freely disseminated to users through LSA SAF website (https://landsaf.ipma.pt). During the different operational phases, these vegetation products have been routinely assessed and some improvements have been made in the upgraded versions. LSA SAF has recently reprocessed the entire MSG archive with the latest version of the several retrieval algorithms in order to obtain a continuous and homogeneous Climate Data Records (CDR) of vegetation products. More than fifteen years (January 2004- present) of homogeneous consistent estimates of FVC, LAI and FAPAR fields are available. The 10-days (FVC, LAI and FAPAR) CDRs are part of the suite of EUMETSAT climate products, which include Fundamental CDRs, Thematic CDRs and Data for operational climate monitoring and are also available from EUMETSAT web page (https://navigator.eumetsat.int/).

The paper presents the following structure. Section 2 is dedicated to describe the algorithms to retrieve FVC, LAI and FAPAR. The suite of LSA SAF SEVIRI/MSG vegetation products are described in Section 3, including expert knowledge and a quality assessment of the consistency between the products. Section 4 provides further insight about the added value and introduces examples of its application in numerous environmental applications.

2. Algorithm Description

2.1. SEVIRI/MSG

The algorithm for retrieving biophysical variables from SEVIRI/MSG data uses as input normalized spectral bidirectional reflectance factor (BRDF) at a fixed angular configuration (i.e., vertical illumination and observation angles of 0°) in order to reduce angular effects. This hotspot geometry leads to a minimum contribution of the shadow proportion and a precise estimation of FVC, coinciding with the complement to unity of the gap fraction at nadir direction [16]. Estimating the FVC with increased sun zenith angles would lead to an overestimation of FVC. The main drawback of this geometry is the significant contribution of illuminated soil background, which constitutes a confounding factor for the retrieval of FVC. Figure 1 outlines the main ingredients for deriving the SEVIRI/MSG products including the stochastic SMM approach for retrieving FVC, a pragmatic method for LAI retrieval from FVC, and a method based on an optimized vegetation index for FAPAR determination.

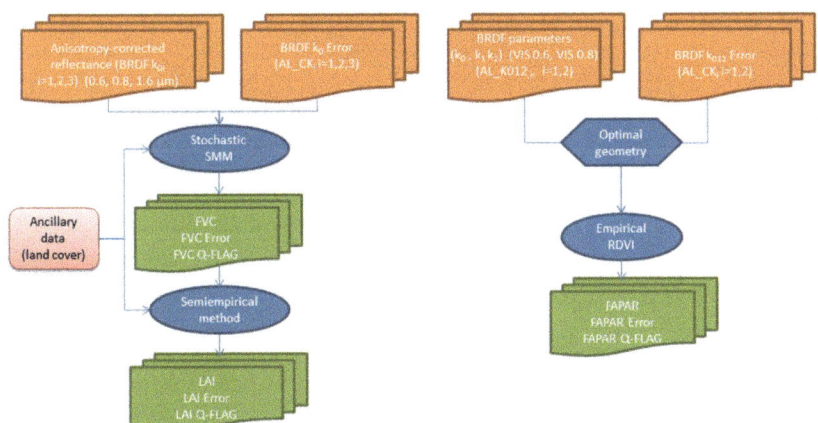

Figure 1. Flow chart of the algorithm for FVC, LAI and FAPAR determination.

The BRDF algorithm applies a semi-empirical reflectance model in order to invert SEVIRI/MSG top-of-canopy reflectance factor values into three parameters (k_0, k_1, k_2). The input of the FVC algorithm is atmospherically corrected BRDF k_0 parameters in the MSG channels 1, 2 and 3, while FAPAR uses all three BRDF parameters (k_0, k_1, k_2) in order to correct surface's reflectance anisotropy and minimize the effect of soil reflectance (see Figure 1). The algorithm to correct atmospheric effects and generate BRDF model parameters is fully described in [33], and in the ATBD of the Albedo MSG product [34].

2.2. FVC Algorithm

Let **r** be the spectrum of each mixture pixel, i.e., a column vector ($r_1, r_2, ..., r_n$), where n is the total number of bands. When multiple scattering can be reasonably disregarded, the spectral reflectance of each pixel can be approximated by a linear mixture of pure spectra **E** (the so-called endmembers) weighted by their corresponding fractional abundances:

$$r = E f + \varepsilon, \tag{1}$$

where **E** [$n \times c$] is the matrix of endmembers, **f** is a vector with the c unknown proportions in the mixture, and ε is the residual vector. The mixing equation is accompanied by two constraints: (1) the normalization constraint says proportions should sum up to one, and (2) the positivity constraint says that no endmember can make a negative contribution. The least-square principle establishes that the unknown parameters are those that minimize the Mahalanobis distance between the pixel **r** and point **E f**:

$$\chi 2 = (r - E f)^T V(r)^{-1} (r - E f) \tag{2}$$

where **V(r)** denotes the error matrix of the observations **r**. The coarse spatial resolution of SEVIRI and the availability of only three spectral bands poses a significant challenge for endmember selection in traditional (deterministic) SMM. The FVC algorithm relies on a novel stochastic SMM which uses a statistical representation of the endmembers to accommodate their variability at a global scale. Each pixel is described by a multiple combination of two pure classes, the target (vegetation) and the background (soil). Both vegetation and soil classes are not treated as deterministic (i.e., using fixed endmembers) but they follow a statistical (multi-modal) distribution, which attempts to capture the variability of soils and vegetation components at a global scale. The main steps of the algorithm are now described (see Figure 2).

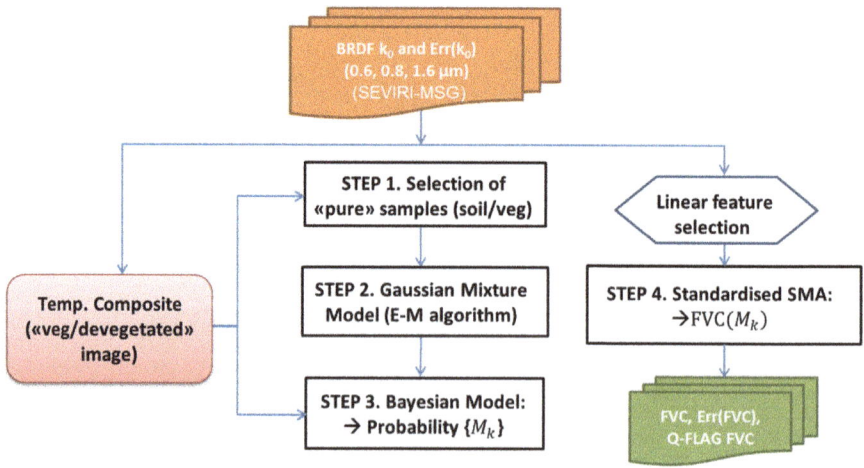

Figure 2. Flow chart of the stochastic SMM algorithm for FVC determination.

Step 1: Selection of a training database

Given the large diversity of vegetation types and underlying soil background that can be found in the SEVIRI disk, a large training database of spectral signatures needs to be generated. To more reliably identify the pure components, two composite k_0 images were generated from all high quality observations (cloud- and snow-free) over a one-year period: a vegetated k_0 image corresponding to the peak of season and a devegetated k_0 image corresponding to the minimum canopy closure. The samples were chosen to be homogeneous over areas higher than SEVIRI spatial resolution, and the classification is based on the 1-km Global Land Cover 2000 (GLC2000) [35]. Training areas of non-vegetated class include desert areas, sparsely vegetated and shrublands. Vegetation areas were identified in crop and forest classes. Samples were further verified using purity methods to filter out possible outliers. In particular, pixels having less than 95% of soil/vegetation were excluded.

Step 2: Gaussian mixture model

At the SEVIRI resolution, the soils and vegetation components usually present a high sub-pixel spatial variability. In order to address this variability, a parametric mixture model weighted sum of Gaussian distributions (or clusters) is adopted [36]. This probabilistic model is applied separately for the target (vegetation) and the background (soil) classes:

$$f_{\text{soil}}(x|\mu_k, \Sigma_k) = \sum_{k=1}^{G_s} \tau_k \phi_k(x|\mu_k, \Sigma_k) \quad (3)$$

$$f_{\text{veg}}(x|\mu_k, \Sigma_k) = \sum_{k'=1}^{G_v} \tau_{k'} \phi_{k'}(x|\mu_{k'}, \Sigma_{k'}), \quad (4)$$

where ϕ_k is class conditional distribution of the k-th Gaussian, with mean μ_k and covariance Σ_k,

$$\phi_k(x|\mu_k, \Sigma_k) = (2\pi)^{-n/2}|\Sigma_k|^{-1/2} \exp\left(-\frac{1}{2}(x-\mu_k)^T \Sigma_k^{-1}(x-\mu_k)\right) \quad (5)$$

τ_k is the probability that an observation belongs to the k-th component ($\tau_k \geq 0$; $\sum_{k=1}^{G_s} \tau_k = 1$), and G_s and G_v are the number of Gaussian components for soil and vegetation, respectively. The algorithm uses the Expectation-Maximization (E-M) approach [37] to estimate the means μ_k and covariances Σ_k of the individual Gaussian components. This simple iterative approach increases the log likelihood of the observed data and usually converges if the data conform reasonably well to the mixture model. The k-means algorithm is used to initialize the E-M parameters (μ_k, Σ_k). Multiple initializations are made to avoid numerical problems local maxima. We assume ellipsoidal unconstrained component covariance matrices and use the Bayesian Information Criterion (BIC) [38,39] to determine an appropriate number of Gaussian components. The value of BIC is the maximized log likelihood with a penalty for the number of parameters. The larger the BIC score, the stronger the evidence for the model. The determination of the number of Gaussian components takes into account not only the BIC score but also the requirement that the distributions must provide a faithful representation of the data. A typical number of 6-7 Gaussians for soil and 3-5 for vegetation has been used to represent the variability of the different SEVIRI geographical areas.

Step 3: Model selection

Each SEVIRI pixel is usually a mixture of several backgrounds and/or vegetation components. The stochastic SMM computes all possible models by taking all possible sets of models $\{M_1, \ldots M_N\}$, with $N = G_s \cdot G_v$. A model M_k is defined a pair of class-conditional distributions for vegetation-soil $M_k \equiv (f_{\text{soil}(k)}, f_{\text{veg}(k')})$. At the SEVIRI resolution, different sets of soil and vegetation may yield to very similar mixtures. Let $p(M_k|r)$ be the posterior probability or likelihood of model M_k given pixel data r, and $\pi(M_K)$ the a priori probability of having the model M_k at a particular pixel. Based on the Bayes theorem most often used in classification problems, the posterior probability assigned to a model M_K is proportional to its likelihood times its prior probability:

$$p(M_k|\mathbf{r}) = \frac{p(\mathbf{r}|M_k) \cdot \pi(M_k)}{\sum_{i=1}^{N} p(\mathbf{r}|M_i) \cdot \pi(M_i)}. \tag{6}$$

Although *a priori* probabilities $\pi(M_k)$ are often unknown and could be assumed to be equal (uninformative prior), they provide a means to inject prior information in the algorithm and, therefore, reduce the misidentification of the models by considering rules on the basis of ancillary data such as land cover classifications. The likelihood of M_k for each individual pixel, $p(\mathbf{r}|M_k)$, can be determined as follows:

$$(\mathbf{r}|M_k) = \int_\mathbf{x} \int_{\mathbf{x}'} \phi_{\text{soil}(k)}(\mathbf{x}|\mu_k, \Sigma_k) \cdot \phi_{\text{veg}(k')}(\mathbf{x}'|\mu_k, \Sigma_{k'}) \Gamma(\mathbf{x}, \mathbf{x}', \mathbf{r}) \, d\mathbf{x}' d\mathbf{x}, \tag{7}$$

where $\Gamma(\mathbf{x}, \mathbf{x}', \mathbf{r})$ is an "accept-reject" function that is the unity when the line joining $\mathbf{x}$ (vegetation) and $\mathbf{x}'$ (soil) intercepts the region in the feature space centered at the mixture $\mathbf{r}$, and is zero otherwise. We considered around $\mathbf{r}$ an envelope volume $V(\mathbf{r})$ given by the radiometric uncertainties attached to the input, i.e., the covariance structure of the k_0 product.

The algorithm solves this integral using Monte Carlo methods, i.e., from endmember spectra drawn from multivariate normal distribution. It basically consists in counting the number of model samples from each possible mixture combination that can give rise to the mixture pixel $\mathbf{r}$, as illustrated in Figure 3 considering a simplified bidimensional problem with 3 vegetation clusters (V_1, V_2, V_3) and 3 soil clusters (S_1, S_2, S_3). The size and orientation in the elliptical probability density contours are determined by the covariance of clusters Σ_k. It is obvious that (S_3, V_1) is the most likely model in the mixture. However, other different model combinations, such as (S_2, V_1) and (S_3, V_2), have non negligible probabilities and may yield to the same mixture spectrum.

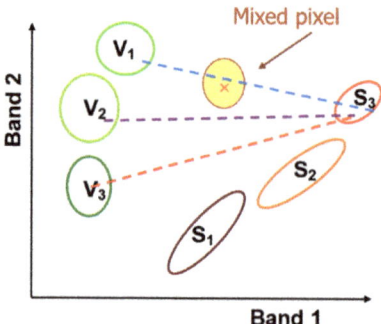

Figure 3. Illustration of the model selection in a bidimensional feature space. Dashed lines join random spectra drawn from soil class S_3 and vegetation classes V_1, V_2 and V_3.

Rather than using a single date, multi-temporal SEVIRI observations $\{t_1, t_2, \ldots t_M\}$ over a seasonal period can be used to more reliably determine the likelihood of model M_K in a mixture pixel $\mathbf{r}$, i.e.,:

$$p(\mathbf{r}|M_k) = \prod_{i=1}^{M} p(\mathbf{r}(t_i)|M_k), \ k = 1, 2, \ldots, N, \tag{8}$$

where $\mathbf{r}(t_i)$ represent the input reflectance at date t_i. The LSA SAF algorithm uses M=2 dates, associated to the maximum and minimum canopy closure of each SEVIRI pixel, making use of the vegetated/devegetated k_0 images generated in step 1:

$$p(\mathbf{r}|M_k) = p(k_0(t_{\text{devegetated}})|M_k) \cdot p(k_0(t_{\text{vegetated}})|M_k), \ k = 1, 2, \ldots, N \tag{9}$$

The simultaneous use of devegetated $k_0(t_{devegetated})$ and vegetated $k_0(t_{vegetated})$ spectral information for extreme situations of the annual growing cycle such as harvested crops and peak of season, are well suited to determine the likelihood of the soil and vegetation components, respectively. Figure 4 allows to better understand the concept. Symbols correspond to pure soil and vegetation samples identified over the Southern Africa SEVIRI geographical area, projected onto the bidimensional k_0 space of SEVIRI channels. Bare areas are predominantly found in sparsely vegetated and open shrublands (GLC2000 classes 19, 14 and 12) whereas purely vegetated areas are mostly found in close forest classes, herbaceous and croplands (GLC2000 classes 1, 2, 13 and 16). In this example, the best suited number of Gaussian components was 3 for vegetation and 6 for soil.

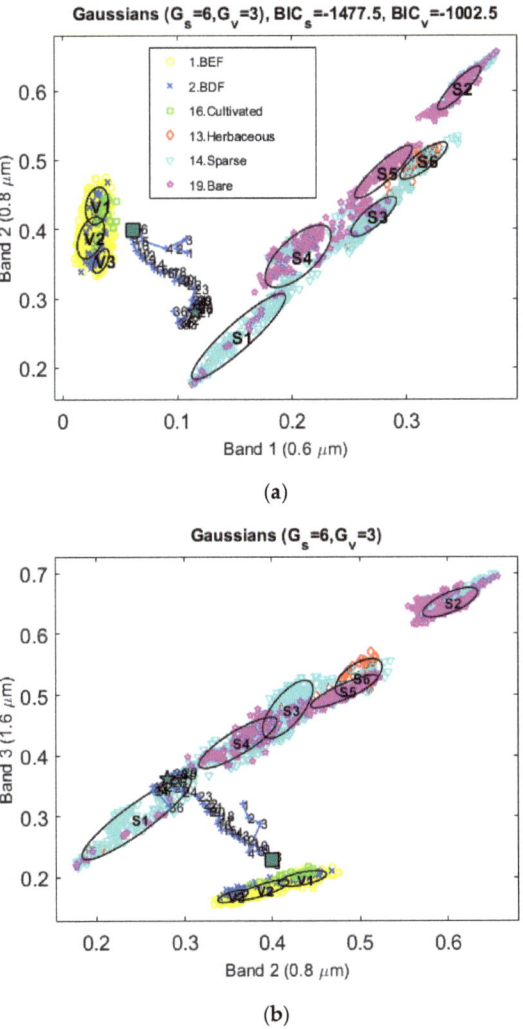

Figure 4. Illustration of the probabilistic mixing model concept over in the k_0 space of SEVIRI channels: (a) channel 1 and 2; (b) channel 2 and 3. Elliptical probability density contours associated with clusters (from V_1 to V_3 for vegetation and from S_1 to S_6 for soil). BDF and BEF correspond to broadleaved deciduous and broadleaved evergreen forest, respectively. BIC_s and BIC_v refer to the values of BIC for the soil and vegetation mixture models, respectively.

The improved multitemporal identification of vegetation and underlying soil background is illustrated for a SEVIRI pixel occupied by open broadleaved deciduous species. Blue symbols correspond to the migration of the pixel signature during year 2015 (numbers from 1 to 36 correspond to decades). The leaf fall is concentrated during the winter season, reaching devegetated period (filled pentagram in Figure 4) at decade 8, when the spectral signature approaches to the soil line. During the development phase, NIR values increase while red and MIR values decrease, and the pixel departs from the soil line, eventually reaching a high canopy closure (filled square in Figure 4a). Combining the probabilities in both extreme periods (Equation (9)), the most likely models for the vegetation and soil are obtained, i.e., vegetation classes V_1 and V_2 with the darkest soil type S_1.

Step 4. Estimation of FVC

The retrieval of vegetation parameters is an ill-posed problem. A feature selection/extraction step, may be useful to transform the original space $r \in \Re^3$ [$(k_0)_{red}$, $(k_0)_{NIR}$, $(k_0)_{SWIR}$] onto a feature space that makes the retrieval model more sensitive to changes in vegetation and minimize the sensitivity to soil background and noise in the inputs. Vegetation retrieval is specially hampered by the influence of undesired soil variability at SWIR wavelengths (1.6 µm), which is particularly large for the soils in Africa (see Figure 4b). This high variability has shown to be a cause for overestimation of FVC in semi-arid regions over dark soils (further insights will be given in Figure 5b). Latest version projects r onto a new features space $w \in \Re^5$, increasing the relative influence of channels 1 and 2 with respect to channel 3 ([$(k_0)_{red}$, $(k_0)_{red}$, $(k_0)_{NIR}$, $(k_0)_{NIR}$, $(k_0)_{SWIR}$]). This strategy has served us to reduce possible biases due to SWIR background reflectance variability.

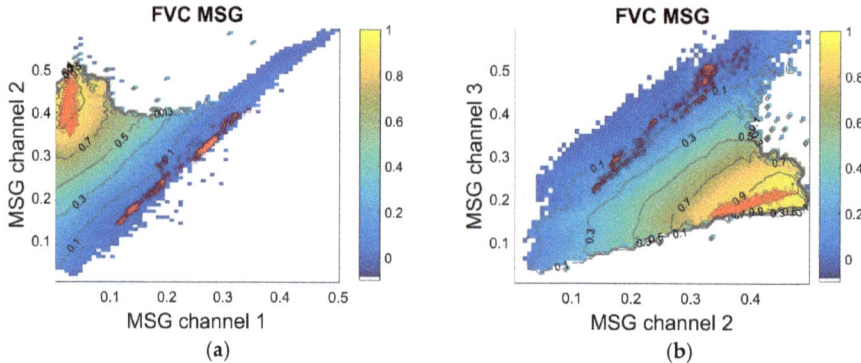

Figure 5. Projection of the SEVIRI/FVC product corresponding to the 15th of June 2015 onto the k_0 feature space of (**a**) channels 1 and 2; (**b**) channels 2 and 3. Red circles correspond to soil and vegetation pure pixels for the Southern Africa (SAfr) region.

In addition, a standardization transform is applied, which transforms the data to a set of variations about the mean value with a mean value of zero and a standard deviation of one:

$$\hat{r} = \frac{r - \mu_r}{\sigma_r}, \tag{10}$$

where $\hat{r}$ is the standardized vector associated to the pixel vector r, with mean and μ_r standard deviation σ_r. Using the standardised endmembers, $\hat{E}_i$ ($i = 1, \ldots, c$), the unmixing is formulated as follows:

$$\hat{r} = \sum_{i=1}^{c} \hat{E}_i \hat{f}_i + \hat{\varepsilon}, \tag{11}$$

where $\hat{f}_i$ is the proportion of such endmember in the standardised coordinates, and $\hat{\varepsilon}$ is the residual vector expressed in standardised units. Although the solution is similar to the conventional SMM, the sum-to-one condition is now expressed as

$$\sum_{i=1}^{c} \frac{\hat{f}_i}{\sigma_{E_i}} = \frac{1}{\sigma_r}. \tag{12}$$

Through this standardization, SMM is less sensitive to the brightness variability within each vegetation-soil component [40]. Standardized signatures are performed for all possible models and image spectra as previous step before computing the fractions of soil and vegetation. In order to compute the fractions, the closed formulation of the method of Lagrange is adopted [36], which provides a fast and unbiased solution. Finally, FVC is estimated as a linear combination of single-model estimates:

$$\text{FVC} = \sum_{k=1}^{N} p(M_k|\mathbf{r}) \cdot \text{FVC}(M_k), \tag{13}$$

where $\text{FVC}(M_k)$ is unmixed abundance of vegetation in the M_k model. The contribution of each model is thus weighted by its *a posteriori* probability $p(M_k|\mathbf{r})$. Although other similar approaches have been developed to address the issue of the spatial variability of endmembers [40–45], the LSA SAF approach decomposes the soil/vegetation into a number of subclasses using non-uniform probabilities for the different models. These probabilities $p(M_k|\mathbf{r})$ vary from pixel to pixel, depending on its likelihood and were pre-computed beforehand in order to increase the robustness of the algorithm and speed up its computational efficiency.

Figure 5 allows understanding better the physical basis of the algorithm and its performance. The results correspond to mean predictions of FVC given a realized value of two SEVIRI channels (considering all possible values of the third channel). We can identify the triangle envelope formed by the mixed pixels which is the physical basis for SMM approaches. The soil/vegetation components are spectrally distinct enough, particularly in the red-NIR space (Figure 5a), where non-vegetation surfaces distribute primarily along the so-called "soil line". Dense canopies situate in the top vertex, although showing a significant spectral variability. The LSA SAF algorithm provides a fair generalization of information from the training data generated in step 1, suggesting that solutions are well constrained. The isolines present realistic values that are consistent with other literature studies [46], and EPS satellite products [24]. The smooth FVC variations with gradual change in reflectance suggests that solutions are stable, without discontinuities that lead to large retrieval errors and are usually found in SMM problems with a large number of endmembers. The algorithm mitigates thus the effects of noise in the input data.

The algorithm uses also SWIR information, which conveys useful information about the vegetation water content and a good sensitivity to FVC (Figure 5b). However, the large variability of SWIR reflectance of soils in the African continent has shown to be confounding factor. In particular, inclusion of SWIR bands causes an overestimation trend for FVC in sparse areas over dark soils in Southern Africa, with values up to 0.15 (see Figure 5b). An effective way to reduce this overestimation is underweighting the SWIR band (step 4 of the algorithm).

2.3. LAI Algorithm

The LSA SAF algorithm uses a pragmatic solution to the radiative transfer problem, which assumes a tractable physical model for interception of solar direct irradiance by leaf canopies [43]. By assuming that leaves are flat with bi-Lambertian properties, the total transmittance, which represents the fraction of incident radiation above the canopy which reaches the soil background level can be expressed as follows [47]:

$$T(\theta_s) = \exp[-b(G(\theta_s)/\mu_s)\text{LAI}], \tag{14}$$

where $\mu_s = \cos\theta_s$, being θ_s the solar zenith angle, $G(\theta_s)$ is the average extinction function [48], $T(\theta_s)$ and b is the backscattered parameter, which can be roughly assumed to be equal to 0.945 for all vegetation types [49]. The fraction of solar radiation intercepted by the vegetation (FIPAR), which coincides with FVC when the sun and the observer are both at zenith, i.e., FVC = FIPAR($\theta_s = 0$) is expressed as:

$$\text{FIPAR}(\theta_s) = 1 - T(\theta_s) = 1 - \exp[-b(G(\theta_s)/\mu_s)\text{LAI}]. \tag{15}$$

The above formulation assumes a random distribution of foliage elements, disregarding the increased probability of light penetration in clumped canopies. A clumping index Ω [50], which accounts for the degree of the deviation of foliage distribution from the random case, can be introduced in Equation (15) to correct possible underestimation of LAI [20]:

$$\text{FVC} = \text{FIPAR}(\theta_s = 0) = 1 - \exp(-b \cdot G(\theta_s = 0) \cdot \Omega \cdot \text{LAI}). \tag{16}$$

It can be assumed for simplicity a value of 0.5 for the leaf projection factor $G(\theta_s)$ considering spherical orientation of the foliage. In order to avoid maximum LAI values in fully vegetated areas exceeding a value about 7, a coefficient a_0 in the range (1.04-1.07) is introduced in Equation (16):

$$\text{FVC} = a_0\{1 - \exp(-0.5 \cdot b \cdot \Omega \cdot \text{LAI})\}. \tag{17}$$

A land cover-dependent clumping index is considered for each of the GLC2000 classes, which corresponds to the maximum values calculated from global POLDER multiangular data [51]. The values range from 0.68 for evergreen forest to 0.83-0.85 for herbaceous, shrub and cultivated areas (see Table S1 in Supplementary Material), leading to a conservative first-order correction of the clumping effect.

2.4. FAPAR Algorithm

The algorithm used for retrieving daily integrated green FAPAR is not novel but it relies on the method proposed by [32], a statistical relationship general enough for global applications based on simulations using the homogeneous SAIL (Scattering by Arbitrary Inclined Leaves) model [52]. The algorithm does not use any prior knowledge on the land cover. Although the SAIL model is widespread in the remote sensing community for the estimation of vegetation biophysical variables, limitations of homogeneous models should be more important in heterogeneous (e.g., savannas and open shrublands) or in canopies showing a complex architecture (e.g., boreal forest). More than 5000 soil/vegetation combinations varying leaf inclination distribution (LIFD), LAI, leaf transmittance, leaf reflectance and soil spectral albedo were considered. The diffuse fraction of incoming radiation was held constant and equal to 0.2, which represents clear sky conditions. For each scenario, red and NIR reflectances with variations of sun and view angles were derived from the SAIL model along with daily-integrated FAPAR, as computed by integration of the instantaneous FAPAR over the day:

$$\text{FAPAR} = \frac{\int_t^{t'} \text{APAR } dt}{\int_t^{t'} \text{PAR } dt}, \tag{18}$$

where t and t' are the time for sunrise and sunset. The FAPAR was integrated over solar angles corresponding to a target located at 45°N latitude and at the equinox.

An optimal geometry based on the criteria of linearity and minimum dispersion between NDVI and daily-integrated FAPAR was found in the solar principal plane (θ_s=45°, θ_v=60°, ϕ=0°). The soil contribution to the canopy reflectance was reduced at this oblique view using a vegetation index, called RDVI (Renormalized Difference Vegetation Index), defined as follows:

$$\text{RDVI} = (\text{NDVI} \cdot \text{DVI})^{1/2} = \frac{\text{NIR} - \text{R}}{\sqrt{\text{NIR} + \text{R}}}, \tag{19}$$

where DVI [16] is the difference vegetation index, defined as follows:

$$\text{DVI} = (k_0)\text{NIR} - (k_0)\text{VIS}. \tag{20}$$

Finally, the RDVI-FAPAR relationship in the optimal geometry is given as follows:

$$\text{FAPAR} = 1.81\ (\text{RDVI})_{\text{opt}} - 0.21, \tag{21}$$

where $(\text{RDVI})_{\text{opt}}$ refers to the RDVI computed in the optimal geometry. The reflectance in the optimal geometry for each spectral channel is estimated as follows [53]:

$$R_{\text{opt}}(\lambda) = k_0(\lambda) - 0.240 k_1(\lambda) + 0.202 k_2(\lambda) \tag{22}$$

2.5. Products Uncertainty Estimation

A per-pixel estimate of the uncertainty for each prediction, namely σ_{FVC}, σ_{LAI} and σ_{FAPAR}, is provided, which propagates uncertainty of inputs and other variables involved in the calculations (see Equations in Supplementary Material).

3. The SEVIRI/MSG Vegetation Products

The LSA SAF generates and disseminates a suite of vegetation products derived from SEVIRI/MSG BRDF composited data for the whole Meteosat disk over land surfaces ideally free of snow and ice cover using the described algorithm at two different time resolutions (see Figure 6), daily (MDLAI, MDFVC, MDFAPAR), and 10-days (MTLAI, MTFVC, MTFAPAR) [54]. The characteristics of SEVIRI based products provided by the LSA SAF are described in Table 1. Daily products use as input the MDAL BRDF parameters computed using a daily rolling compositing approach with a characteristic 5-day compositing period, whereas the 10-days product use as input the MTAL BRDF product based on a composite over a 30-day period. We refer for further details about the recursive temporal composition to Geiger et al. [33]. Using the latest version of the daily (v3.1) and 10-days (v1.2) algorithms, released in January 2016, the LSA SAF system generates and disseminated the products in Near-Real-Time (NRT) with a time lag of about six hours. NRT users are also encouraged to use EUMETCast (https://navigator.eumetsat.int), i.e., the primary distribution means for EUMETSAT image data and derived products.

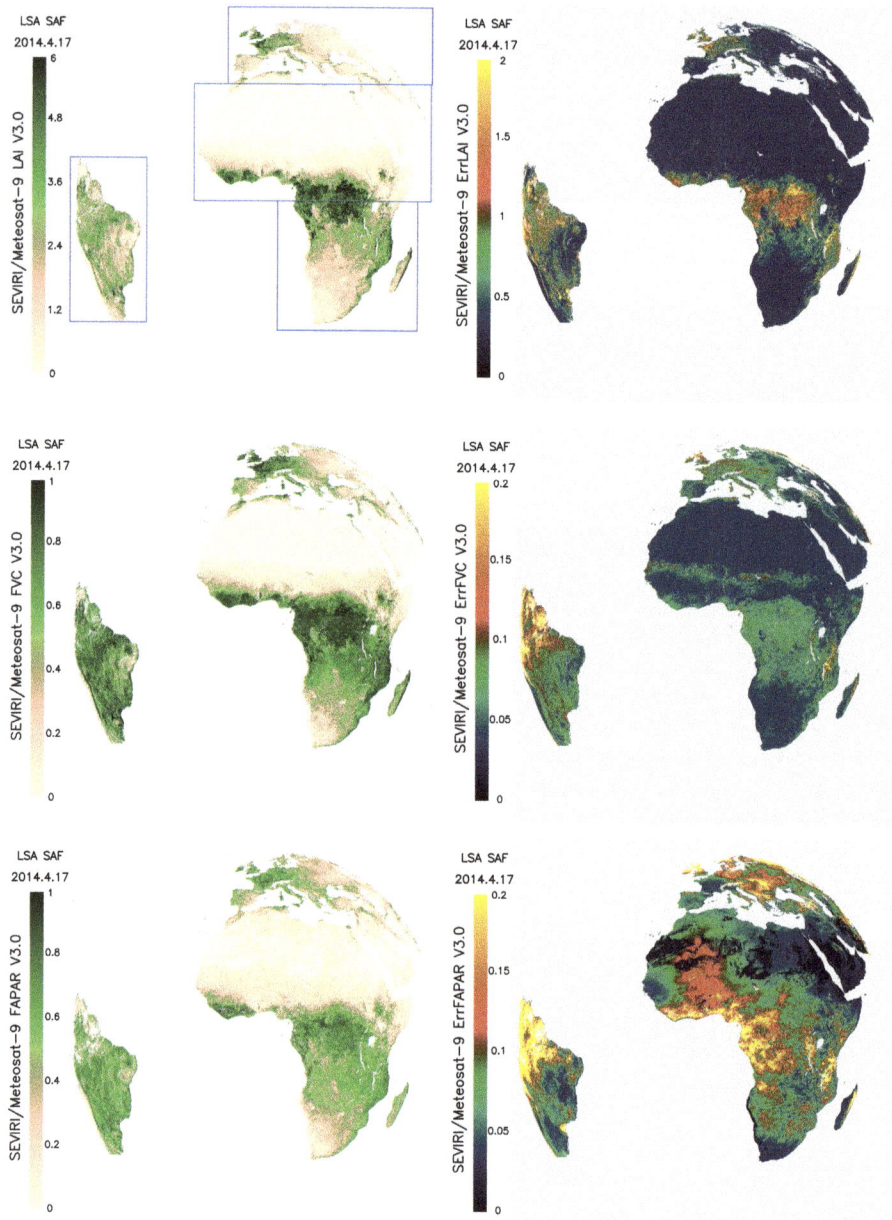

Figure 6. MSG Daily LAI (top), FVC (middle) and FAPAR (bottom) LSA SAF product composition corresponding to the 17th of April 2014 products (left panels) and their respective error estimates (right panels). Location of the four LSA SAF geographical areas is also provided. Rectangles in the LAI field are associated to Euro, NAfr, SAfr and Same SEVIRI geographical regions for illustrating purposes.

Table 1. Product Requirements for MSG vegetation products, in terms resolution and accuracy.

Product	Identifier	Distribution	Temporal Resolution	Spatial Resolution	Target Accuracy
MDFVC	LSA-421	NRT	1-day	MSG pixel	Max [0.075,15%]
MTFVC	LSA-422	NRT	10-days	MSG pixel	Max [0.075,15%]
MTFVC-R	LSA-450	CDR([1])	10-days	MSG pixel	Max [0.075,15%]
MDLAI	LSA-423	NRT	1-day	MSG pixel	Max [0.5,20%]
MTLAI	LSA-424	NRT	10-days	MSG pixel	Max [0.5,20%]
MTLAI-R	LSA-451	CDR([2])	10-days	MSG pixel	Max [0.5,20%]
MDFAPAR	LSA-425	NRT	1-day	MSG pixel	Max [0.075,15%]
MTFAPAR	LSA-426	NRT	10-days	MSG pixel	Max [0.075,15%]
MTFAPAR-R	LSA-452	CDR([3])	10-days	MSG pixel	Max [0.075,15%]

[1] http://doi.org/10.15770/EUM_SAF_LSA_0003. [2] http://doi.org/10.15770/EUM_SAF_LSA_0004. [3] http://doi.org/10.15770/EUM_SAF_LSA_00035.

The algorithms were reprocessed to generate the full archive of CDRs for 10-days vegetation products for FVC (MTFVC-R), LAI (MTLAI-R) and FAPAR (MTFAPAR-R) over the period 2004-2015, which complement the datasets available in the LSA SAF website since 2016 onwards. Both NRT products and homogeneous CDR 10-days vegetation products are available from the website (https://landsaf.ipma.pt/en/products/vegetation/). A similar dataset of reprocessed daily products has been produced as internal products for the period 2004-2015. Although the daily products are not foreseen to be distributed as CDR, they may be made available upon request.

The algorithm is firstly applied separately at four SEVIRI geographic regions: Europe (Euro), Northern Africa (NAfr), Southern Africa (SAfr), and South America (SAme) (see location in Figure 6). However, since the latest version (v3.1) onwards, products are disseminated for the whole Meteosat disk. The projection and spatial resolution correspond to the characteristics of Level 1.5 (standard geolocation) MSG/SEVIRI instrument data. Information on geo-location and data distribution is available at the LSA SAF web-site: http://landsaf.ipma.pt.

Each product is delivered in the Hierarchical Data Format version 5 (HDF5), which contains three separate datasets: a biophysical product, its respective error estimate and a Quality Flag (QF) field. The error dataset is a pixel-wise estimate of the observation uncertainty, which is indicative of the quality of the retrieval error (see details of calculation in Supplementary Material). These retrieval errors have shown to be within the range of the typical differences found between satellite products. For examples the uncertainty ranges typically between 0.05 and 0.10 for FVC, between 0.5 and 1.0 for LAI and between 0.05 and 0.20 for FAPAR. The FAPAR product presents a lower quality because it uses as input the three BRDF parameters (k_0, k_1, k_2), with k_2 presenting large uncertainties and noisy profiles on a short time scale, mainly in Western Africa.

The algorithms to retrieve LSA SAF vegetation require ideally free of snow and ice cover observations. Since snow events in previous days, though influencing the signal, may be unidentified BRDF product, an empirical criterion was used to identify and mask out pixels with residual, by combining the SEVIRI channels 1 (0.6 μm) and 3 (1.6 μm):

$$k_0(\lambda_1) - k_0(\lambda_3) > 0$$
$$\text{or } k_0(\lambda_1) > k_{0,\max}(\lambda_1) + 0.06$$
$$\text{or } (k_0(\lambda_1) > k_{0,\max}(\lambda_1) + 0.02 \text{ and } k_0(\lambda_3) < k_{0,\min}(\lambda_3))$$

where $k_{0,\max}(\lambda_1)$ and $k_{0,\min}(\lambda_3)$ refer to the values of the k_0 devegetated image in channel 1 and 3, respectively. The QF records relevant information at the pixel level about the quality and processing status of retrievals (e.g., unrealistic input ranges, traces of inland water, traces of snow and retrieval values out of maximum physical range). The QF codes and several empirical thresholds used to blind problematic areas are shown in Tables S2 and S3 in Supplementary Material, respectively, as in the Product User Manual (available from http://landsaf.ipma.pt).

Figure 7 shows quality information of the SEVIRI/MSG FVC, LAI and FAPAR products, using the mean error of the products along the year 2014. Green color refers to consolidate regions with optimal quality. Reliable areas with medium quality are shown in cyan color. Orange color refers to regions with low quality. Finally, red color refers to generally unusable areas, presenting large view zenith angles and frequent snow cover (e.g., Europe during wintertime). FAPAR is generally unusable in areas with persistent cloud occurrence (e.g., western Africa).

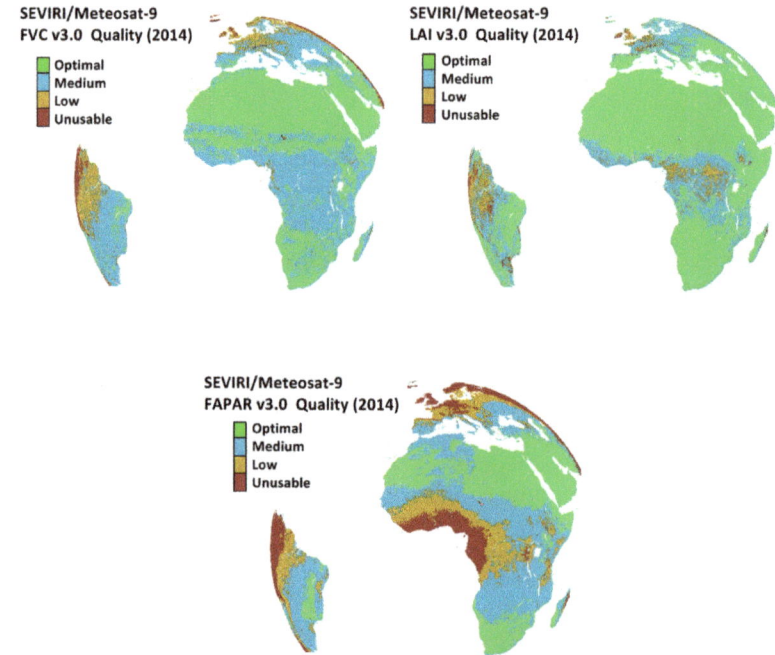

Figure 7. Quality of the MSG daily FVC, LAI and FAPAR products based on the information provided by the mean values of its theoretical uncertainty along the year 2014. The levels of accuracy stand for: Optimal (Err(FVC/FAPAR)<0.05; Err(LAI)<0.5); Medium (0.05<Err(FVC/FAPAR)<0.10; 0.5<Err(LAI)<1.0); Low (0.10<Err(FVC/FAPAR)<0.12; 1.0<Err(LAI)<1.5); Poor (Err(FVC/FAPAR)>0.15; Err(LAI)>1.5).

The seasonal variations in the quality and coverage of the daily MSG FVC product along 2014 are depicted in Figure 8. Optimal or medium quality retrievals are obtained for north and south Africa continental zones with a negligible percentage of poor quality or missing values. An acceptable performance is also found over Europe from April through September, with a significant quality reduction during late autumn and winter. The worst performance is observed for south America regions, generally due to decreased accuracy and a larger percentage of missing values. Both daily MSG LAI and FAPAR products show similar performance stages with slightly lower rates of high quality values.

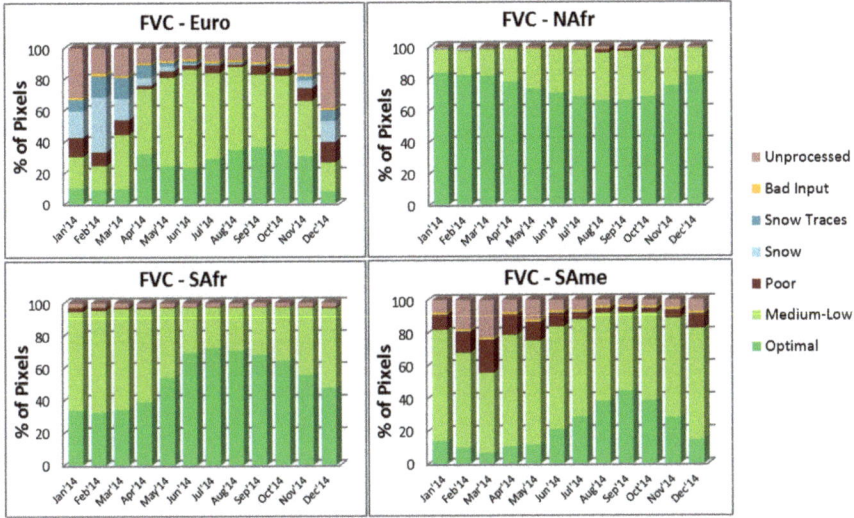

Figure 8. Monthly fraction of valid pixels for daily MSG FVC product during the entire 2014 year over the four SEVIRI geographical areas.

A detailed assessment of the MSG CDR products is provided in a validation report [55], showing good spatial and temporal consistency as compared to currently available validated vegetation products and no presence of artefacts. Noteworthy good inter-annual precision and stability of the long-term time series was found. Based on the validation results, performed over a representative network of sites over global conditions, CDR MSG vegetation products reached the operational status, what means that the products fulfilled all defined requirements and are suitable for distribution to users. Positive results were found for all the considered criteria, evaluated by comparisons with similar products such as Copernicus Global Land Monitoring Service GEOV1 based on SPOT/VGT data and NASA MODIS/TERRA products. Accuracy assessment against ground references provided by the Committee on Earth Observation Satellite (CEOS) through its Working Group in Calibration and Validation (WGCV) showed improved accuracy as compared to similar products and up to 84% of validation samples within target requirements [55].

Internal Consistency between the LSA SAF Products

Temporal profiles of the 10-days MSG vegetation products were verified through a comparison with the daily vegetation products. A very good agreement was found for the three products (see examples in Figure 9). Although one main drawback of the daily FAPAR product with regard to other operational products was the shaky temporal profiles on a daily basis, the 10-days (MTFAPAR) improves notably the smoothness of daily FAPAR.

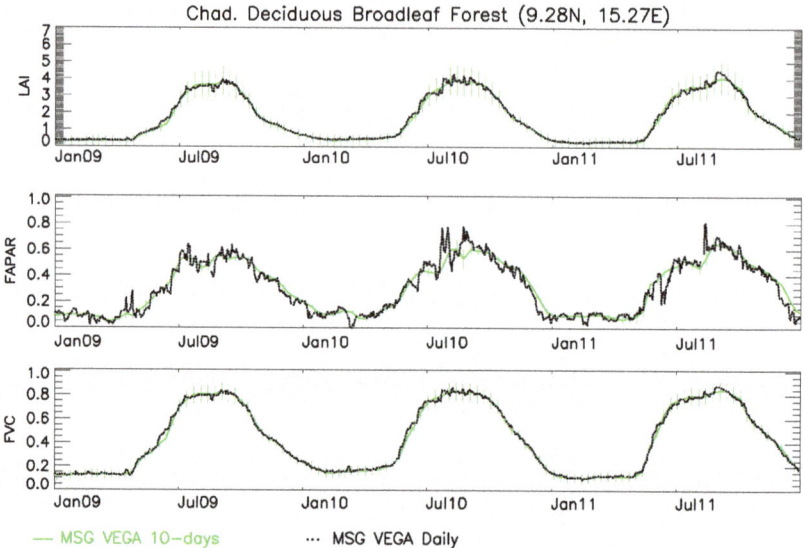

Figure 9. Time series of LAI, FAPAR and FVC daily and 10-days products at a Deciduous Broadleaf Forest site.

The aim of this section is also to analyse the consistency among the suite of LSA SAF vegetation products (i.e., FAPAR and LAI) and albedo fields. The energy absorbed by the ground below vegetation (FGROUND) can be estimated by combining the FAPAR and an independent LSA SAF field from the albedo SEVIRI product, the bi-hemispherical reflectance integrated over the photosynthetically active spectral region (BHRPAR) [33]. From energy conservation, the fraction of PAR absorbed by the ground beneath the canopy is given by [56]:

$$\text{FGROUND} = 1 - \text{BHRPAR} - \text{FAPAR}, \qquad (23)$$

Following previous works [56], FGROUND can be expressed as the downward PAR flux density, F, times the soil absorptance, $1-\alpha$, where α is the reflectance of the canopy background and F can be expressed via the Beer's law [57]:

$$\text{FGROUND} \approx \frac{1-\alpha}{1-\alpha r^*} e^{[-G(\theta_s)\text{LAI}/\cos(\theta_s)]}, \qquad (24)$$

where r* is the probability that photon entering through the lower canopy boundary will be reflected back by the vegetated layer [58,59]. Figure 10 shows the relationship between MSG LAI and FGROUND for two SEVIRI regions, Europe (Euro), and southern Africa (SAfr), as derived from BHRPAR and FAPAR SEVIRI/MSG products using Equation (24). The distribution can be well approximated by the exponential function, FGROUND = $Ae^{-B\cdot\text{LAI}}$ for different periods, which indicates a strong consistency between LAI and FGROUND at the resolution of SEVIRI products.

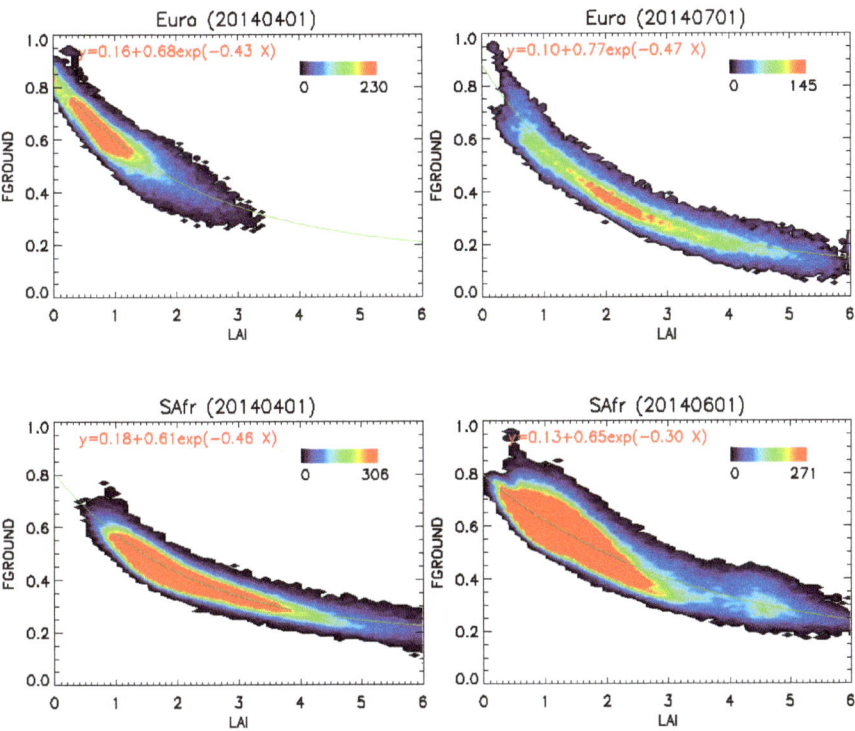

Figure 10. Joint probability density plots between FGROUND and MSG daily LAI at several different periods of the year. Top Figures correspond to the Euro SEVIRI zone, whereas bottom Figures correspond to the SAfr SEVIRI region.

4. Potential Applications of SEVIRI Vegetation Products

Different studies have assessed the quality of SEVIRI vegetation products [60–62]. MSG/SEVIRI LAI and FAPAR products were included in a detailed scientific analysis for quality information provision within the Evaluation and Quality Control (EQC) of climate data products derived from satellite and in situ observations to be catalogued within the Copernicus Climate Change Service (C3S) Climate Data Store (CDS) [63]. The products have been extensively used as input for environmental applications [64,65]. Particularly, the MSG LAI product found applicability to examine the feedback of land surface in response to vegetation changes [66], and in a hydrological model for flood early warning systems [67]. Gessner et al. [61] reported that the MSG LAI product is well suited to capture critical periods of growing cycle during the rainy season in West Africa as compared to existing global LAI products derived from polar orbit instruments. The use of MSG FVC and LAI products improved evapotranspiration estimates with respect to ECOCLIMAP-I in a land surface model [68]. Daily MSG LAI products were used for automatic derivation of detailed phenological information at the African continent [69]. The MSG LAI was fused with VGT/SPOT LAI to fill frequent gaps in VGT product during the rainy season in West Africa [70]. Further case studies and product applications of the SEVIRI/MSG vegetation products are provided in the LSA SAF web page (https://landsaf.ipma.pt/).

The SEVIRI/MSG vegetation products are routinely used as input in the operational line of several LSA SAF products. Daily MSG FVC and LAI are required inputs to update information about the state of vegetation for the derivation of the LSA SAF evapotranspiration and (latent and sensible) heat flux products [71]. The MSG FVC information on the pixel fraction of vegetation cover is required to operationally retrieve thermal Land Surface Emissivity [72]. The daily MSG FAPAR is also included as input on the Monteith's LUE concept used for the computation of the 10-days Gross Primary

Production (MGPP) product recently developed in LSA SAF [73,74]. The remaining of this section provides further insights about potential use of the LSA SAF vegetation products, considering three illustrative examples of land processes applications.

4.1. Application 1: Monitoring of Seasonal Cycle and Phenology

The enhanced data quality of geoestationary SEVIRI products can be demonstrated by comparing time series of FVC and polar-orbiting satellite sensors such as MODIS and SPOT/VGT across the cloud-prone equatorial regions, such as Amazonian, West Africa and Central Africa. As illustrated in a representative site in the western part of Equatorial Africa (Figure 11), the quality of polar-orbiting satellite products is severely hampered by cloud occurrence. The presence of missing values for long periods and unrealistic short-term variations reduces the ability of polar orbit products to capture the seasonal cycle.

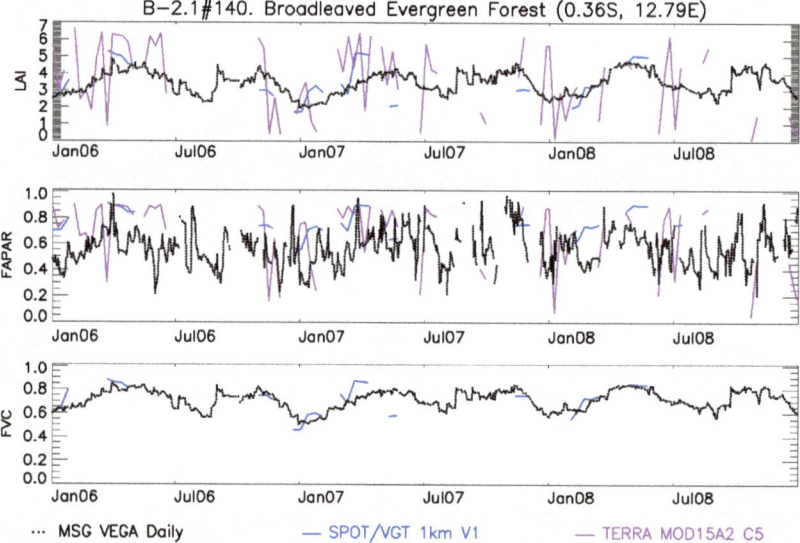

Figure 11. Comparison between LSA SAF products and equivalent MODIS and SPOT/VGT over a three-year period representative example at a site in Gabon.

Conversely, temporal profiles of MSG FVC are consistent from year to year showing realistic smooth variations, even during periods with high cloud coverage during the growing season. These findings are in line with previous studies [61,75]. This is mainly because the frequent sampling of its geostationary platform allows to accumulate enough cloud-free observations along a single day.

4.2. Application 2: Interrelation between Vegetation and Rainfall

Accumulated precipitation data derived from 10-days rainfall estimates (RFE 2.0) at a spatial resolution of 0.1° were considered to explore the performance of the SEVIRI FVC product over the Africa continent. The Rainfall Estimator (RFE) is produced by NOAA Climate Prediction Center (NOAA/CPC) merging information from polar-orbiting microwave measurements, geostationary infrared satellite estimates and gridded rainfall gauge measurements [76,77]. As illustrated in Figure 12, the SEVIRI FVC is an indicator of spatial and temporal rainfall variability, showing a strong temporal correlation with rainfall measurements (Pearson coefficient of determination r ranging from 0.7 to 0.9) in most of Africa.

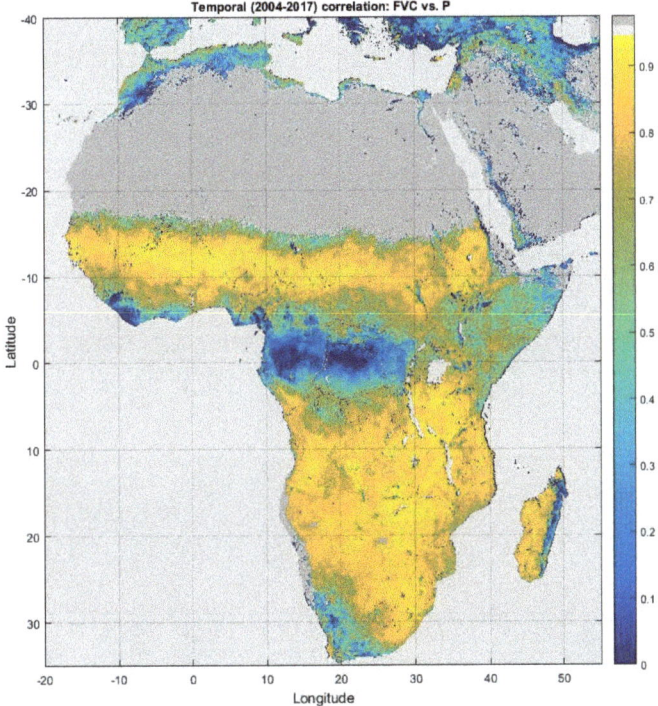

Figure 12. Pearson correlation coefficient between time series of FVC and 3-month accumulated precipitation for the 2004-2017 period.

Figure 13 shows an example of the relationships between FVC and 3-month accumulated rainfall for an herbaceous biome located in southern Africa. A clear interrelation is found between inter-annual variations of both magnitudes, proving the FVC product to be a proper indicator to detect sensitive areas. In particular, the rainfall deficit occurred during the rainy seasons of 2005, 2013, 2015 and 2016 produced a severe impact on the FVC during the growing season. Although not all regions present similar response to rainfall precipitation, NRT monitoring of FVC anomalies can be used to diagnose in below-normal rainfall values in sensitive areas, and inputted in early warning system of water stress initiation and severe drought events.

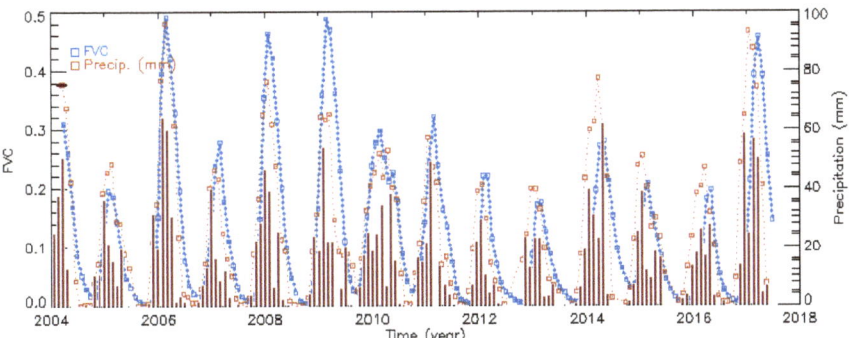

Figure 13. Time course of FVC and 3-month accumulated precipitation for an herbaceous region situated in Bostwana region (-23.1° N, 26.4° E).

4.3. Application 3: The Detection of Inter-Annual Vegetation Trends over the Period 2004-2017

A trend analysis over Africa to report long-term variations due to subtle changes over the period 2004-2017 has been performed. A multiresolution analysis (MRA) wavelet transform (WT) has been applied to the 10-days MSG FVC product [78]. Figure 14 shows the derived normalized trend of inter-annual changes over Africa for the period 2004-2017. In general, no trend is observed over central Africa (light blue and green colors) with some distinguishable negative trend areas (dark blue colors) localized at eastern (Horn of Africa, region 1), southern Mozambique (region 2) and western Namibia (region 5). The impact on vegetation is particularly evident in these regions since they have been affected by a precipitation deficit due to El Niño-Southern Oscillation (ENSO) events [79–81]. Conversely, positive trend values (yellow colors) are mostly located in the grassland biomes of the Sahel zone (region 3) and a few patches in east of Namibia and west of Bostwana (region 4). The Sahel results agree with the widely reported re-greening behavior in this area for the last two decades [82,83] whereas the behavior of region 4 is possibly explained by exceptionally wet years in southwestern Africa since circa 2005.

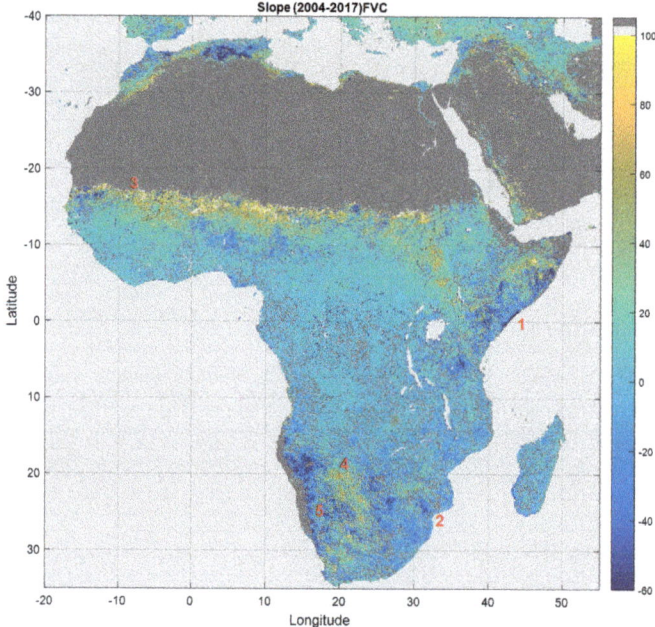

Figure 14. Trend of inter-annual changes over Africa derived by applying the MRA-WT method to 10-days FVC time series (2004-2017).

5. Summary and Conclusions

This paper describes the algorithm currently used in the LSA SAF operational system to produce in NRT daily and 10-days basis global FVC, LAI and FAPAR biophysical products from SEVIRI/MSG data. A novel algorithm has been proposed to retrieve FVC, which is also the base for retrieving LAI using a semi-empirical method. It relies on a stochastic spectral mixture model which characterizes the large variability of soil background and target vegetation using statistical (Gaussian mixture) models with a finite number of classes, and determines a posteriori probabilities of the background/soil classes making use of the pixel multitemporal trajectory. The FAPAR algorithm uses statistical relationships with BRDF in an optimal angular geometry, general enough for global applications. This study evidences a good consistency among the suite of SEVIRI/MSG (daily and 10-days) vegetation products, as well as among FAPAR, LAI and BHRPAR, enabling a realistic partitioning of incoming solar radiation between the canopy and the ground below the canopy.

Main advantages and relevant information derived from this study are summarized below:

1. NRT daily (MDFVC, MDLAI, MDFAPAR) and 10-days (MTFVC, MTLAI, MTFAPAR) products are generated and disseminated from LSA SAF since January 2004 over the geostationary Meteosat disk offering almost fifteen years of an alternative dataset to the user community.
2. The 10-days (MTFVC-R, MTLAI-R and MTFAPAR-R) CDRs are provided as a suite of EUMETSAT climate products data records estimated consistently along the years using the latest versions of the whole processing chain algorithms. The 10-days products could be suitable for a community of users that requires observations representative of a 30-day period with at frequency of 10 days (e.g., numerical weather and climate models, and flood forecasting systems).
3. The daily SEVIRI/MSG timeliness of the distribution of the observations and its smaller compositing period avoids possible shifts regarding the actual state of the vegetation (e.g., for an early estimate of key phenological parameters and seasonal production).
4. The absence of gaps and the high temporal frequency and continuity of the products over Africa offer major potentials for NRT monitoring of land cover dynamics for applications that require frequent observations such as agriculture, and food management.
5. The SEVIRI/MSG vegetation products have demonstrated its suitability to accurately resolving long term changes in large regions, allowing improving the understanding of interactions between land surface and climate.

The MSG programme is expected to provide observations for the next years to ensure long-term operations. In order to improve the coverage of the LSA SAF service, an ongoing work is the clone of MSG algorithms to produce similar LSA SAF products based on the Indian Ocean Data Coverage (IODC) observations. A priority work in LSA SAF is the adaptation of the algorithms to take advantage of enhanced capabilities in terms of spatial and spectral resolution of the future EUMETSAT sensors. The algorithm will be adapted to the enhanced characteristics of the Flexible Combined Imager (FCI)/Meteosat Third Generation satellites (MTG), allowing a more accurate characterization of the soil and vegetation signatures and improved screening of snow pixels. The FCI based products will ensure the continuity of the service during the lifetime of the MTG program.

Supplementary Materials: The following are available online at http://www.mdpi.com/2072-4292/11/18/2103/s1, Figure S1: Theoretical FAPAR error as a function of input k_0 and k_2 errors for a given k_1 error of 0.01. Two different cases have been considered: Low FAPAR values (a) and high FAPAR values (b)., Table S1: Cover-dependent clumping index values for LAI algorithm based on the GLC2000 land cover classification based on values obtained in [51], Table S2: VEGA products QF information. The default missing value for the product fields is −10. The associated error estimate fields for unprocessed pixels take different negative values, depending on the identified problem (default missing value = −10). Main identified problems in the VEGA products and empirical thresholds used to blind problematic areas. Note that although the missing value for the product fields is unique (−10), associated error estimate fields for unprocessed pixels take different negative values, depending on the identified problem (default missing value = −10).

Author Contributions: Conceptualization, F.J.G-H. & F.C.; Algorithm design, F.J.G-H; writing—original draft preparation, F.J.G-H; Formal analysis, writing, review & editing, all authors.

Funding: This research was funded by LSA SAF (EUMETSAT) and ESCENARIOS (CGL2016-75239-R) projects.

Acknowledgments: LSA SAF (EUMETSAT) and ESCENARIOS (CGL2016-75239-R) projects are acknowledged.

Conflicts of Interest: The authors declare no conflict of interest.

References

1. WMO. *A Global Framework for Climate Services-Empowering the Most Vulnerable*; WMO: Geneva, Switzerland, 2011.
2. Hewitt, C.; Mason, S.; Walland, D. The Global Framework for Climate Services. *Nat. Clim. Chang.* **2012**, *2*, 831–832. [CrossRef]
3. Dowell, M.; Lecomte, P.; Husband, R.; Schulz, J.; Mohr, T.; Tahara, Y.; Eckman, R.; Lindstrom, E.; Wooldridge, C.; Hilding, S.; et al. Strategy Towards an Architecture for Climate Monitoring from Space. 2013. Available online: http://www.wmo.int/pages/prog/sat/documents/ARCH_strategy-climate-architecture-space.pdf (accessed on 1 June 2019).

4. Trigo, I.F.; Dacamara, C.C.; Viterbo, P.; Roujean, J.-L.; Olesen, F.; Barroso, C.; Camacho-de-Coca, F.; Carrer, D.; Freitas, S.C.; García-Haro, J.; et al. The Satellite Application Facility for Land Surface Analysis. *Int. J. Remote Sens.* **2011**, *32*, 2725–2744. [CrossRef]
5. Schmetz, J.; Pili, P.; Tjemkes, S.; Just, D.; Kerkmann, J.; Rota, S.; Ratier, A. An introduction to Meteosat Second Generation (MSG). *Bull. Am. Meteorol. Soc.* **2002**, *83*, 977–992. [CrossRef]
6. Liang, S. *Comprehensive Remote Sensing*; Elsevier: Amsterdam, The Netherlands, 2017.
7. Chase, T.N.; Pielke, R.; Kittel, T.; Nemani, R.; Running, S. Sensitivity of a general circulation model to global changes in leaf area index. *J. Geophys. Res.* **1996**, *101*, 7393–7408. [CrossRef]
8. Buermann, W.; Dong, J.; Zeng, X.; Myneni, R.B.; Dickinson, R.E. Evaluation of the utility of satellite-based leaf area index data for climate simulation. *J. Clim.* **2001**, *14*, 3536–3550. [CrossRef]
9. Leuning, R.; Zhang, Y.Q.; Rajaud, A.; Cleugh, H.; Tu, K. A simple surface conductance model to estimate regional evaporation using MODIS leaf area index and the Penman–Monteith equation. *Water Resour. Res.* **2008**, *44*. [CrossRef]
10. Pagani, V.; Guarneri, T.; Busetto, L.; Ranghetti, L.; Boschetti, L.; Movedi, E.; Campos-Taberner, M.; García-Haro, F.J.; Katsantonis, D.; Stavrakoudis, D.; et al. A high resolution, integrated system for rice yield forecast at district level. *Agric. Syst.* **2018**, *168*, 181–190. [CrossRef]
11. Gilardelli, C.; Stella, T.; Confalonieri, R.; Ranghetti, L.; Campos-Taberner, M.; García-Haro, F.J.; Boschetti, M. Downscaling rice yield simulation at sub-field scale using remotely sensed LAI data. *Eur. J. Agron.* **2018**, *103*, 108–116. [CrossRef]
12. López-Lozano, R.; Duveiller, G.; Seguini, L.; Meroni, M.; Garcia-Condado, S.; Hooker, J.; Leo, O. Towards regional grain yield forecasting with 1-km resolution EO biophysical products: Strengths and limitation at pan-European level. *Agric. For. Meteorol.* **2015**, *206*, 12–32. [CrossRef]
13. Barlage, M.; Zeng, X. The effects of observed fractional vegetation cover on the land surface climatology of the community land model. *J. Hydrometeorol.* **2004**, *5*, 823–830. [CrossRef]
14. Bojinski, S.; Verstraete, M.; Peterson, T.; Richter, C.; Simmons, A.; Zemp, M. The concept of essential climate variables in support of climate research, applications, and policy. *Bull. Am. Meteorol. Soc.* **2014**, *95*, 1431–1443. [CrossRef]
15. CTOS. *Implementation Plan for the Global Observing System for Climate in Support of the UNFCCC (2010 Update)*; GCOS Rep. 138; CTOS: Geneva, Switzerland, 2010; p. 186. Available online: https://library.wmo.int/doc_num.php?explnum_id=3851 (accessed on 9 September 2019).
16. GCOS-200. *The Global Observing System for Climate: Implementation Needs*; Rep. 200; World Meteorological Organization; GCOS: Geneva, Switzerland, 2016; p. 315. Available online: https://library.wmo.int/doc_num.php?explnum_id=3417 (accessed on 9 September 2019).
17. Roujean, J.-L.; Lacaze, R. Global mapping of vegetation parameters from POLDER multiangular measurements for studies of surface-atmosphere interactions: A pragmatic method and validation. *J. Geophys. Res.* **2002**, *107*, ACL6:1–ACL6:14. [CrossRef]
18. Knyazikhin, Y.; Glassy, J.; Privette, J.L.; Tian, Y.; Lotsch, A.; Zhang, Y.; Wang, Y.; Morisette, J.T.; Votava, P.; Myneni, R.B.; et al. *MODIS Leaf Area Index (LAI) and Fraction of Photosynthetically Active Radiation Absorbed by Vegetation (FPAR) Product (MOD15) Algorithm Theoretical Basis Document*; NASA Goddard Space Flight Center: Greenbelt, MD, USA, 1999; Volume 20771.
19. Hu, J.; Tan, B.; Shabanov, N.; Crean, K.A.; Martonchik, J.V.; Diner, D.J.; Knyazikhin, Y.; Myneni, R.B. Performance of the MISR LAI and FPAR algorithm: A case study in Africa. *Remote Sens. Environ.* **2003**, *88*, 324–340. [CrossRef]
20. Gobron, N.; Pinty, B.; Verstraete, M.; Govaerts, Y. The MERIS Global Vegetation Index (MGVI): Description and preliminary application. *Int. J. Remote Sens.* **1999**, *20*, 1917–1927. [CrossRef]
21. Bacour, C.; Baret, F.; Béal, D.; Weiss, M.; Pavageau, K. Neural network estimation of LAI, fAPAR, fCover and LAI×Cab, from top of canopy MERIS reflectance data: Principles and validation. *Remote Sens. Environ.* **2006**, *105*, 313–325. [CrossRef]
22. Gobron, N.; Mélin, F.; Pinty, B.; Verstraete, M.M.; Widlowski, J.-L.; Bucini, G. A Global Vegetation Index for SeaWiFS: Design and Applications. In *Remote Sensing and Climate Modeling: Synergies and Limitations SE-1*; Beniston, M., Verstraete, M.M., Eds.; Springer: New York, NY, USA, 2001; Volume 7, pp. 5–21.
23. Baret, F.; Hagolle, O.; Geiger, B.; Bicheron, P.; Miras, B.; Huc, M.; Berthelot, B.; Niño, F.; Weiss, M.; Samain, O.; et al. LAI, fAPAR and fCover CYCLOPES global products derived from VEGETATION: Part 1: Principles of the algorithm. *Remote Sens. Environ.* **2007**, *110*, 275–286. [CrossRef]

24. Baret, F.; Weiss, M.; Lacaze, R.; Camacho, F.; Makhmara, H.; Pacholcyzk, P.; Smets, B. GEOV1: LAI and FAPAR essential climate variables and FCOVER global time series capitalizing over existing products. Part1: Principles of development and production. *Remote Sens. Environ.* **2013**, *137*, 299–309. [CrossRef]
25. García-Haro, F.J.; Campos-Taberner, M.; Muñoz-Marí, J.; Laparra, V.; Camacho, F.; Sánchez-Zapero, J.; Camps-Valls, G. Derivation of global vegetation biophysical parameters from EUMETSAT Polar System. *ISPRS J. Photogramm. Remote Sens.* **2018**, *139*, 57–74. [CrossRef]
26. Gitelson, A.A. Wide dynamic range vegetation index for remote quantification of biophysical characteristics of vegetation. *J. Plant Physiol.* **2004**, *161*, 165–173. [CrossRef] [PubMed]
27. Widlowski, J.L.; Taberner, M.; Pinty, B.; Bruniquel-Pinel, V.; Disney, M.; Fernandes, R.; Gastellu-Etchegorry, J.P.; Gobron, N.; Kuusk, A.; Lavergne, T. Third Radiation Transfer Model Intercomparison (RAMI) exercise: Documenting progress in canopy reflectance models. *J. Geophys. Res. Atmos.* **2007**, *112*. [CrossRef]
28. Weiss, M.; Baret, F.; Myneni, R.B.; Pragnere, A.; Knyazikhin, Y. Investigation of a model inversion technique to estimate canopy biophysical variables from spectral and directional reflectance data. *Agronomie* **2000**, *20*, 3–22. [CrossRef]
29. Jacquemoud, S.; Verhoef, W.; Baret, F.; Bacour, C.; Zarco-Tejada, P.J.; Asner, G.P.; François, C.; Ustin, S.L. PROSPECT + SAIL models: A review of use for vegetation characterization. *Remote Sens. Environ.* **2009**, *113*, S56–S66. [CrossRef]
30. Fisher, J.I.; Mustard, J.F.; Vadeboncoeur, M.A. Green leaf phenology at Landsat resolution: Scaling from the field to the satellite. *Remote Sens. Environ.* **2006**, *100*, 265–279. [CrossRef]
31. Filipponi, F.; Valentini, E.; Nguyen Xuan, A.; Guerra, C.A.; Wolf, F.; Andrzejak, M.; Taramelli, A. Global MODIS Fraction of Green Vegetation Cover for Monitoring Abrupt and Gradual Vegetation Changes. *Remote Sens.* **2018**, *10*, 653. [CrossRef]
32. Roujean, J.-L.; Breon, F.-M. Estimating PAR absorbed by vegetation from bidirectional reflectance measurements. *Remote Sens. Environ.* **1995**, *51*, 375–384. [CrossRef]
33. Geiger, B.; Carrer, D.; Franchisteguy, L.; Roujean, J.L.; Meurey, C. Land surface albedo derived on a daily basis from Meteosat second generation observations. *IEEE Trans. Geosci. Remote Sens.* **2008**, *46*, 3841–3856. [CrossRef]
34. Geiger, B.; Carrer, D.; Hautecoeur, O.; Franchistéguy, L.; Roujean, J.-L.; Catherine Meurey, X.C.; Jacob, G.; Algorithm Theoretical Basis Document (ATBD). Land Surface Albedo PRODUCTS: LSA-103 (ETAL). Available online: Ref:SAF/LAND/MF/ATBD_ETAL/1.3,25November2016,41pp (accessed on 12 July 2019).
35. Bartholome, E.; Belward, A.S. GLC2000: A new approach to global land cover mapping from earth observation data. *Int. J. Remote Sens.* **2005**, *26*, 1959–1977. [CrossRef]
36. Bishop, C.M. *Neural Networks for Pattern Recognition*; Oxford University Press: Oxford, UK, 1995.
37. McLachlan, G.J.; Krishnan, T. *The EM Algorithm and Extensions*; Wiley: New York, NY, USA, 1997.
38. Stone, M. Comments on model selection criteria of Akaike and Schwartz. *J. R. Stat. Soc.* **1979**, *41*, 276–278.
39. Fraley, C.; Raftery, A.E. Model-Based Clustering, Discriminant Analysis, and Density Estimation. *J. Am. Stat. Assoc.* **2002**, *97*, 611–631. [CrossRef]
40. Garcia-Haro, F.J.; Sommer, S.; Kemper, T. A new tool for variable multiple endmember spectral mixture analysis (VMESMA). *Int. J. Remote Sens.* **2005**, *26*, 2135–2162. [CrossRef]
41. Roberts, D.A.; Gardner, M.; Church, R.; Ustin, S.; Scheer, G.; Green, R.O. Mapping chaparral in the Santa Monica Mountains using multiple endmember spectral mixture models. *Remote Sens. Environ.* **1998**, *65*, 267–279. [CrossRef]
42. Bateson, C.A.; Asner, G.P.; Wessman, C.A. Endmember Bundles: A New Approach to Incorporating Endmember Variability into Spectral Mixture Analysis. *IEEE Trans. Geosci. Remote Sens.* **2000**, *38*, 1083–1094. [CrossRef]
43. Asner, G.P.; Heidebrecht, K.B. Spectral unmixing of vegetation, soil and dry carbon cover in arid regions: Comparing multispectral and hyperspectral observations. *Int. J. Remote Sens.* **2002**, *23*, 3939–3958. [CrossRef]
44. Song, C. Spectral mixture analysis for subpixel vegetation fractions in the urban environment: How to incorporate endmember variability? *Remote Sens. Environ.* **2005**, *95*, 248–263. [CrossRef]
45. Somers, B.; Asner, G.P.; Tits, L.; Coppin, P. Endmember variability in spectral mixture analysis: A review. *Remote Sens. Environ.* **2011**, *115*, 1603–1616. [CrossRef]
46. Huete, A.R. A soil-adjusted vegetation index (SAVI). *Remote Sens. Environ.* **1988**, *25*, 295–309. [CrossRef]
47. Roujean, J.-L. A tractable physical model of shortwave radiation interception by vegetative canopies. *J. Geophys. Res.* **1996**, *101*, 9523–9532. [CrossRef]
48. Ross, J. *The Radiation Regime and Architecture of Plant Stands*; Springer Science & Business Media: Berlin, Germany, 2012; Volume 3.

49. Roujean, J.-L.; Tanré, D.; Bréon, F.-M.; Deuzé, J.-L. Retrieval of land surface parameters from airborne polder bidirectional reflectance distribution function during hapex-sahel. *J. Geophys. Res. Atmos.* **1997**, *102*, 11201–11218. [CrossRef]
50. Nilson, T. A theoretical analysis of the frequency of gaps in plant stands. *J. Agric. Meteorol.* **1971**, *8*, 25–38. [CrossRef]
51. Chen, J.M.; Menges, C.H.; Leblanc, S.G. Global mapping of foliage clumping index using multi-angular satellite data. *Remote Sens. Environ.* **2005**, *97*, 447–457. [CrossRef]
52. Verhoef, W. Light scattering by leaf layers with application to canopy reflectance modeling: The SAIL model. *Remote Sens. Environ.* **1984**, *16*, 125–141. [CrossRef]
53. Roujean, J.-L.; Leroy, M.; Deschamps, P.-Y. A bidirectional reflectance model of the Earth's surface for the correction of remote sensing data. *J. Geophys. Res. Atmos.* **1992**, *97*, 20455–20468. [CrossRef]
54. Garcia-Haro, F.J.; Gilabert, M.A.; Melia, J. Linear spectral mixture modelling to estimate vegetation amount from optical spectral data. *Int. J. Remote Sens.* **1996**, *17*, 3373–3400. [CrossRef]
55. Camacho, F.; García-Haro, F.J.; Fuster, B.; Sanchez-Zapero, J. MSG/SEVIRI Vegetation Parameters (VEGA) Validation Report. SAF/LAND/UV/VR_VEGA_MSG, v3.1. 2018, p. 109. Available online: https://landsaf.ipma.pt/en/products/vegetation/ (accessed on 9 September 2019).
56. García-Haro, F.J.; Camacho-de Coca, F.; Meliá, J.; Martínez, B. Operational derivation of vegetation products in the framework of the LSA SAF project. In Proceedings of the 2005 EUMETSAT Meteorological Satellite Conference, Dubrovnik, Croatia, 19–23 September 2005.
57. Diner, D.J.; Braswell, B.H.; Davies, R.; Gobron, N.; Hu, J.; Jin, Y.; Kahn, R.A.; Knyazikhin, Y.; Loeb, N.; Muller, J.-P.; et al. The value of multiangle measurements for retrieving structurally and radiatively consistent properties of clouds, aerosols, and surfaces. *Remote Sens. Environ.* **2005**, *97*, 495–518. [CrossRef]
58. Hu, J.; Su, Y.; Tan, B.; Huang, D.; Yang, W.; Schull, M.; Bull, M.A.; Martonchik, J.V.; Diner, D.J.; Knyazikhin, Y.; et al. Analysis of the MISR LAI/FPAR product for spatial and temporal coverage, accuracy and consistency. *Remote Sens. Environ.* **2007**, *107*, 334–347. [CrossRef]
59. Knyazikhin, Y.; Marshak, A. Mathematical aspects of BRDF modeling: Adjoint problem and Green's function. *Rem. Sens. Rev.* **2000**, *18*, 263–280. [CrossRef]
60. Wang, Y.; Buermann, W.; Stenberg, P.; Smolander, H.; Häme, T.; Tian, Y.; Hu, J.; Knyazikhin, Y.; Myneni, R.B. A new parameterization of canopy spectral response to incident solar radiation: Case study with hyperspectral data from pine dominant forest. *Remote Sens. Environ.* **2003**, *85*, 304–315. [CrossRef]
61. Gessner, U.; Niklaus, M.; Kuenzer, C.; Dech, S. Intercomparison of leaf area index products for a gradient of sub-humid to arid environments in West Africa. *Remote Sens.* **2013**, *5*, 1235–1257. [CrossRef]
62. Camacho, F.; García-Haro, F.J.; Sánchez-Zapero, J.; Fuster, B.; Validation Report MSG/SEVIRI Vegetation Parameters (VEGA). SAF/LAND/UV/VR_VEGA_MSG, Issue 3.1. 2017. Available online: http://www.landsaf.meteo.pt (accessed on 9 September 2019).
63. Nightingale, J.; Mittaz, J.P.; Douglas, S.; Dee, D.; Ryder, J.; Taylor, M.; Old, C.; Dieval, C.; Fouron, C.; Duveau, G.; et al. Ten Priority Science Gaps in Assessing Climate Data Record Quality. *Remote Sens.* **2019**, *11*, 986. [CrossRef]
64. Martínez, B.; Camacho, F.; Verger, A.; García-Haro, F.J.; Gilabert, M.A. Inter-comparison and quality assessment of MERIS, MODIS and SEVIRI fAPAR products over the Iberian Peninsula. *Int. J. Appl. Earth Obs. Geoinf.* **2013**, *21*, 463–476. [CrossRef]
65. Verger, A.; Camacho, F.; García-Haro, F.J.; Meliá, J. Prototyping of Land-SAF leaf area index algorithm with VEGETATION and MODIS data over Europe. *Remote Sens. Environ.* **2009**, *113*, 2285–2297. [CrossRef]
66. Mamadou, O.; Galle, S.; Cohard, J.M.; Peugeot, C.; Kounouhewa, B.; Biron, R.; Zannou, A.B. Dynamics of water vapor and energy exchanges above two contrasting Sudanian climate ecosystems in Northern Benin (West Africa). *J. Geophys. Res.-Atmos.* **2016**, *121*, 11–269. [CrossRef]
67. Koriche, S.A.; Rientjes, T.H.M. Application of satellite products and hydrological modelling for flood early warning. *Phys. Chem. Earth* **2016**, *93*, 12–23. [CrossRef]
68. Ghilain, N.; Arboleda, A.; Sepulcre-Canto, G.; Batelaan, O.; Ardo, J.; Gellens-Meulenberghs, F. Improving evapotranspiration in a land surface model using biophysical variables derived from MSG/SEVIRI satellite. *Hydrol. Earth Syst. Sci.* **2012**, *16*, 2567–2583. [CrossRef]
69. Guan, K.; Medvigy, D.; Wood, E.F.; Caylor, K.K.; Li, S.; Jeong, S.J. Deriving vegetation phenological time and trajectory information over Africa using SEVIRI daily LAI. *IEEE Trans. Geosci. Remote Sens.* **2014**, *52*, 1113–1130. [CrossRef]

70. Klein, C.; Bliefernicht, J.; Heinzeller, D.; Gessner, U.; Klein, I.; Kunstmann, H. Feedback of observed interannual vegetation change: A regional climate model analysis for the West African monsoon. *Clim. Dyn.* **2017**, *48*, 2837–2858. [CrossRef]
71. Arboleda, A.; Ghilain, N.; Meulenberghs, F. First Product User Manual for MET&DMET (v2) and new LE&H products. Available online: SAF/LAND/RMI/PUM/ET&SF/1.1,2018,35pp (accessed on 12 July 2019).
72. Trigo, I.F.; Peres, L.F.; DaCamara, C.C.; Freitas, S.C. Thermal land surface emissivity retrieved from SEVIRI/METEOSAT. *IEEE Trans. Geosci. Remote Sens.* **2008**, *46*, 307–315. [CrossRef]
73. Martínez, B.; Sánchez-Ruiz, S.; Gilabert, M.A.; Moreno, A.; Campos-Taberner, M.; García-Haro, F.J. Retrieval of daily gross primary production over Europe and Africa from an ensemble of SEVIRI/MSG products. *Int. J. Appl. Earth Obs.* **2018**, *65*, 124–136. [CrossRef]
74. Martínez, B.; Gilabert, M.A.; Sánchez-Ruiz, S.; Campos-Taberner, M.; García-Haro, F.J.; Brüemmer, C.; Carrara, A.; Feig, G.; Grünwald, T.; Mammarella, I.; et al. Evaluation of the LSA-SAF Gross Primary Production product derived from SEVIRI/MSG data (MGPP). *ISPRS J. Photogramm. Remote Sens.* **2019**. in review.
75. García-Haro, F.J.; Camacho, F.; Verger, A.; Meliá, J. Current status and potential applications of the LSA-SAF suite of vegetation products. In Proceedings of the 29th EARSeL Symposium, Chania, Greece, 15–18 June 2009.
76. Xie, P. *CPC RFE Version 2.0. NOAA/CPC Training Guide*; Drought Monitoring Centre: Nairobi, Kenya, 2001.
77. Laws, K.B.; Janowiak, J.E.; Huffman, G.J. Verification of rainfall estimates over Africa using RFE, NASA MPA-RT, and CMORPH. In Proceedings of the Combined Preprints CD-ROM, 84th AMS Annual Meeting, Paper P2.2 in 18th Conference on Hydrology, Seattle, WA, USA, 11–15 January 2004; p. 6.
78. Martínez, B.; Gilabert, M.A. Vegetation dynamics from NDVI time series analysis using the wavelet transform. *Remote Sens. Environ.* **2009**, *113*, 1823–1842. [CrossRef]
79. Loewenberg, S. Humanitarian response inadequate in Horn of Africa crisis. *Lancet* **2011**, *378*, 555–558. [CrossRef]
80. Archer, E.R.M.; Landman, W.A.; Tadross, M.A.; Malherbe, J.; Weepener, H.; Maluleke, P.; Marumbwa, F.M. Understanding the evolution of the 2014–2016 summer rainfall seasons in southern Africa: Key lessons. *Clim. Risk Manag.* **2017**, *16*, 22–28. [CrossRef]
81. Qu, C.; Hao, X.; Qu, J.J. Monitoring Extreme Agricultural Drought over the Horn of Africa (HOA) Using Remote Sensing Measurements. *Remote Sens.* **2019**, *11*, 902. [CrossRef]
82. Brandt, M.; Mbow, C.; Diouf, A.A.; Verger, A.; Samimi, C.; Fensholt, R. Ground and satellite-based evidence of the biophysical mechanisms behind the greening Sahel. *Glob. Chang. Biol.* **2015**, *21*, 1610–1620. [CrossRef] [PubMed]
83. Nicholson, S.E.; Funk, C.; Fink, A.H. Rainfall over the African continent from the 19th through the 21st century. *Glob. Planet. Chang.* **2017**, *165*, 114–127. [CrossRef]

© 2019 by the authors. Licensee MDPI, Basel, Switzerland. This article is an open access article distributed under the terms and conditions of the Creative Commons Attribution (CC BY) license (http://creativecommons.org/licenses/by/4.0/).

Article

Lead-Induced Changes in Fluorescence and Spectral Characteristics of Pea Leaves

Marlena Kycko [1,*], Elżbieta Romanowska [2] and Bogdan Zagajewski [1]

[1] Department of Geoinformatics, Cartography and Remote Sensing, Faculty of Geography and Regional Studies, University of Warsaw, 00-927 Warsaw, Poland
[2] University of Warsaw, Faculty of Biology, Department of Molecular Plant Physiology, 02-096 Warsaw, Poland
* Correspondence: marlenakycko@uw.edu.pl; Tel.: +48-225-521-507

Received: 22 June 2019; Accepted: 9 August 2019; Published: 12 August 2019

Abstract: Chlorophyll fluorescence parameters can provide useful indications of photosynthetic performance in vivo. Coupling appropriate fluorescence measurements with other noninvasive techniques, such as absorption spectroscopy or gas exchange, can provide insights into the limitations to photosynthesis under given conditions. Chlorophyll content is one of the dominant factors influencing the conditions of a vegetation growing season, and can be tested using both fluorescence and remote sensing methods. Hyperspectral remote sensing and recording the narrow range of the spectrum can be used to accurately analyze the parameters and properties of plants. The aim of this study was to analyze the influence of lead ions (Pb, 5 mM $Pb(NO_3)_2$) on the growth of pea plants using spectral properties. Hyperspectral remote sensing and chlorophyll fluorescence measurements were used to assess the physiological state of plants seedlings treated by lead ions during the experiment. The plants were growing in hydroponic cultures supplemented with Pb ions under various conditions (control, complete Knop + phosphorus (+P); complete Knop + phosphorus (+P) + Pb; Knop (-P) + Pb, distilled water + Pb) affecting lead uptake via the root system. Spectrometric measurements allowed us to calculate the remote sensing indices of vegetation, which were compared with chlorophyll and carotenoids content and fluorescence parameters. The lead contents in the leaves, roots, and stems were also analyzed. Spectral characteristics and vegetation properties were analyzed using statistical tests. We conclude that: (1) pea seedlings grown in complete Knop (with P) and in the presence of Pb ions were spectrally similar to the control plants because lead was not transported to the shoots of plants; (2) lead most influenced plants that were grown in water, according to the highest lead content in the leaves; and (3) the effects of lead on plant growth were confirmed by remote sensing indices, whereas fluorescence parameters identified physiological changes induced by Pb ions in the plants.

Keywords: fluorescence; in vivo; spectrometry; ASD Field Spec; lead ions; remote sensing indices

---

1. Introduction

The use of chlorophyll a fluorescence measurements to examine stress in algae and plants is now widespread in physiological and ecophysiological studies [1]. Fluorescence can be a powerful tool to study photosynthetic performance, especially when coupled with other noninvasive measurements such as absorption spectroscopy, gas analyses, and infrared thermometry. Many environmental stresses affect CO_2 assimilation. Changes in fluorescence induced by illumination of dark-adapted leaves are qualitatively correlated with their photosynthetic rates [2,3]. Photosynthetic CO_2 fixation is a process significantly affected by heavy metals in a number of plant species [4]. The mechanism(s) of heavy metal toxicity on photosynthesis is still a matter of speculation, but it almost certainly involves electron transport in light reactions [5] and enzyme activity in the dark reactions [6].

The ions determine the proper development and functioning of organisms [2,3], but they can also cause stress in plant cells, which produces disturbances in the functioning of processes including, among others, photosynthesis [4,7]. Burzynski and Kłobus [8] reported the inhibitory effect of metals on growth and biomass of plants as well as changes in the metabolic processes, which reduce photosynthetic activity. The effects are dependent on the operation of the ion, its concentration and duration of action, and the environment in which the plant grows [9]. The most damaging heavy metals (HMs) are cadmium (Cd), lead (Pb), mercury (Hg), nickel (Ni), and arsenic (As). Pb causes a reduction in chlorophyll content [10] and assimilation of CO_2, and inhibits the respiratory intensity [10,11]. Pb also affects the absorption of ions from the soil including, among others, iron and magnesium [12,13]. Pb is known to induce a broad range of toxic effects, including those that are morphological, physiological, and biochemical in origin. This metal impairs plant growth, root elongation, seed germination, seedling development, transpiration, chlorophyll production, and cell division [14]. A high Pb content in plants thus contributes to a reduction in the water absorption capacity [12]. Heavy metal ions also affect the formation of reactive oxygen species (ROS), which cause changes in signal transduction, change the properties of the membranes, and affect gene expression, damaging photosystem II (PSII) [15,16]. Stress-induced decreases in stomatal conductance, carbon metabolism, and transport processes can all decrease PSII efficiency [7]. Changes in chlorophyll a fluorescence measured in vivo enables the estimation of the physiological state of the plant and illustrates how absorbed light energy is used.

Stress induces changes in light absorption and can result in loss of excitation energy in the form of heat (non-photochemical quenching, NPQ) [17]. NPQ protects the photosynthetic membrane against photodamage, leading to early senescence and reduced plant growth and fitness [18]. NPQ is assume to be zero in the dark adapted state, because Fv' = Fv and Fm' = Fm. When plants are exposed to light, we can estimate the non-photochemical quenching from maximal fluorescence with an adaptation to darkness (F_m) to maximal fluorescence without adaptation to darkness (F_m') and monitors the apparent rate constant for non-radiative decay (heat loss) from PSII and its antennae for understanding the protective action of NPQ [19]. Plants have various responses to the noxious effects of lead, such as selective metal uptake, metal binding to the root surface, metal binding to the cell wall, and induction of antioxidants. The published findings on the evaluation of photochemical processes and physiological do not allow for the unambiguous determination of the response of photosynthetic apparatus to stress caused by lead [14,20,21]. Pb was found to induce changes in the absorption and dissipation of energy within PSII, whereas small changes were observed in electron transport [22]. Heavy metals, such as cadmium, nickel, and lead, are phytotoxic [23]. The literature also confirms a variable response to Pb as a stress factor, depending on the species. For example, corn and soybean showed varying degrees of photosynthesis sensitivity to lead [24].

Remote and non-invasive detection of changes in vegetation due to contamination with lead is possible using remote sensing techniques (spectrometric measurements, and multispectral and hyperspectral imaging). Using hyperspectral techniques is becoming increasingly important in the study of vegetation stress [25,26]. In response to the HMs presence, remote sensing indices show a decrease in photosynthetic active radiation (PAR), chlorophyll and carotenoids contents, a shift of red-edge spectral bands to the visible part of the spectrum (VIS), and changes in plant-water accumulation in short-wave infrared (SWIR). A limited growth, changes of canopy structures, decrease of chlorophyll content (leaf chlorosis), and changes in cell water content is a consequence of pollution [27].

The development of technologies and retrieval algorithms to evaluate fluorescence has progressed with model developments [28–30]. The state of vegetation contaminated with heavy metals may be tested using remote sensing indices [31] and by changes in the spectral reflectance curves, mainly the red edge [32]. The spectral reflectance and thereby the relevant spectral features offer an easy and simple method to assess plant health and metal-concentration. ASD FieldSpec was used for monitoring the phytoaccumulation of arsenic into above-ground parts of barley [33] and paddy plants (measurement of field spectral reflectance was one basis of airborne or spaceborne remote sensing monitoring). Ren et al. [34] also studied these spectral changes to relate them with the phytoaccumulation of metals [35,36]. Most of the studies used the various vegetation indices (ratios or linear combinations of two or more spectral wavelengths [37]) and red-edge position (a sharp transition between red and NIR wavelengths; positively related to chlorophyll concentration [38]) as indications of plant stress. The impact of heavy metals was observed in the spectral reflectance curve, where, under the influence of, for example Cd and Ni [39], the red edge decreased. Zagajewski et al. [39] observed that stressed plants (expressed as lower chlorophyll content) showed an increase in the fluorescence at F690 (red) and a decrease in the intensity of fluorescence at F740 (far-red) with decreasing intensity at 710 nm. Different fluorescence features are affected by leaf chlorophyll content in the antenna and the F690/F740 ratio appears to be one of the most indicative of chlorophyll variations [40]. Changes in the state of vegetation caused by the impact of heavy metals were also observed in multispectral images, where decreases in crop biomass and chlorophyll content were observed, as were changes in the red edge during the measurements [31]. Many observations indicate that the interactions between chlorophyll and carotenoids seem to be crucial for excess energy dissipation [41]. Remote sensing technologies provide complementary capacity for measuring and interpreting fluorescence in the context of physiological processes in field- [42,43], airborne- [44,45], and satellite-level measurements [46].

Despite extensive studies conducted on the effects of various heavy metals on different photosynthetic parameters of plants, our knowledge is still incomplete. Various changes in physiological and biochemical events occurring at different sites during photosynthesis are tightly linked so that even slight inhibition of one may result in a series of processes limiting gas exchange in the plants. Fluorescence assessment in the laboratory or growth chamber involves a suite of measurement devices, including fluorescence microscopes, spectrophotometers spectrofluorometers, fluorometers, and others. These have allowed studies on scales ranging from isolated photosystems to small vegetation canopies. Many researchers underlined that in the future, laboratory-scale or controlled-environment trials can support sun-induced fluorescence (SIF) remote sensing activities, such as by the elucidation of confounding factors for interpreting SIF changes, or by identifying ancillary data types needed for airborne- or space-based missions [47].

The aim of this study was to combine devices, such as a fluorometer and spectrometer, for detection of the influence of lead on photosynthetic parameters in pea plants (*Pisum sativum*) growing in laboratory conditions. Laboratory and field experiments provide new insights into fluorescence–photosynthesis linkages, stress effects, and relationships between other photosynthetic parameters. We investigated the relationship between fluorescence parameters: the maximum quantum efficiency of PSII in the light, measured as a Fv'/Fm' ratio (variable fluorescence without adaptation (Fv') to maximal fluorescence without adaptation to darkness (F_m')) and non-photochemical quenching of fluorescence (NPQ) and hyperspectral remote sensing indices. We also report the effect of lead ions on the growth of plants in various conditions influencing Pb uptake from the root medium. These data provide a novel insight into the relationships between the methods used for measurements of heavy metal effects on the plants.

2. Materials and Methods

2.1. Plant Material

Pea (*P. sativum* L.) seedlings were grown hydroponically in aerated:

(1) Knop's solution [1] containing (g·L^{-1}) 0.8 CaNO$_3$·4H$_2$O, 0.2 KNO$_3$, 0.2 KH$_2$PO$_4$, 0.2 MgSO$_4$·7H$_2$O, and 0.028 EDTA-Fe (EthyleneDiamineTetraacetic) enriched with A–Z microelement nutrients (Control);
(2) Knop solution [1] with 5 mM Pb(NO$_3$)^2 (lead-treated plants; Knop + Pb);
(3) Knop solution [1] without phosphorus with 5mM Pb(NO$_3$)$_2$ (described below as Knop (-P) +Pb);
(4) distilled water with 5 mM Pb(NO$_3$)$_2$ (described below as H$_2$O + Pb).

Plants were grown in a growth chamber with a 14 h photoperiod and a day/night regime at 25/20 °C at an irradiance of 50 µmol photons·m^{-2}·s^{-1} (Figure 1). Pb ions (5 mM Pb(NO$_3$)$_2$) were added to the growth medium (see above) when seedlings were two weeks old. Lead was introduced into the leaves via the transpiration stream. The leaves from three-week-old plants were used for experiments.

Figure 1. Pea plants used in experiments. From left to right: control plants in Knop's solution; Knop medium without phosphorus; Knop solution with 5 mM Pb(NO$_3$)$_2$ (lead-treated plants, Pb); and distilled water with 5 mM Pb(NO$_3$)$_2$ (photo: M. Kycko).

2.2. In Vivo Measurements of Chlorophyll a Fluorescence

Chlorophyll a fluorescence of pea leaves was measured at room temperature with a fluorometer (FMS 1, Hansatech, Norfolk U.K.) run by Modfluor software (Hansatech, Norfolk U.K.). Leaves were adapted to darkness for 30 minutes prior to measurements, and then were used in these assays with an actinic radiation of 60–1100 µmol photons·m^{-2}s^{-1} and the saturation radiation of 4500 µmol photons·m^{-2}s^{-1}. The standard amber modulating beam had a center frequency of 594 nm. The photochemical quenching coefficient (qP), non-photochemical quenching (NPQ), and quantum efficiency of PSII electron transport in the light (Fv'/Fm') were measured at steady-state photosynthesis. The procedure developed by Genty at al. [48] was followed. All results are represented as means ± standard error (SE) from 8 independent series of experiments (10 measurements each).

2.3. Determination of Lead Content

After harvesting, leaves, stalks, and roots were washed with redistilled water, then samples were oven-dried at 85 °C and ground to powder. The powder was washed in a muffle furnace at 550 °C for 4 h and the residue was brought to a standard volume with 1 M HNO$_3$. The Pb concentration of the extract was determined by an Atomic Absorption Spectrophotometer (AAS; PerkinElmer, model 3300; PerkinElmer, Waltham, MA, USA). The Pb ion content, expressed in mg·g^{-1} dry weight of the tissue tested, was determined on the basis of indications obtained from the corrected concentration of lead using Equation (1):

$$A = Cs \times V/m \tag{1}$$

where

A is the Pb ion content in mg·g^{-1} dry weight,
Cs is the corrected concentration Pb (mg·L^{-1}),
V is the volume of the sample for analysis (L), and
m is the dry weight of a sample (g).

2.4. Laboratory Measurements

To determine the effect of lead on pea plants, we measured the spectral characteristics of the plant vegetation with a ASD FieldSpec 3 spectrometer (ASD Inc., Longmont, CO, USA) with the ASD PlantProbe (ASD Inc., Longmont, CO, USA), and chlorophyll a fluorescence using a Hansatech FS-1 fluorometer (Hansatech, Norfolk U.K.); and the growth parameters including lengths and fresh and dry weights of the seedlings. On the basis of the dry weight, we determined the water content in plants and the lead content in the leaves by AAS.

Several randomly chosen pea seedlings, including control and plants grown in the presence of lead in different types of hydroponic medium, were used for measurements. The measurements were recorded using the contact probe (ASD PlantProbe) on fragments of the tested pea plant for each growth environment. The spectral characteristics of the plants were registered in the range from 350 to 2500 nm. The study consisted of calibrating the spectrometer (25 independent measurements per calibration) using a white pattern of spectralon (SG 33,151 Zenith Lite Reflectance Target and calibration screen P/N A122634 Leaf clip) and appropriate measurements of vegetation (the reported value of one record is the average of 25 measurements). Table 1 lists the measurements recorded for every group of plants.

Table 1. ASD FieldSpec spectrometric measurements for control and lead-treated plants.

	Control	Knop + 5 mM $Pb(NO_3)_2$	Knop (-P) + 5 mM $Pb(NO_3)_2$	H_2O + 5 mM $Pb(NO_3)_2$
Number of spectrometric measurements	60	40	40	40

The data obtained from spectrometric measurements were saved in *.asd format and then exported to the ASD FieldSpec View in *.txt ASCII format and imported into an Excel spreadsheet and the Statistica 13 software (TIBCO Software Inc., Palo Alto, CA, USA) for statistical analyses. The reflectance values were averaged and the standard deviations were computed. For particular research elements, spectral reflectance curves were drawn. The program also calculated the selected remote sensing vegetation indices:

(1) The content and structure of chlorophyll: modified normalized difference vegetation index 705 (mNDVI705) [49], red edge position index (REPI) [50], ratio analysis of reflectance spectra—chlorophyll a (RARSa) [51], RARS—chlorophyll b (RARSb) [51], and RARS—carotenoids (RARSc) [51];
(2) The amount of light used in photosynthesis: structural independent pigment index (SIPI) [52];
(3) Nitrogen content: normalized difference nitrogen index (NDNI) [53];
(4) Amount of carbon: normalized difference lignin index (NDLI) [53] and plant senescence reflectance index (PSRI) [54];
(5) Amount of carotenoids: carotenoid reflectance index (CRI 1) [55], anthocyanin reflectance index (ARI 1) [56], red/green ratio (RGR), and anthocyanins/chlorophyll [57]; and
(6) Water content of the plant cover: normalized difference water index (NDWI) [58], water index (WI) [59], normalized water index-2 (NWI-2) [60], and disease water stress index (DSWI) [61].

Then, to determine the overall physiological activity, changes in the functioning photosynthetic parameters of the plants [62] and their condition [17], the chlorophyll fluorescence measurements were used. This measurement allows for the determination of the physiological state of photosynthetic apparatus on the basis of the energy used in photochemical reactions [62]. The leaf was exposed to a short pulse of high light (4500 µmol photons·$m^{-2}s^{-1}$): Q_A is the maximally reduced fluorescence level and the Fm is the maximal fluorescence level observed. On the basis of these two parameters, we calculated the maximum quantum efficiency of PSII (Fv/Fm) in plants adapted to darkness and after actinic light was used. Measurement cycles were performed using light intensities of 60, 120, 180, 240, 420, 600, 820 and 1020 µmol photons·$(m^{-2}s^{-1})$. We measured the fluorescence maximum (Fm') for the plant adapted to relevant light conditions and components that decreased in fluorescence (NPQ).

The growth rate of seedlings (the length of roots and the stalk), fresh and dry weights of leaves, and the roots of four randomly selected plants from each of the investigated variants were measured (Figure 2). The dry weight was determined after drying plant material at 85 °C for 24 h until a constant weight was obtained.

Figure 2. A plant treated with 5 mM Pb(NO$_3$)$_2$ on the left side and control pea plant on the right side (photo: I. Kaźmierczak).

The pigment contents in the leaves were measured in acetone. The leaves (approx. 0.1–0.2 g) were cut into small pieces and then mixed with a pestle in a mortar cooled to about 0 °C with the addition the purified sand and CaCO$_3$ in 2 mL cold pure acetone. The extracts were filtered and washed with 80% acetone in a schott funnel (Schott AG, Mainz, Germany) and refilled with 80% acetone to a constant volume. Measurements were recorded with a UV-VIS 160A Shimadzu spectrophotometer (Shimadzu, Kyoto, Japan) at wavelengths of 663.2, 646.8, and 470 nm relative to 80% acetone as described by Arnon [49]. Pigment content was analyzed based on the fresh weight (FW) of the leaves (mg Chl/FW). The concentrations of the pigments (in µg/mL) was calculated with Equations (2)–(4) [38]:

$$\text{Chlorophyll a} = 12.25 \times A663.2 - 2.79 \times A646.8 \quad (2)$$

$$\text{Chlorophyll b} = 21.5 \times A646.8 - 5.1 \times A663.2 \quad (3)$$

$$\text{Carotenoids} = (1000 \times A470 - 1.82 \times \text{Chl a} - 85.02 \times \text{Chl b})/198 \quad (4)$$

The statistical analyses were conducted basing on the Statistica 13 software and used included univariate ANOVA and Kruskal–Wallis one-way ANOVA by ranks. Univariate ANOVA for independent groups (called one-way ANOVA), proposed by Fisher [63], was used to verify the hypothesis of equality of means to test variables in several (k > 2) populations. The Kruskal–Wallis ANOVA is used for nonparametric data to study the effect of factors (independent variables) on the dependent variable [64]. To assess significant differences in the studied indices, we used the Mann–Whitney U test. This test is used to verify hypotheses about the insignificance of the differences between the variable medians studied in two populations; in this case, this was used to verify the differences between the controls and those treated with lead. The assessment of the significance of variance was performed at a significance level of 0.001. This analysis allowed the separation of statistically significant intervals of the spectrum and remote sensing vegetation indicators for the plants treated with lead. In the analysis, we considered all the samples and a statement of the trial with the trial of control. The characteristic spectral ranges are marked on the graphs of spectral reflection of plants.

3. Results

Lead toxicity on photosynthesis has often been examined with various species but rarely or not at all with ecotypes or populations all [65]. The generally accepted view is that heavy metals cause a reduction in photosynthesis in leaves primarily through induction of stomatal closure rather than a direct effect on light or dark reactions [20]. The lead treatment of pea plants resulted in statistically significant changes in the condition of the peas. The ANOVA performed for all the samples indicated that the presented ranges of the electromagnetic spectrum (Figures 3 and 4) were statistically significant at the significance level of 0.001 (the most sensitive to the effects of heavy metals in the case of lead). In the figures, the marked grey ranges are the sensitive range for pigments and water content (the statistically significant changes depending on the medium in which the plants were grown: control; Knop + Pb; Knop (-P) + Pb; distilled water + Pb; at the $p < 0.001$). In the case of plants treated with the lead, the red-edge is noticeably, indicating damage to the cell structures of plants (Figure 3). In the plants growing in distilled water with lead, we perceived that the changes were noticed for 480–515 nm and 540–720 nm (responsible for absorption of light by chlorophyll. For Knop + Pb and control plants the following spectral ranges were significantly important: 710–735, 830–950, 995–1145, 1385–1570, 1760–1902, and 1980–2500 nm (responsible for water absorption). In contrast to ranges obtained for plants from Knop (-P) + Pb and control plants, also 850–1400, 1570–1760 and 1800–1830 nm indicated water absorption features (Figure 4). The differences occurred in the ranges: 714–723, 858–951, 995–1145, 1386–1389, 1570–1579, and 1801–1829 nm (marked with a black dashed line in Figure 4).

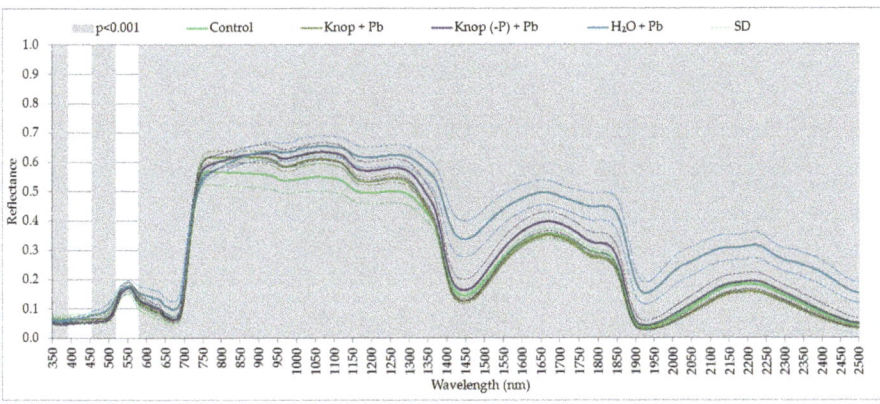

Figure 3. The statistically significant spectral ranges ($p < 0.001$) for the different treatments of pea plants at different wavelengths.

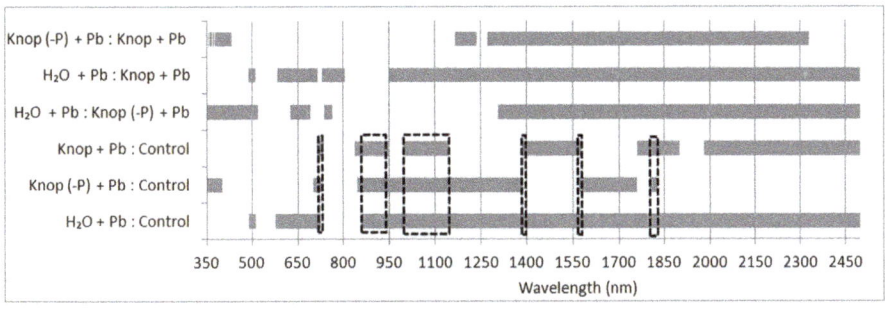

Figure 4. The statistically significant spectral ranges ($p < 0.001$) for pairs of the different treatments of the measurements.

The calculated remote sensing indicators of Pb-treated plants showed significant differences in comparison with the reference plants. Statistical analysis was performed using the Mann–Whitney U test at the 0.001 level, which confirmed the significance of the observed changes (marked in red in Table 2). We observed a statistically significant ($p < 0.001$) decrease in the value of most of the indicators compared with the control. These changes were clearly visible in the indicators describing the contents of chlorophyll in the plants (mNDVI705, REPI1, RARSa, and RARSb), the amount of light used in photosynthesis (SIPI), nitrogen (NDNI), the amount of carbon (NDLI) and carotenoids (CRI1, ARI1), and the water content in vegetation cover (WBI, NDWI, MSI, and NDII). Overall analysis of all samples at a level of significance of 0.001 indicated that the most sensitive indicator to changes caused in the plant due to lead is the water content as indicated by NDWI, WI, NWI-2, and DSWI. The values of these indicators decreased by 5–34%, 1–5%, 23–43%, and 5–27%, respectively, thereby indicating the water stress of the plants. In the H_2O + Pb-treated plants, we also observed a significant decrease in the value of chlorophyll b (RARSb) by about 35% in comparison to the control, a statistically significant difference in the indicators describing the dry parts of plants, and an increase of about 57% in the amount of anthocyanins (ARI1) in relation to the control. In the plants from the Knop (-P) + Pb treatments, we also observed significant decreases in the value of water (generally about 1–34% in relation to the control), a statistically significant difference in the indicators describing the dry parts of plants and chlorophyll content (mNDVI705, a difference about 19% in relation to the control).

Table 2. Mean values of the remote sensing indices with the standard deviation of the different measurements indices comparing Knop + Pb, Knop (-P) + Pb, and H_2O + Pb treatments with the control. * indicates a statistical significant difference α = 0.001; ±SD).

Plant Characteristic	Index	H_2O + Pb	Knop (-P) + Pb	Knop + Pb	Control
Chlorophyll content	mNDVI 705	0.41 ± 0.11 *	0.46 ± 0.07 *	0.57 ± 0.04	0.57 ± 0.09
	REPI1	715.15 ± 1.48 *	715.91 ± 1.79 *	719.94 ± 1.9	719.23 ± 1.94
	RARSc	5.52 ± 0.9	8.79 ± 0.46	8.38 ± 0.89	7.83 ± 0.45
	RARSa	5.98 ± 0.91 *	9.49 ± 0.78	9.99 ± 0.39	9.18 ± 0.42
	RARSb	5.31 ± 0.4 *	7.91 ± 0.45	8.62 ± 0.1	8.21 ± 0.54
Amount of light used in photosynthesis	SIPI	1.06 ± 0.07 *	1.025 ± 0.02 *	0.99 ± 0.005	0.99 ± 0.01
Amount of nitrogen	NDNI	0.14 ± 0.05 *	0.22 ± 0.02	0.22 ± 0.02	0.21 ± 0.04
Amount of carbon	NDLI	0.05 ± 0.01	0.06 ± 0.01 *	0.06 ± 0.01 *	0.05 ± 0.01
	PSRI	0.001 ± 0.03	0.0004 ± 0.01 *	−0.0145 ± 0.01	−0.0133 ± 0.01
Amount of carotenoids and other pigments	CRI 1	2.94 ± 0.07 *	6.065 ± 0.05 *	5.46 ± 0.26	5.54 ± 0.27
	ARI 1	1.02 ± 0.06 *	0.44 ± 0.06 *	−0.66 ± 0.04	−0.78 ± 0.02
	RGR	1.32 ± 0.05 *	1.05 ± 0.08	0.98 ± 0.04	1.07 ± 0.07
Water content in the plant	NDWI	−0.003 ± 0.02 *	0.04 ± 0.03 *	0.066 ± 0.01 *	0.06 ± 0.01
	WI	0.99 ± 0.02 *	1.03 ± 0.02	1.05 ± 0.01 *	1.04 ± 0.01
	NWI-2	0.015 ± 0.01 *	−0.006 ± 0.02 *	−0.027 ± 0.003 *	−0.02 ± 0.004
	DSWI	1.27 ± 0.22 *	1.66 ± 0.22	1.91 ± 0.1 *	1.74 ± 0.1

The results represent the fluorescence parameter values obtained for the different variants of pea leaves, using light intensities of 60, 120, 180, 240, 420, 600, 820, and 1020 µmol photons·m^{-2}s^{-1} (Figure 5). When leaves of high- or low-light-grown plants are exposed to other environmental stresses, a pronounced and sustained increase in energy dissipation may be induced, which could largely account for the change in photochemical efficiency. It is not clear whether the growth light conditions can modify the response of plants to heavy metals. We measured the quantum efficiency of PSII (Fv'/Fm') and non-photochemical quenching of fluorescence (NPQ) in the light. The results show that the the maximum quantum efficiency of PSII (Fv'/Fm') was significantly reduced in the presence of lead (Figure 5). At the higher light intensities above 400 µmol photons·m^{-2}s^{-1}, we observed slightly decreased photosynthetic efficiency (by about 4% to 8%), which was also confirmed by a significant difference in SIPI values, indicating the amount of light used in the photosynthesis process for peas growing in H_2O + Pb and Knop (-P) + Pb medium was statistically significantly different compared with the control. Maximal photosynthetic efficiency (Fv'/Fm' of ~0.78) was observed for control plants. We observed that in the Pb-treated plants in the range of light intensity from 60 to 600 µmol photons·m^{-2}s^{-1}, heat emission (NPQ) significantly reduced, but increased significantly at higher light intensities. This may indicate participation of carotenoids in the NPQ process. This was confirmed by a significant difference in the values for remote sensing CRI1 and ARI1 indices, which indicate the amount of carotenoids and anthocyanins, respectively. For peas grown in H_2O + Pb and Knop (-P)

+ Pb medium, these indices showed a statistically significant change compared with the control. In control plants, NPQ was higher than in Pb-treated plants; thus, heat emission protects photosynthesis components and systems when light absorption is maximal.

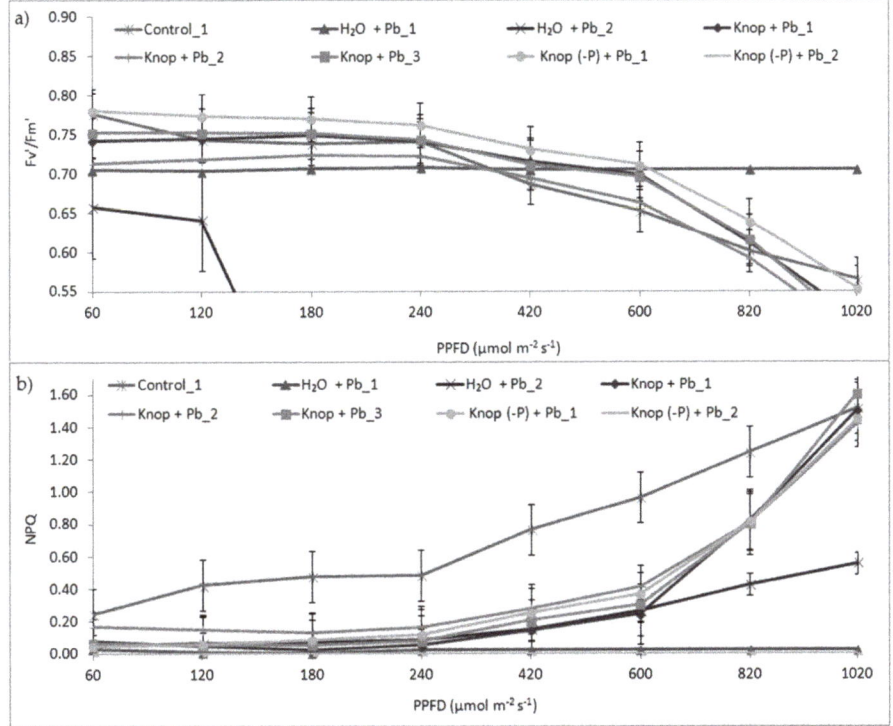

Figure 5. Relationship between photosynthetic photon flux density (PPFD) and the maximum quantum efficiency of PSII (Fv'/Fm'; (**a**)) in the light adapted state and non-photochemical fluorescence quenching (NPQ; (**b**)) for the different variants of the measurement pea leaves. Data are means ± SE ($n \geq 5$).

The maximum quantum efficiency of PSII in the light (Fv'/Fm'; Figure 5) for all Pb-treated plants was low and decreased significantly as PPFD increased from 420 to 1020 µmol·photon m^{-2}s^{-1}, When the Pb ions were added to the distilled water, we observed a strong decrease in the F'/Fm' values, even in lowest PPDF, or no change at all. This result suggests that PSII operating efficiency is inhibited, which demonstrates that the ability to oxidize QA decreases and irreversible damage of PSII occurs. Generally, decreases in quantum efficiency of PSII is indicated by increases in NPQ. NPQ was measured after illuminating the leaf at increasing light intensities to investigate the light responses of this parameter. NPQ increased for control plants from the lowest PPDF, which demonstrates that both the energy processing and the dissipating capacities increased, in turn indicating the better performance of photosynthetic apparatus. In Pb-treated plants, the NPQ was suppressed by Pb ions, even at low PPDF; thus, the photoprotective mechanisms was not present in these plants. The changes caused by lead, as well as impeding the proper course of photosynthesis, were visible in the leaves, Deposits of lead were visible in the wall of the vascular bundles (Figure 6). The data also indicate that the water thermal dissipation of excessive excitation energy was at the lowest level in Pb-plants.

Figure 6. Leaves of pea plants grown in Knop medium with addition of 5mM Pb(NO$_3$)$_2$. Deposits of lead are visible in the wall of vascular bundles (photo: I. Kaźmierczak).

The plants under stress conditions related to the presence of lead ions in the medium were characterized by a much lower height, a longer but less branched root system, a smaller number of trichomes, and clearly visible deposits of lead in the vascular tissues of the leaves (Figure 6). Also, the fresh and dry masses of Pb-treated plants were lower than those of the control (Table 3). The lead contents (median value) in various parts of the tested plants (30 plants from each environment) are presented in Table 3.

Table 3. The content of lead in the roots, stalks, and leaves of peas for different treatments ($\mu g \cdot g^{-1}$ dry weight).

Treatment	Root	Stalk	Leaf
Control	6	133 ± 10	0
Knop + 5 mM Pb(NO$_3$)$_2$	15,260 ± 330	6430 ± 142	68 ± 8
H$_2$O + 5 mM Pb(NO$_3$)$_2$			13,260 ± 245
Knop -P (without phosphate) + 5 mM Pb(NO$_3$)$_2$			1730 ± 75

The content of lead in control plants was very low, which might indicate the minor contamination of reagents used (Table 3). The highest amounts of lead content (an average of 15.3 mg Pb g^{-1} dry weight) was found in the roots of plants growing in complete Knop medium supplemented with 5 mM Pb(NO$_3$)$_2$, whereas the lowest content was found in the leaves of these plants, resulting from the reduced transport of the lead ions to the leaves. The addition of Pb to the water increased this metal concentration in the leaves approximately 200 times competed to that in Knop + Pb plants. This may be related to a lack of nutrients (salts present in Knop medium) in distilled water, which resulted in increased transport of lead ions from the medium. The Pb concentration in the leaves from plants grown in the Knop medium without phosphate and with 5 mM Pb(NO$_3$)$_2$ was increased approximately 25 times compared to that in the leaves of plants grown with the water with Pb ions. The results indicate that the lead content in the leaves can be decreased by phosphorus in the growth medium. Plants grown on Knop (-P) + Pb accumulated more lead in the leaves than the plants grown in complete medium with Pb, which creates less insoluble salts, which facilitate the transport of metal ions to the leaves. The obtained results indicate the importance of the selected growth media on degree of accumulation of heavy metals in the plant leaves. The leaves and roots of vegetables have a particular predisposition for accumulating toxic metals such as lead and therefore can be used for biomonitoring of the environment, mainly as a tool for assessing the extent of soil contamination. The presence of phosphorus in Knop media reduces the uptake of Pb ions by pea leaves. This indicates that P ions help reduce Pb toxicity by limiting its availability for plants. This result is also confirmed by the differences in the values of the remote sensing indicators.

The lead had negative effects on plant vegetative growth. Root and stalk growth was sensitive to lead and was inhibited approximately 30–40% for all investigated plants compared to control plants (Table 4). Our results show that although the accumulation of lead was not uniform among organs, sensitivity to heavy metal of growth processes was similar. Lead may induce unknown signal(s) that are transmitted from root to shoot. Table 5 presents the effect of lead on the fresh weight of stalks, leaves, and roots of the investigated plants. The fresh weight of the organs markedly decreased in all plants treated with lead compared to the control. The largest decrease (approximately 70%) was observed for all organs in pea seedlings grown in distilled H_2O + Pb. For plants grown in Knop + Pb, the decrease was highest in the roots (about 50%) according to highest accumulation of lead. Thus, changes in the growth and biomass accumulation depends on the photochemical activity and thus on the rate of photosynthesis.

Table 4. The effect of lead on the length (cm) of stalks and roots. The length of control plants was set to 100%.

Treatment	Stalk		Root	
Control	14.58 ± 1.76	100%	18.35 ± 1.18	100%
Knop + 5 mM $Pb(NO_3)_2$	10.08 ± 1.15	70%	13.05 ± 1.65	71%
Knop -P (without phosphate) + 5 mM $Pb(NO_3)_2$	8.98 ± 0.94	61%	12.45 ± 1.42	68%
H_2O + 5 mM $Pb(NO_3)_2$	9.08 ± 1.94	62%	11.53 ± 1.50	63%

Table 5. The effect of lead on a fresh mass (g) of selected plant organs (calculated for one plant). The mass of control plants was set to 100%.

Treatment	Stalk		Root		Leaf	
Control	0.54 ± 0.17	100%	1.19 ± 0.43	100%	1.27 ± 0.31	100%
Knop + 5 mM $Pb(NO_3)_2$	0.36 ± 0.05	67%	0.63 ± 0.10	53%	0.82 ± 0.10	64%
Knop -P (without phosphate) + 5 mM $Pb(NO_3)_2$	0.24 ± 0.05	44%	0.50 ± 0.07	42%	0.69 ± 0.09	54%
H_2O + 5 mM $Pb(NO_3)_2$	0.14 ± 0.03	26%	0.31 ± 0.16	26%	0.29 ± 0.08	23%

We also observed that the reduction in plant fresh mass and inhibition of growth elongation by Pb ions was accompanied by water loss of plants organs (data not presented). The greatest depletion of water content was noted in the leaves of plants grown in water with Pb ions (approximately 40%), whereas the smallest occurred in the leaves of plants grown in Knop + Pb compared to the control plants. This was also confirmed by the highest difference in the indicators from the group describing water content in vegetation cover (WBI, NDWI, MSI, and NDII); these differences ranged from 1% to 42%.

Leaves of pea seedlings grown in Knop (with and without P) with Pb ions exhibited reduction in chlorophyll a content compared with control seedlings (Figure 7). Growth in water and $Pb(NO_3)_2$ treatment increased the chlorophyll content significantly (about 70%, Figure 7). The plants treated with Pb ions showed lower chlorophyll a/b ratios than control plants (2.7) and was lowest (1.3) in plants grown in water with Pb. The plants grown in hydroponic cultures with Knop medium, in both variants (with and without phosphorus), contained less carotenoids relative to the control plants (Figure 7). Carotenoids (Car) were less affected by lead compared to chlorophyll (Chl); however, the Chl/Car ratio increased significantly after Pb treatment: 5.9 for control plants but approximately 10 for plants grown in water with Pb. Remote sensing indicators describing the content of chlorophyll (such as RARSa and RARSb; Table 6) also confirmed a decrease of about 35% (in the case of both indicators) between the growth environment, H_2O + Pb, and control (Table 6). However, the amount of dyes, such as anthocyanins or carotenoids reflected by remote sensing vegetation indicators ARI1 and CRI1, also confirmed the fluorescence results. For the H_2O + Pb growth environment, the difference in CRI1 was 47% and 575 for ARI1 compared with the control.

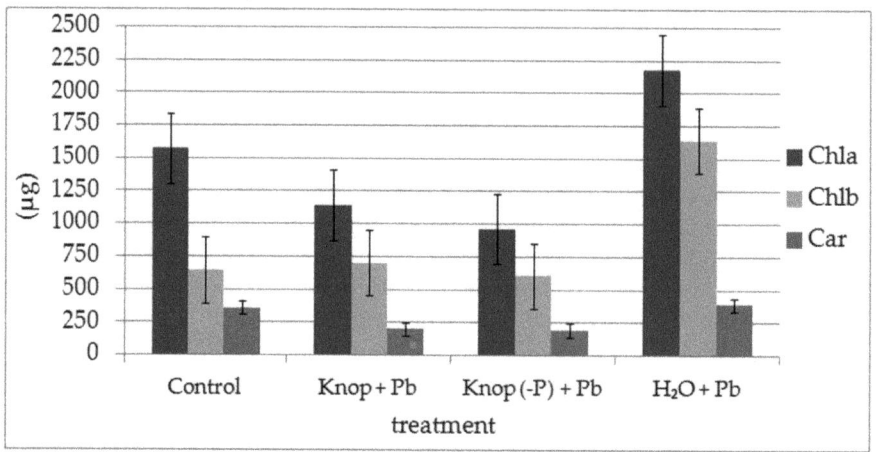

Figure 7. The photosynthetic pigments (chlorophyll a and b and carotenoids) contents in the pea leaves, expressed as micrograms per gram fresh weight.

Table 6. The ratio of Chla/b and chlorophyll to carotenoids (Chl/Car) in the pea leaves contaminated with lead ion, $Pb(NO_3)_2$, calculated as in Arnon [49] and indicators of remote sensing (RARS, RARSb, and RARSc).

Treatment	Chl a/b	Chl/Car	RARSa/RARSb	(RARSa + RARSb)/RARSc
Control	2.7	5.9	1.12	2.22
Knop + 5 mM $Pb(NO_3)_2$	1.8	7.4	1.16	2.22
Knop -P (without phosphate) + 5 mM $Pb(NO_3)_2$	1.6	6.7	1.20	1.98
H_2O + 5 mM $Pb(NO_3)_2$	1.3	10.0	1.13	2.05

In summary, the plants grown in distilled water with Pb ions were most sensitive to the toxic effects of lead. This result was also confirmed by the differences in the remote sensing indicator values, where the effects in peas could be ranked depending on the medium in which they were grown as follows from the worst to the best condition: H_2O + Pb, Knop (-P) + Pb, Knop + Pb, control. Although under these conditions lead is accumulated to a large degree in the leaves with only a small amount in the roots, both the elongation growth and mass the organs were inhibited. The maximum quantum efficiency of PSII in the light and the Chl a/b ratio were very low for these plants, indicating the inhibition of photosynthesis. The Fv/Fm ratio is used for monitoring in vivo effects of stress because it is linearly correlated with the quantum yield of light limited O_2 evolution and the number of PSII centers. In contrast, the plants grown on complete Knop + Pb ions showed the least response to the presence of the toxic metal due to the accumulation of Pb in the roots and the formation of insoluble phosphate salts. The indicators that showed statistically significant differences and thereby damage to plant cells caused by the action of lead were indicators determining the chlorophyll content: mNDVI705, RARSa, and RARSb; the amount of light used in photosynthesis: SIPI; the amount of carotenoids and other pigments: CRI1 and ARI1; and water content: NDWI, WI, NWI-2, and DSWI. These results indicate that the type of growth medium has a considerable influence on the absorption and accumulation of toxic metals in different parts of the plant, thereby affecting their function and modifying growth.

4. Discussion

Coupling fluorescence measurements with other non-invasive techniques, such as absorption spectroscopy in model laboratory conditions, we obtained more comprehensive results showing changes in the rate of growth, water content, photosynthetic activity, and others induced by lead under given conditions that change the transport of Pb ions to the leaves. Hence, irradiance during growth is responsible for changes in the antioxidant enzymes activities by affecting the rate of photosynthesis and related processes. The presented results confirm that the effect of Pb ions on photosynthetic parameters in pea seedlings can be detected using hyperspectral and fluorescence techniques. Despite the visual differences and accumulation of lead in the leaves, we confirm that the chlorosis of leaves may be a consequence of the inhibition of chlorophyll synthesis, as well as the increase in its degradation [66,67]. We previously observed [14] that Pb ions stimulated the respiration rate in pea leaves in plants grown in high light conditions, which was accompanied by an increase in ATP production in mitochondria. We also found that the inhibition of photosynthesis by Pb ions was higher in low- than in high-light-grown plants as a consequence of decreasing electron transport in the photosynthesis light reaction [68]. Lead is absorbed by plants mainly through the root system and in minor amounts through the leaves. At lethal concentrations, this barrier is broken and lead may enter vascular tissues. Lead in plants may form deposits of various sizes, present mainly in intercellular spaces, cell walls, and vacuoles. The negative effects of lead on plant vegetative growth mainly result from the following factors: distortion of chloroplast ultrastructure, obstructed electron transport, inhibition of Calvin cycle enzymes, impaired uptake of essential elements such as Mg and Fe, and induced deficiency of CO_2 resulting from stomatal closure [69,70].

Important for research on the impact of lead and heavy metals on peas is the environment in which the plant grows, both in terms of the soil and the nutrient medium. In our study, pea seedlings had the lowest values in comparison to the control when grown in H_2O + Pb medium. In the literature, we found an example where detached leaves of 14-day-old dark-grown pea seedlings were immersed with their cut ends either in water (control) or in 20 mM $Pb(NO_3)_2$ solution [71]. They were exposed to continuous illumination for 24 and 48 h. Values of Fv, Fm, and Fv/Fm were reduced by Pb^{2+}. Here, with H_2O + Pb treatment, the Fv'/Fm' value decreased about two-fold. The decrease in the chlorophyll a fluorescence parameters occurred in parallel with the strong inhibition of the biosynthesis of chlorophyll a and b but less of the carotenoids by this metal [71]. We also confirmed that the difference in the chlorophyll value between the H_2O + Pb and control treatments was significant (Chl a was 700 µg/g FW; Chl b was 1000 µg/g FW), whereas the difference in the carotenoids value was similar or less by only 20 µg/g FW. Pb^{2+} drastically reduced photosynthesis but had a stimulatory action on evolution in darkness (DR) after 24 h and only somewhat inhibited DR after 48 h exposure of leaves to this metal [71].

Fluorescence indicators confirmed the negative effect of lead ions on pea seedlings through a decline in the value of Fv/Fm [72]. Fluctuations were observed in PSII electron transport during photosynthesis induction. When the light intensity increased, the maximum quantum efficiency of PSII (Fv/Fm) decreased, as was also observed by Baker et al. [19]. The decrease in the Fv/Fm value was also noted for oats plants (*Avena sativa*) growing on contaminated places with excess Cu and Pb ions where the half-rise time ($t_{1/2}$) of Chl a fluorescence decreased, suggesting that the amount of active pigments decreased and the functional Chl antennae size of the photosynthetic apparatus was smaller compared with control plants [73]. In pea plants, after adding lead to the water, we observed that the greatest amount of lead was transported to the leaves, which was also noted for spinach plants treated with lead acetate for 15 days; 45% of Pb ions were accumulated in roots and 55% in the leaves [74]. More than 90% of heavy metals are generally adsorbed into cell walls. The oldest leaves accumulated the highest concentrations of heavy metals. Heavy metals have different mobility in plant tissues and translocation is probably regulated by the carrier proteins of the vascular tissues [75,76]. Decreases in the protein content in pea roots was observed after the addition of cadmium ions [67,77]. Roots accumulate larger amounts of metals than shoots [78]. Permanent stomatal closure, structural damage of chloroplasts, increased synthesis of ethylene (induce senescence), and imbalance in water relationships might be acting synergistically on heavy metal exposed plants, resulting in impairment of photosynthetic functions [79].

Carotenoids were less affected by lead compared to chlorophyll; however, the Chl/Car ratio increased significantly after Pb treatment: 5.9 for control plants 5.9 but and about 10 for plants grown in water with Pb. An increase in the Chl/Car ratio is characteristic for stress conditions because carotenoids are involved in photoprotective mechanisms. This is particularly important when photochemical quenching activity is exceeded, leading to photoinhibition [80]. Carotenoids protect chloroplasts (1) by modulating NPQ, (2) by mediating direct quenching of chlorophyll triples, and (3) by scavenging the ROS generated during photosynthesis. Our results indicate that the main function of carotenoids is scavenging the ROS because the NPQ values were low in Pb-treated plants. The reduced chlorophyll content due to heavy metals toxicity in different plant species has been well documented [81]. The observed increases in chlorophyll content in the leaves of plants grown in water + Pb may be related to the lead transported to the young plants inhibiting the activity of chlorophyllase, so the content of chlorophyll may be even higher in plants treated with lead [10].

The obtained results can be compared with previously reported data of tests of plants under stress conditions. Zagajewski [82], using multispectral spectrometer SPZ5 (24 spectral bands in the visible and near-infrared range from 400 to 1025 nm; Research Space Centre, PAS, Warsaw, Poland), noted the presence of lead in the contaminated grasses with 128 ppm Pb^{2+}. He observed not only a decrease in the light absorption used in the process of photosynthesis (VIS), but also changes in cellular structures. Heavy metals reduced the chlorophyll content, which is closely related to the inhibition of photosynthesis and, as a consequence, causes a decrease in biomass production [51]. Here, this was confirmed by the reduced LAI index value, and the roots and the leaf index determined by the ratio of fresh to dry mass of the leaves [51]. A hyperspectral method was used to assess concentrations of heavy metals in common cane growing along the Le An River in China [83], where the spatial distribution of metal concentration was examined. The stress status of plants was determined by using three indicators: NDVI, red edge position (REP), and continuum removal normalized band depth (CRNBD).

The correlation of remote sensing indices with chlorophyll content has enabled the creation of chlorophyll content models. On the basis of the proposed model, significant differences were observed in the level of heavy metals between the top and bottom of the plant stand corresponding to the chlorophyll gradients within. The concentrations of copper, zinc, and lead were highest in the lower parts of plants, but the coefficient of variation revealed differences in the values between heavy metals used. For Cu ions, the highest value attained was 0.61, 0.53 for Pb, and Zn was lowest at 0.22. The results also showed that the HMs content correlates negatively with chlorophyll level, whereas the vegetation indices showed a positive correlation [83]. Remote sensing of vegetation indices in the research areas contaminated with heavy metals were used, among others, by Elbe in Germany [84]. The aim of this study was to assess the pollution of the river floodplain ecosystems using ground spectrometric measurements (ASD FieldSpec Pro FR) and laboratory measurements. Remote sensing vegetation indices were able to assess the vitality of the plants, including biomass, chlorophyll, water, cellulose, and lignin contents. The water status of plants was excluded as a stressor because the coefficient of variation was small for WBI and MSI. The indicators NPCI, PRI, NDVI, and SIPI showed higher variability over time. In the case of the nitrogen content (NDNI) and lignin (NDLI), small changes were identified. Especially indicators NPCI, PRI, REP, and CR1730 were proven to be sensitive to the stress caused by heavy metals.

The acquired results are also confirmed by the differences in the values of the remote sensing indicators, where the effect on peas can be ranked from the worst to best condition depending on the medium in which they were grown as follows: H_2O + Pb, Knop + Pb, Knop (-P) + Pb, and control. Different environments significantly affect the impact of stress on plant growth caused by lead on its vital functions. The single-metal sorption uptake capacity of the biomass for Pb was slightly inhibited by the presence of the other heavy metals in the system. The presence of Cu and Cd cations in separate systems reduced the biomass uptake capacity of Pb by only 6% [85]. Two plant species, spinach (*Spinacia oleracea*) and wheat (*Triticum aestivum*), were grown under hydroponic conditions and stressed with lead nitrate, $Pb(NO_3)_2$, at three concentrations (1.5, 3, and 15 mM) [86]. Lead was accumulated in a dose-dependent manner in both plant species, which resulted in reduced growth and lower uptake of all mineral ions tested. Total amounts and concentrations of most mineral ions (Na, K, Ca, P, Mg, Fe, Cu, and Zn) decreased, although Mn concentrations increased, as its uptake decreased less relative to the whole plant's growth. The deficiency in mineral nutrients was correlated with a strong decrease in the chlorophyll a and b and proline contents in both species, but these effects were less pronounced in spinach than in wheat. In contrast, the effects of lead on soluble proteins differed between species; they reduced in wheat at all lead concentrations, whereas they increased in spinach, where their value peaked at 3 mM Pb. In wheat seedlings, concentrations of chlorophylls a and b were already significantly lowered at 1.5 mM Pb, and this effect was even more pronounced at 3 and 15 mM Pb. Lead stress resulted in a heavy reduction in chlorophyll due to both chloroplast disorganization, the reduction in the amount of thylakoids and grana, direct inhibition of chlorophyll synthesis, as well as changes of chlorophyll structure due to replacement of key nutrients (Mg, Fe, and Cu) by lead [87,88]. Lead exposure resulted in dose-dependent damage to both plant species. In wheat plants exposed to 15 mM Pb, growth inhibition was clear, whereas spinach fresh and dry weights decreased by only 28% and 29%, respectively, at 15 mM Pb, when compared with controls.

The spectral reflectance curves obtained for pea plants in the near and mid-infrared indicate a high sensitivity to Pb. Plants treated with lead ions are the most sensitive to light in the red edge range, which also has been confirmed for other metals as Ni, Cd, and Cu [31,40]. An experimental design was established with three Pb concentrations (0, 2.5 × 207.2, 5.0 × 207.2 mg Pb per 1000 g soil) for 30, 50, 65, 80, and 90 days. The differences in the shape of the spectral reflection curve were visible mainly in the red edge range, which was found here (e.g., 714–723 nm). The highest impact of lead (the difference between the spectral characteristics of the concentrations tested) was observed on day 50 and 80 by a flattening of the edges of red [34]. The derivative reflectance calculated from canopy-level CASI airborne imagery showed a peak at the 700–730 nm region, which was experimentally shown to be related to stress conditions and potentially caused by fluorescence emission and chlorophyll content changes in vegetation under stress [89,90]. Here, the ranges in the electromagnetic spectrum were statistically significant for condition, stress influence was the range: 714–723 nm (plant stress), 858–951 nm (biomass, plant density), 995–1145 (biophysical quantity and yield), and 1386–1389, 1570–1579, and 1801–1829 nm (lignin, cellulose, plant litter and water). Similar results were also observed by Zagajewski [82] for grasses 6 and 11 weeks after application of lead. Spectral reflectance curves can also indicate disturbances in the cell structures. The largest differences in reflectance, especially in near-infrared, were shown for the grasses 11 weeks after application of lead.

Lead causes large changes in plant vitality, inducing cellular structures by creating deposits in the cell walls. This may cause developmental disorders, as well as impact the water status of plants and then photosynthesis. In analyzing the spectrum bands for pea and common cane plants growing in the presence of lead, lead induced changes in the spectrum differently for both plants. In the case of the pea plants, the range of the electromagnetic spectrum, describing the structure of the cells and the water content in the plants, indicated the high sensitivity of the pea plants to lead ions. Whereas for cane plants grown on the reservoir flotation [91], changes occurred in the edge of the red (400–800 nm), the near infrared (from 1100 to 1400 nm), and in the mid-infrared.

Numerous studies that have tested the physiological responses to excess levels of heavy metal ions indicate that plants have evolved various mechanisms to cope with heavy-metal-induced stress. Lead is highly reactive and inactivates various enzymes; hence, it impairs plant growth and development [71]. Thus, using noninvasive techniques, such as fluorescence, absorption spectroscopy, or gas exchange, is important for monitoring plant photosynthesis under given conditions even for a long time. The transferability of laboratory-based results to field situations is a subject of discussion. Laboratory results might not mirror in situ behavior due to differences in growing environments, sampling protocols, and operating conditions, but they can provide information about the correlation between investigated processes under controlled conditions. Thus, further comparative work is warranted, which will allow for further development, such as the novel ground-based spectrometer system, PhotoSpec, for measuring SIF in the red (670–732 nm) and far-red (729–784 nm) wavelength range as well as canopy reflectance (400–900 nm) to calculate vegetation indices, such as the normalized difference vegetation index (NDVI), the enhanced vegetation index (EVI), and the photochemical reflectance index (PRI) [92].

5. Conclusions

The obtained results confirm a significant decrease in the vitality of the pea plants treated with lead. Using fluorescence techniques to assess the degree of inhibition of plant metabolism by lead ions provides useful information about the actual physiological state of vegetation in the presence of this heavy metal. These changes are visible in both the spectral reflection curve shape and the remote sensing vegetation indices. Combined with remote sensing measurements and the quantitative survey of selected parameters (e.g., photosynthetic pigments), they are one of the most accurate techniques for describing photosynthesis and evaluating the function of the photosynthetic apparatus in the presence of stress factors. Uptake of heavy metals, their distribution, and accumulation result in permanent structural changes in the plants, which was confirmed by the results obtained from spectrometric measurements. Their effect changes in terms of the content of plant components such as water and photosynthetically active pigments, confirming an adverse effect on the photosynthetic apparatus of the plant (fluorescence measurements). One of the main symptoms induced by lead on the plants is accelerated aging, visible in the decrease in chlorophyll content and the lowering of the water content in tissues (the drying of plants). Lead exposure in plants strongly limits the development and sprouting of seedlings. Plant biomass can also be restricted by high doses of lead exposure.

The disruption of plant water status after lead treatment has been addressed in many studies. Results of such exposures show a decrease in transpiration as well as reduction in the moisture content. Lead reduces plant cell wall plasticity, and thereby influences the cell turgor pressure. The obtained results confirm that hyperspectral techniques can be used to detect the effect of lead as a stress factor on plant vegetation. The contribution of this research is the identification of noticeable changes in the physical properties of pea plants treated with Pb ions and grown in the laboratory conditions, where visible changes were observed in the cell structure and water content as well as changes leading to the death of plants. This information provides the basis for the assumption that the hyperspectral data and fluorescence measurements are useful for the analysis of vegetation contaminated with heavy metals and monitoring of contaminated areas. The determination of the content of heavy metals accumulated in vegetables and physiological effects on the plants can provide information on the source and extent of the impact of pollution emissions.

Author Contributions: The experiment was designed and conducted by E.R. and M.K.; both researchers analyzed the data. All authors prepared the manuscript.

Funding: The publishing costs were financed from by grant No. 500-D119-12-1190000 awarded by the Polish Ministry of Science and Higher Education.

Acknowledgments: The authors express thanks to Anna Robak and Ignacy Kaźmierczak for their participation in laboratory research. The analyses were conducted in the frame of the COST Action: "Optical synergies for spatiotemporal SENsing of Scalable ECOphysiological traits" (SENSECO). Sincere thanks to Echo Chi (editor), Kelly O'Keefe (English editor) and anonymous Reviewer, who allowed us to improve the manuscript.

Conflicts of Interest: The authors declare no conflict of interest.

References

1. Mohr, H.; Schopfer, P. Physiology of Movement. In *Plant Physiology*; Springer: Berlin/Heidelberg, Germany, 1995; pp. 497–538. [CrossRef]
2. Fryer, M.J.; Andrews, J.R.; Oxborough, K.; Blowers, D.A.; Baker, N.R. Relationship between CO_2 Assimilation, Photosynthetic Electron Transport, and Active O2 Metabolism in Leaves of Maize in the Field during Periods of Low Temperature. *Plant Physiol.* **1998**, *116*, 571–580. [CrossRef] [PubMed]
3. Hendrickson, L.; Furbank, R.T.; Chow, W.S. A Simple Alternative Approach to Assessing the Fate of Absorbed Light Energy Using Chlorophyll Fluorescence. *Photosynth. Res.* **2004**, *82*, 73–81. [CrossRef] [PubMed]
4. Romanowska, E. Gas Exchange Functions in Heavy Metal Stressed Plants. In *Physiology and Biochemistry of Metal Toxicity and Tolerance in Plants*; Springer: Dordrecht, The Netherlands, 2002; pp. 257–285. [CrossRef]
5. Rashid, A.; Camm, E.L.; Ekramoddoullah, A.K.M. Molecular mechanism of action of Pb^{2+} and Zn^{2+} on water oxidizing complex of photosystem II. *FEBS Lett.* **1994**, *350*, 296–298. [CrossRef]
6. Yadav, S.K. Heavy metals toxicity in plants: An overview on the role of glutathione and phytochelatins in heavy metal stress tolerance of plants. *S. Afr. J. Bot.* **2010**, *76*, 167–179. [CrossRef]
7. Schreiber, U. Pulse-Amplitude-Modulation (PAM) Fluorometry and Saturation Pulse Method: An Overview. In *Chlorophyll a Fluorescence*; Springer: Dordrecht, The Netherlands, 2007; pp. 279–319. [CrossRef]
8. Burzyński, M.; Kłobus, G. Changes of photosynthetic parameters in cucumber leaves under Cu, Cd, and Pb stress. *Photosynthetica* **2004**, *42*, 505–510. [CrossRef]

9. Woźniak, A.; Drzewiecka, K.; Kęsy, J.; Marczak, Ł.; Narożna, D.; Grobela, M.; Motała, R.; Bocianowski, J.; Morkunas, I. The Influence of Lead on Generation of Signalling Molecules and Accumulation of Flavonoids in Pea Seedlings in Response to Pea Aphid Infestation. *Molecules* **2017**, *22*, 1404. [CrossRef] [PubMed]
10. Parys, E.; Romanowska, E.; Siedlecka, M.; Poskuta, J.W. The effect of lead on photosynthesis and respiration in detached leaves and in mesophyll protoplasts of *Pisum sativum*. *Acta Physiol. Plant.* **1998**, *20*, 313. [CrossRef]
11. Romanowska, E.; Igamberdiev, A.U.; Parys, E.; Gardestrom, P. Stimulation of respiration by Pb^{2+} in detached leaves and mitochondria of C_3 and C_4 plants. *Physiol. Plant.* **2002**, *116*, 148–154. [CrossRef]
12. Nas, F.S.; Ali, M. The effect of lead on plants in terms of growing and biochemical parameters: A review. *MOJ Ecol. Environ. Sci.* **2018**, *3*, 265–268. [CrossRef]
13. Romanowska, E.; Wasilewska, W.; Fristedt, R.; Vener, A.V.; Zienkiewicz, M. Phosphorylation of PSII proteins in maize thylakoids in the presence of Pb ions. *J. Plant Physiol.* **2012**, *169*, 345–352. [CrossRef]
14. Romanowska, E.; Wróblewska, B.; Drozak, A.; Siedlecka, M. High light intensity protects photosynthetic apparatus of pea plants against exposure to lead. *Plant Physiol. Biochem.* **2006**, *44*, 387–394. [CrossRef] [PubMed]
15. Aro, E.-M.; Suorsa, M.; Rokka, A.; Allahverdiyeva, Y.; Paakkarinen, V.; Saleem, A.; Battchikova, N.; Rintamäki, E. Dynamics of photosystem II: A proteomic approach to thylakoid protein complexes. *J. Exp. Bot.* **2005**, *56*, 347–356. [CrossRef] [PubMed]
16. Miles, C.D.; Brandle, J.R.; Daniel, D.J.; Chu-Der, O.; Schnare, P.D.; Uhlik, D.J. Inhibition of Photosystem II in Isolated Chloroplasts by Lead. *Plant Physiol.* **1972**, *49*, 820–825. [CrossRef] [PubMed]
17. Baker, N.R. Chlorophyll Fluorescence: A Probe of Photosynthesis in vivo. *Annu. Rev. Plant Biol.* **2008**, *59*, 89–113. [CrossRef] [PubMed]
18. Ruban, A.V. Nonphotochemical Chlorophyll Fluorescence Quenching: Mechanism and Effectiveness in Protecting Plants from Photodamage. *Plant Physiol.* **2016**, *170*, 1903–1916. [CrossRef] [PubMed]
19. Baker, N.R.; Oxborough, K. Chlorophyll Fluorescence as a Probe of Photosynthetic Productivity. In *Chlorophyll a Fluorescence*; Springer: Dordrecht, The Netherlands, 2007; pp. 65–82. [CrossRef]
20. Clijsters, H.; Van Assche, F. Inhibition of photosynthesis by heavy metals. *Photosynth. Res.* **1985**, *7*, 31–40. [CrossRef]
21. Kalaji, H.M.; Loboda, T. Photosystem II of Barley seedlings under cadmium and lead stress. *Plant Soil Environ.* **2007**, *53*, 511–516. [CrossRef]
22. Lazár, D.; Jablonsky, J. On the approaches applied in formulation of a kinetic model of photosystem II: Different approaches lead to different simulations of the chlorophyll a fluorescence transients. *J. Theor. Biol.* **2009**, *257*, 260–269. [CrossRef]
23. Shamshad, S.; Shahid, M.; Rafiq, M.; Khalid, S.; Dumat, C.; Sabir, M.; Murtaza, B.; Farooq, A.B.U.; Shah, N.S. Effect of organic amendments on cadmium stress to pea: A multivariate comparison of germinating vs. young seedlings and younger vs. older leaves. *Ecotoxicol. Environ. Saf.* **2018**, *151*, 91–97. [CrossRef]
24. Bazzaz, F.A.; Rolfe, G.L.; Carlson, R.W. Effect of cadmium on photosynthesis and transpiration of excised leaves of corn and sunflower. *Physiol. Plant.* **1974**, *32*, 373–376. [CrossRef]
25. Zagajewski, B.; Kycko, M.; Tømmervik, H.; Bochenek, Z.; Wojtuń, B.; Bjerke, J.W.; Kłos, A. Feasibility of hyperspectral vegetation indices for the detection of chlorophyll concentration in three high Arctic plants: *Salix polaris*, *Bistorta vivipara*, and *Dryas octopetala*. *Acta Soc. Bot. Pol.* **2018**, *87*, 3604. [CrossRef]
26. Kycko, M.; Zagajewski, B.; Lavender, S.; Romanowska, E.; Zwijacz-Kozica, M. The Impact of Tourist Traffic on the Condition and Cell Structures of Alpine Swards. *Remote Sens.* **2018**, *10*, 220. [CrossRef]
27. Clevers, J.G.P.W. Beyond NDVI: Extraction of Biophysical Variables from Remote Sensing Imagery. In *Remote Sensing and Digital Image Processing*; Springer: Dordrecht, The Netherlands, 2014; pp. 363–381. [CrossRef]
28. Farquhar, G.D.; von Caemmerer, S.; Berry, J.A. A biochemical model of photosynthetic CO_2 assimilation in leaves of C_3 species. *Planta* **1980**, *149*, 78–90. [CrossRef]
29. Guanter, L.; Zhang, Y.; Jung, M.; Joiner, J.; Voigt, M.; Berry, J.A.; Frankenberg, C.; Huete, A.R.; Zarco-Tejada, P.; Lee, J.-E.; et al. Global and time-resolved monitoring of crop photosynthesis with chlorophyll fluorescence. *Proc. Natl. Acad. Sci. USA* **2014**, *111*, E1327–E1333. [CrossRef]
30. Norton, A.J.; Rayner, P.J.; Koffi, E.N.; Scholze, M. Assimilating solar-induced chlorophyll fluorescence into the terrestrial biosphere model BETHY-SCOPE v1.0: Model description and information content. *Geosci. Model Dev.* **2018**, *11*, 1517–1536. [CrossRef]

31. Kancheva, R.; Georgiev, G. Spectrally-based quantification of plant heavy metal-induced stress. In *Remote Sensing for Agriculture, Ecosystems, and Hydrology XIV*; SPIE: Bellingham, WA, USA, 2012; Volume 8531, p. 85311D. [CrossRef]
32. Rajewicz, P.A.; Atherton, J.; Alonso, L.; Porcar-Castell, A. Leaf-level spectral fluorescence measurements: Comparing methodologies for broadleaves and needles. *Remote Sens.* **2019**, *11*, 532. [CrossRef]
33. Rathod, P.H.; Brackhage, C.; Van der Meer, F.D.; Müller, I.; Noomen, M.F.; Rossiter, D.G.; Dudel, G.E. Spectral changes in the leaves of barley plant due to phytoremediation of metals—Results from a pot study. *Eur. J. Remote Sens.* **2015**, *48*, 283–302. [CrossRef]
34. Ren, H.-Y.; Zhuang, D.-F.; Pan, J.-J.; Shi, X.-Z.; Wang, H.-J. Hyper-spectral remote sensing to monitor vegetation stress. *J. Soils Sediments* **2008**, *8*, 323–326. [CrossRef]
35. Dunagan, S.C.; Gilmore, M.S.; Varekamp, J.C. Effects of mercury on visible/nearinfrared reflectance spectra of mustard spinach plants (*Brassica rapa* P.). *Environ. Pollut.* **2007**, *148*, 301–311. [CrossRef]
36. Su, Y.; Sridhar, B.B.M.; Han, F.X.; Monts, D.L.; Diehl, S.V. Effect of bioaccumulation of Cs and Sr natural isotopes on foliar structure and plant spectral reflectance of Indian mustard (*Brassica juncea*). *Water Air Soil Pollut.* **2007**, *180*, 65–74. [CrossRef]
37. Bannari, A.; Morin, D.; Bonn, F.; Huete, A. A review of vegetation indices. *Remote Sens. Rev.* **1995**, *13*, 95–120. [CrossRef]
38. Kycko, M.; Zagajewski, B.; Lavender, S.; Dabija, A. In Situ Hyperspectral Remote Sensing for Monitoring of Alpine Trampled and Recultivated Species. *Remote Sens.* **2019**, *11*, 1296. [CrossRef]
39. Zagajewski, B.; Tømmervik, H.; Bjerke, J.; Raczko, E.; Bochenek, Z.; Kłos, A.; Jarocińska, A.; Lavender, S.; Ziółkowski, D. Intraspecific Differences in Spectral Reflectance Curves as Indicators of Reduced Vitality in High-Arctic Plants. *Remote Sens.* **2017**, *9*, 1289. [CrossRef]
40. Kancheva, R.; Borisova, D.; Iliev, I. Chlorophyll fluorescence as a quantitative measure of plant stress. In *New Developments and Challenges in Remote Sensing*; Bochenek, Z., Ed.; Millpress: Rotterdam, The Netherlands, 2007; ISBN 978-90-5966-053-3.
41. Holleboom, C.-P.; Walla, P.J. The back and forth of energy transfer between carotenoids and chlorophylls and its role in the regulation of light harvesting. *Photosynth. Res.* **2014**, *119*, 215–221. [CrossRef]
42. Rossini, M.; Fava, F.; Cogliati, S.; Meroni, M.; Marchesi, A.; Panigada, C.; Giardino, C.; Busetto, L.; Migliavacca, M.; Amaducci, S.; et al. Assessing canopy PRI from airborne imagery to map water stress in maize. *ISPRS J. Photogramm. Remote Sens.* **2013**, *86*, 168–177. [CrossRef]

43. Cierniewski, J.; Kazmierowski, C.; Krolewicz, S.; Piekarczyk, J.; Wrobel, M.; Zagajewski, B. Effects of different illumination and observation techniques of cultivated soils on their hyperspectral bidirectional measurements under field and laboratory conditions. *IEEE J. Sel. Top. Appl. Earth Obs. Remote Sens.* **2014**, *7*, 2525–2530. [CrossRef]
44. Pinto, F.; Damm, A.; Schickling, A.; Panigada, C.; Cogliati, S.; Müller-Linow, M.; Balvora, A.; Rascher, U. Sun-induced chlorophyll fluorescence from high-resolution imaging spectroscopy data to quantify spatio-temporal patterns of photosynthetic function in crop canopies. *Plant. Cell Environ.* **2016**, *39*, 1500–1512. [CrossRef]
45. Colombo, R.; Celesti, M.; Bianchi, R.; Campbell, P.K.E.; Cogliati, S.; Cook, B.D.; Corp, L.A.; Damm, A.; Domec, J.-C.; Guanter, L.; et al. Variability of sun-induced chlorophyll fluorescence according to stand age-related processes in a managed loblolly pine forest. *Glob. Change Biol.* **2018**, *24*, 2980–2996. [CrossRef]
46. Drusch, M.; Moreno, J.; Del Bello, U.; Franco, R.; Goulas, Y.; Huth, A.; Kraft, S.; Middleton, E.M.; Miglietta, F.; Mohammed, G.; et al. The Fluorescence Explorer Mission Concept—ESA's Earth Explorer 8. *IEEE Trans. Geosci. Remote Sens.* **2017**, *55*, 1273–1284. [CrossRef]
47. Rascher, U.; Alonso, L.; Burkart, A.; Cilia, C.; Cogliati, S.; Colombo, R.; Damm, A.; Drusch, M.; Guanter, L.; Hanus, J.; et al. Sun-induced fluorescence—A new probe of photosynthesis: First maps from the imaging spectrometer HyPlant. *Glob. Change Biol.* **2015**, *21*, 4673–4684. [CrossRef]
48. Genty, B.; Briantais, J.-M.; Baker, N.R. The relationship between the quantum yield of photosynthetic electron transport and quenching of chlorophyll fluorescence. *Biochim. Biophys. Acta Gen. Subj.* **1989**, *990*, 87–92. [CrossRef]
49. Arnon, D.I. Cooper enzymes in isolated chloroplasts. Polypenyloxidase in *Beta vulgaris*. *Plant Physiol.* **1949**, *24*, 1–15. [CrossRef]
50. Dawson, T.P.; Curran, P.J. Technical note A new technique for interpolating the reflectance red edge position. *Int. J. Remote Sens.* **1998**, *19*, 2133–2139. [CrossRef]
51. Chappelle, E.W.; Kim, M.S.; McMurtrey, J.E. Ratio analysis of reflectance spectra (RARS): An algorithm for the remote estimation of the concentrations of chlorophyll A, chlorophyll B, and carotenoids in soybean leaves. *Remote Sens. Environ.* **1992**, *39*, 239–247. [CrossRef]
52. Peñuelas, J.; Baret, F.; Filella, I. Semi-Empirical Indices to Assess Carotenoids/Chlorophyll-a Ratio from Leaf Spectral Reflectance. *Photosynthetica* **1995**, *31*, 221–230.
53. Fourty, T.; Baret, F. On spectral estimates of fresh leaf biochemistry. *Int. J. Remote Sens.* **1998**, *19*, 1283–1297. [CrossRef]
54. Merzlyak, M.N.; Gitelson, A.A.; Chivkunova, O.B.; Rakitin, V.Y. Non-destructive optical detection of pigment changes during leaf senescence and fruit ripening. *Physiol. Plant.* **1999**, *106*, 135–141. [CrossRef]
55. Gitelson, A.A.; Zur, Y.; Chivkunova, O.B.; Merzlyak, M.N. Assessing Carotenoid Content in Plant Leaves with Reflectance Spectroscopy. *Photochem. Photobiol.* **2002**, *75*, 272–281. [CrossRef]
56. Gitelson, A.A.; Merzlyak, M.N.; Chivkunova, O.B. Optical Properties and Nondestructive Estimation of Anthocyanin Content in Plant Leaves. *Photochem. Photobiol.* **2001**, *74*, 38. [CrossRef]
57. Fuentes, D.A.; Gamon, J.A.; Qiu, H.; Sims, D.A.; Roberts, D.A. Mapping Canadian boreal forest vegetation using pigment and water absorption features derived from the AVIRIS sensor. *J. Geophys. Res. Atmos.* **2001**, *106*, 33565–33577. [CrossRef]
58. Gao, B. NDWI—A normalized difference water index for remote sensing of vegetation liquid water from space. *Remote Sens. Environ.* **1996**, *58*, 257–266. [CrossRef]
59. Sims, D.A.; Gamon, J.A. Relationships between leaf pigment content and spectral reflectance across a wide range of species, leaf structures and developmental stages. *Remote Sens. Environ.* **2002**, *81*, 337–354. [CrossRef]
60. Babar, M.A.; Reynolds, M.P.; van Ginkel, M.; Klatt, A.R.; Raun, W.R.; Stone, M.L. Spectral Reflectance Indices as a Potential Indirect Selection Criteria for Wheat Yield under Irrigation. *Crop Sci.* **2006**, *46*, 578. [CrossRef]
61. Galvão, L.S.; Epiphanio, J.C.N.; Breunig, F.M.; Formaggio, A.R. *Biophysical and Biochemical Characterization and Plant Species Studies*; Thenkabail, P.S., Lyon, J.G., Huete, A., Eds.; CRC Press: Boca Raton, FL, USA, 2018; ISBN 9780429431180. [CrossRef]
62. Evans, E.H.; Brown, R.G. New trends in photobiology. *J. Photochem. Photobiol. B Biol.* **1994**, *22*, 95–104. [CrossRef]
63. Fisher, R.A. *Statistical Methods for Research Workers; Landmark Writings in Western Mathematics: Case Studies*, 11th ed.; Grattan-Guinness, I., Ed.; Elsevier: Amsterdam, The Netherlands, 1925; pp. 1640–1940.

64. StatSoft. *StatSoft Manual, Internetowy Podręcznik Statystyki*; Statistica: Krakow, Poland, 2012.
65. Sharma, P.; Dubey, R.S. Lead Toxicity in Plants. *Braz. J. Plant Physiol.* **2005**, *17*, 1–19. [CrossRef]
66. Myśliwa-Kurdziel, B.; Prasad, M.N.V.; Strzałtka, K. Photosynthesis in heavy metal stressed plants. In *Heavy Metal Stress in Plants*; Springer: Berlin/Heidelberg, Germany, 2004; pp. 146–181. [CrossRef]
67. Bavi, K.; Kholdebarin, B.; Moradshahi, A. Effect of cadmium on growth, protein content and peroxidase activity in pea plants. *Pak. J. Bot.* **2011**, *43*, 1467–1470.
68. Romanowska, E.; Wróblewska, B.; Drożak, A.; Zienkiewicz, M.; Siedlecka, M. Effect of Pb ions on superoxide dismutase and catalase activities in leaves of pea plants grown in high and low irradiance. *Biol. Plant.* **2008**, *52*, 80–86. [CrossRef]
69. Sengar, R.S.; Gautam, M.; Sengar, R.S.; Sengar, R.S.; Garg, S.K.; Sengar, K.; Chaudhary, R. Lead Stress Effects on Physiobiochemical Activities of Higher Plants. In *Reviews of Environmental Contamination and Toxicology*; Springer: New York, NY, USA, 2008; pp. 73–93. ISBN 9780387784434. [CrossRef]
70. Pourrut, B.; Shahid, M.; Dumat, C.; Winterton, P.; Pinelli, E. Lead Uptake, Toxicity, and Detoxification in Plants. In *Reviews of Environmental Contamination and Toxicology*; Springer: New York, NY, USA, 2011; pp. 113–136. ISBN 9781441998590. [CrossRef]
71. Łukaszek, M.; Poskuta, J.W. Development of photosynthetic apparatus and respiration in pea seedlings during as influenced by toxic concentration of lead. *Acta Physiol. Plant.* **1998**, *20*, 35. [CrossRef]
72. Joshi, M.K.; Mohanty, P. Chlorophyll a Fluorescence as a Probe of Heavy Metal Ion Toxicity in Plants. In *Chlorophyll a Fluorescence*; Springer: Dordrecht, The Netherlands, 2007; pp. 637–661. [CrossRef]
73. Moustakas, M.; Lanaras, T.; Symeonidis, L.; Karataglis, S. Growth and some photosynthesis characteristics of field grown Avena sativa under copper and lead stress. *Photosynthetica* **1994**, *30*, 389–396.
74. Ernst, W.H.O.; Verkleij, J.A.C.; Schat, H. Metal tolerance in plants. *Acta Bot. Neerl.* **1992**, *41*, 229–248. [CrossRef]
75. Gallego, S.M.; Pena, L.B.; Barcia, R.A.; Azpilicueta, C.E.; Iannone, M.F.; Rosales, E.P.; Zawoznik, M.S.; Groppa, M.D.; Benavides, M.P. Unravelling cadmium toxicity and tolerance in plants: Insight into regulatory mechanisms. *Environ. Exp. Bot.* **2012**, *83*, 33–46. [CrossRef]
76. Kłos, A.; Ziembik, Z.; Rajfur, M.; Dołhańczuk-Śródka, A.; Bochenek, Z.; Bjerke, J.W.; Tømmervik, H.; Zagajewski, B.; Ziółkowski, D.; Jerz, D.; et al. Using moss and lichens in biomonitoring of heavy-metal contamination of forest areas in southern and north-eastern Poland. *Sci. Total Environ.* **2018**, *627*, 438–449. [CrossRef]
77. Usman, K.; Al-Ghouti, M.A.; Abu-Dieyeh, M.H. The assessment of cadmium, chromium, copper, and nickel tolerance and bioaccumulation by shrub plant Tetraena qataranse. *Sci. Rep.* **2019**, *9*, 5658. [CrossRef]
78. Cannata, M.G.; Carvalho, R.; Bertoli, A.C.; Augusto, A.S.; Bastos, A.R.R.; Carvalho, J.G.; Freitas, M.P. Effects of Cadmium and Lead on Plant Growth and Content of Heavy Metals in Arugula Cultivated in Nutritive Solution. *Commun. Soil Sci. Plant Anal.* **2013**, *44*, 952–961. [CrossRef]
79. Prasad, M.N.V.; Strzałka, K. Impact of heavy metals on photosynthesis. In *Heavy Metal Stress in Plants*; Springer: Berlin/Heidelberg, Germany, 1999; pp. 117–138. [CrossRef]
80. Külheim, C.; Ågren, J.; Jansson, S. Rapid regulation of light harvesting and plant fitness in the field. *Science* **2002**, *297*, 91–93. [CrossRef]
81. Pandey, N.; Pathak, G.C. Nickel alters antioxidative defense and water status in green gram. *Ind. J. Plant Physiol.* **2006**, *11*, 113–118.
82. Zagajewski, B. Remote Sensing Measurements of Lead Concentration in Plants. *Misc. Geogr.* **2000**, *9*, 267–282. [CrossRef]
83. Chen, H. *The Possibility of Assessing Heavy Metal Concentrations in Reed along le an River (China) Using Hyperspectral Data*; International Institute for Geo-information Science and Earth Observation ITC: Enschede, The Netherlands, 2008; pp. 1–54.
84. Götze, C.; Jung, A.; Henrich, V.; Merbach, I.; Gläßer, C. Spectrometric analyses in comparison to the physiological condition of heavy metal stressed floodplain vegetation in a standardised experiment. In Proceedings of the 6th EARSeL Workshop on Imaging Spectroscopy SIG, Tel Aviv, Israel, 16–19 March 2009.
85. Hawari, A.H.; Mulligan, C.N. Effect of the presence of lead on the biosorption of copper, cadmium and nickel by anaerobic biomass. *Process Biochem.* **2007**, *42*, 1546–1552. [CrossRef]
86. Lamhamdi, M.; El Galiou, O.; Bakrim, A.; Nóvoa-Muñoz, J.C.; Arias-Estévez, M.; Aarab, A.; Lafont, R. Effect of lead stress on mineral content and growth of wheat (Triticum aestivum) and spinach (Spinacia oleracea) seedlings. *Saudi J. Biol. Sci.* **2013**, *20*, 29–36. [CrossRef]

87. Haider, S.; Kanwal, S.; Uddin, F.; Azmat, R. Phytotoxicity of Pb: II. Changes in Chlorophyll Absorption Spectrum due to Toxic Metal Pb Stress on Phaseolus mungo and Lens culinaris. *Pak. J. Biol. Sci.* **2006**, *9*, 2062–2068. [CrossRef]
88. Akinci, I.E.; Akinci, S.; Yilmaz, K. Response of tomato (*Solanum lycopersicum* L.) to lead toxicity: Growth, element uptake, chlorophyll and water content. *Afr. J. Agric. Res.* **2010**, *5*, 416–423.
89. Zarco-Tejada, P.J.; Miller, J.R.; Mohammed, G.H.; Noland, T.L.; Sampson, P.H. Estimation of chlorophyll fluorescence under natural illumination from hyperspectral data. *Int. J. Appl. Earth Obs. Geoinf.* **2001**, *3*, 321–327. [CrossRef]
90. Zarco-Tejada, P.J.; Miller, J.R.; Mohammed, G.H.; Noland, T.L.; Sampson, P.H. Vegetation Stress Detection through Chlorophyll + Estimation and Fluorescence Effects on Hyperspectral Imagery. *J. Environ. Qual.* **2002**, *31*, 1433. [CrossRef]
91. Zagajewski, B.; Lechnio, J.; Sobczak, M. *Wykorzystanie Teledetekcji Hiperspektralnej w Analizie Roślinności Zanieczyszczonej Metalami Ciężkimi*; Teledetekcja Środowiska: Warszawa, Poland, 2007; Volume 37, pp. 82–100.
92. Grossmann, K.; Frankenberg, C.; Magney, T.S.; Hurlock, S.C.; Seibt, U.; Stutz, J. PhotoSpec: A new instrument to measure spatially distributed red and far-red solar-induced chlorophyll fluorescence. *Remote Sens. Environ.* **2018**, *216*, 311–327. [CrossRef]

© 2019 by the authors. Licensee MDPI, Basel, Switzerland. This article is an open access article distributed under the terms and conditions of the Creative Commons Attribution (CC BY) license (http://creativecommons.org/licenses/by/4.0/).

Article

Estimating Forest Leaf Area Index and Canopy Chlorophyll Content with Sentinel-2: An Evaluation of Two Hybrid Retrieval Algorithms

Luke A. Brown, Booker O. Ogutu and Jadunandan Dash *

School of Geography and Environmental Science, University of Southampton, Highfield, Southampton SO17 1BJ, UK
* Correspondence: j.dash@soton.ac.uk

Received: 18 June 2019; Accepted: 23 July 2019; Published: 25 July 2019

Abstract: Estimates of biophysical and biochemical variables such as leaf area index (LAI) and canopy chlorophyll content (CCC) are a fundamental requirement for effectively monitoring and managing forest environments. With its red-edge bands and high spatial resolution, the Multispectral Instrument (MSI) on board the Sentinel-2 missions is particularly well-suited to LAI and CCC retrieval. Using field data collected throughout the growing season at a deciduous broadleaf forest site in Southern England, we evaluated the performance of two hybrid retrieval algorithms for estimating LAI and CCC from MSI data: the Scattering by Arbitrarily Inclined Leaves (SAIL)-based L2B retrieval algorithm made available to users in the Sentinel Application Platform (SNAP), and an alternative retrieval algorithm optimised for forest environments, trained using the Invertible Forest Reflectance Model (INFORM). Moderate performance was associated with the SNAP L2B retrieval algorithm for both LAI (r^2 = 0.54, RMSE = 1.55, NRMSE = 43%) and CCC (r^2 = 0.52, RMSE = 0.79 g m^{-2}, NRMSE = 45%), while improvements were obtained using the INFORM-based retrieval algorithm, particularly in the case of LAI (r^2 = 0.79, RMSE = 0.47, NRMSE = 13%), but also in the case of CCC (r^2 = 0.69, RMSE = 0.52 g m^{-2}, NRMSE = 29%). Forward modelling experiments confirmed INFORM was better able to reproduce observed MSI spectra than SAIL. Based on our results, for forest-related applications using MSI data, we recommend users seek retrieval algorithms optimised for forest environments.

Keywords: artificial neural networks; canopy chlorophyll content; INFORM; leaf area index; SAIL

1. Introduction

Estimates of biophysical and biochemical variables such as leaf area index (LAI) and canopy chlorophyll content (CCC) provide vital information on the condition, structure and function of vegetation canopies. They are a key input into climate and numerical weather prediction models, and characterising their spatial and temporal dynamics is crucial in understanding biogeochemical fluxes between the biosphere and atmosphere [1–3]. Of particular importance for modelling terrestrial carbon exchange are forest environments, which cover approximately 30% of the terrestrial surface and account for approximately 50% of its gross primary productivity [4,5]. In addition to carbon sequestration, forests provide a range of ecosystem services, acting as a source of food, fibre, fuel and timber [6]. If we are to effectively monitor and manage these resources, estimates of the biophysical and biochemical variables that describe their status are a fundamental requirement.

Estimating LAI and CCC in the field is a time-consuming and labour-intensive process that is often constrained by logistical challenges and financial resources. Advances in indirect, non-destructive field-based techniques have provided increases in efficiency [7–9], but nevertheless, field observations are insufficient to characterise spatial and temporal variability in LAI and CCC at regional to global scales. Satellite remote sensing data are particularly advantageous in this respect, offering consistent, repeat observations with global coverage. Over the last two decades, methods of retrieving vegetation biophysical and biochemical variables from optical satellite remote sensing have been established, typically making use of radiative transfer models (RTMs) that simulate canopy reflectance as a function of biophysical and biochemical properties [10–12].

Retrieval approaches that have proven popular for generating operational products include the inversion of RTMs using look-up-tables (LUTs) [13,14], in addition to hybrid methods, which use RTM outputs to train machine learning algorithms such as artificial neural networks (ANNs) [15–17]. When compared to LUT inversion, hybrid methods are typically more computationally efficient [10,18–20]. This is because large LUTs are required to achieve high retrieval accuracies, slowing inversion, which involves searching the LUT on a per-pixel basis [10,11,21]. In contrast, hybrid methods such as ANNs are able to accurately represent complex non-linear relationships, and provide an input to output mapping that can be applied almost instantly [21–23]. Thus, they can be quickly and automatically applied to large data sets, while providing comparable retrieval accuracies [12,18,19]. The principle drawbacks of ANNs are their 'black box' nature, the need to specify and tune the network architecture, and their tendency to perform unpredictably when inputs strongly deviate from their training data [10,20].

With its red-edge bands and high spatial resolution (10 m to 60 m), the Multispectral Instrument (MSI) on-board the Sentinel-2 missions is well-suited to vegetation biophysical and biochemical variable retrieval [24]. In particular, its spectral sampling opens up opportunities for the retrieval of CCC, to which the position of the red-edge is highly sensitive [25–29]. A considerable body of work related to the retrieval of vegetation biophysical and biochemical variables using MSI data has been published in recent years, although the majority of studies have been restricted to agricultural environments containing crop canopies. Despite the importance of forest environments, few studies have focussed on forest biophysical and biochemical variable retrieval from MSI data [28,30–32]. Of these, the majority have relied on simulated MSI data that cannot fully represent the observational characteristics of the instrument in orbit (i.e., they make use of resampled airborne, field spectroradiometer or RTM data, with different viewing/illumination geometries and atmospheric characteristics).

Although estimates of LAI and CCC are not produced operationally by the Sentinel-2 ground segment, a retrieval algorithm is provided in the freely available Sentinel Application Platform (SNAP), enabling users to generate so-called 'L2B' products. Developed by Weiss and Baret [33], the algorithm is based on the hybrid retrieval approach, making use of a series of ANNs. Each is pre-trained using the coupled Scattering by Arbitrarily Inclined Leaves (SAIL) and Leaf Optical Properties Spectra (PROSPECT) RTMs [34,35]. While the SNAP L2B retrieval algorithm is described as generic, providing reasonable retrieval accuracies over all vegetation types [33], previous work has demonstrated poor performance when RTMs such as SAIL, which describe the canopy as a turbid medium, are applied over forest environments [23]. Because of its ease of use and integration within the image processing software, it is expected that many users will adopt the SNAP L2B retrieval algorithm as a first port of call. Although good performance has been demonstrated over crop canopies [36–39], evaluation of its performance over forest canopies would enable users to more explicitly assess its fitness for purpose for non-agricultural applications, particularly when compared to retrieval algorithms specifically optimised for forest environments.

Using data collected throughout the first full year of the Sentinel-2 mission during a series of field campaigns, we assess the performance of the SNAP L2B retrieval algorithm for estimating LAI and CCC over a deciduous broadleaf forest site in Southern England. We also develop and evaluate an alternative hybrid retrieval algorithm optimised for forest environments, trained using the Invertible Forest Reflectance Model (INFORM) [40]. The objectives of the paper are to:

- Determine the retrieval accuracy that can be expected when the SNAP L2B retrieval algorithm (which is based on SAIL and not optimised for forest environments) is used for LAI and CCC retrieval over deciduous broadleaf forest.
- Evaluate the extent to which a retrieval algorithm trained using INFORM (and thus optimised for forest environments) will improve LAI and CCC retrieval accuracy.

2. Materials and Methods

2.1. Study Site

The study site (50.8498°N, 1.5741°W) covers a 1 km × 1 km area in the New Forest National Park, Hampshire, United Kingdom, and is comprised of deciduous broadleaf forest (Figure 1). Lying approximately 40 m above sea level, the site is a unique example of ancient and ornamental woodland, dating back to the 17th century. The dominant species are beech (*Fagus sylvatica*) (45%), oak (*Quercus robur*) (40%) and silver birch (*Betula pendula*) (5%), while the dominant soil type is a dark-grey clay loam. Statistics of key stand attributes are listed in Table 1. The site has previously been used in the validation of a range of satellite-derived vegetation biophysical and biochemical variables [41,42].

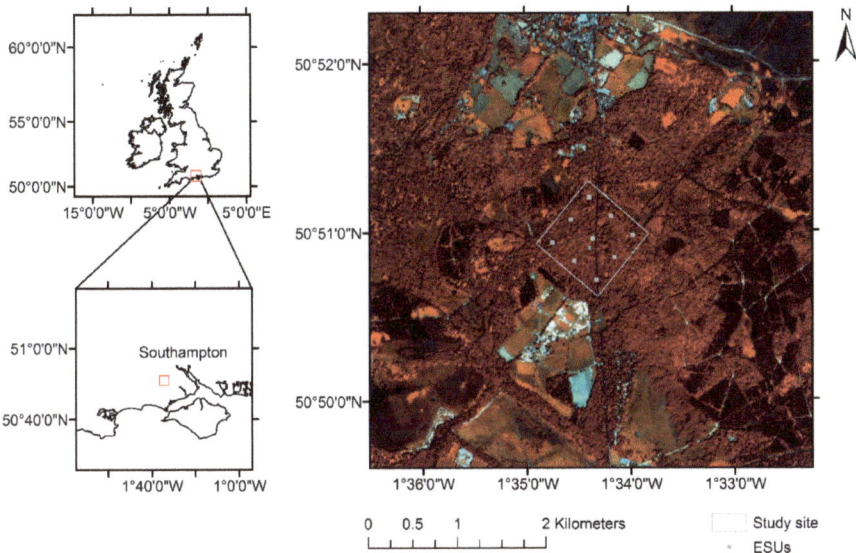

Figure 1. Location of the study site within the United Kingdom. The background image is an MSI false colour composite acquired on 19 July 2016.

Table 1. Statistics of key stand attributes after Cantarello and Newton [43] and Mountford et al. [44].

Stand Attribute	Minimum	Maximum	Mean	Standard Deviation
Canopy height (m)	10	34	-	-
Diameter at breast height (cm)	25.9	80.5	43.4	12.9
Crown diameter (m)	5.76	14.29	8.49	3.73
Stem density (ha^{-1})	72	536	256	106

2.2. Field Data Collection and Processing

Field data were collected on sixteen dates throughout 2016, enabling the phenological cycle to be captured between day of year (DOY) 101 and 307 (Figure 2). A total of nine ESUs of 40 m × 40 m were established over the study site in a regular grid pattern, enabling spatial variability in LAI and CCC to be characterised (Figure 1). All nine ESUs were sampled on each measurement date. The ESU dimensions were selected according to Justice and Townshend [45], to enable the known 10 m positional uncertainty in the investigated 20 m MSI data to be accounted for [46]. The location of each ESU was determined using a handheld Garmin eTrex H global positioning system (GPS) device, which has a reported positional error of less than 10 m [47]. Within each ESU, sampling was carried out at five points arranged in a cross pattern, following the approach adopted in the BigFoot project [48]. The four peripheral points were located approximately 20 m from the central point. Despite the fact that several species are present in the forest, all selected ESUs were dominated by a single species.

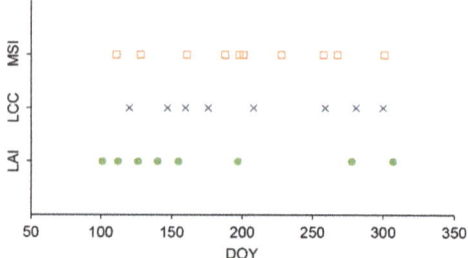

Figure 2. Field data collection of leaf area index (LAI) and leaf chlorophyll concentration (LCC), in addition to MSI data availability throughout the study period.

Leaf area index (LAI) was derived using digital hemispherical photography (DHP). Images were acquired using a Nikon Coolpix 4500 digital camera equipped with an FC-E8 fisheye lens, calibrated using the procedures described by Weiss and Baret [49]. To calculate LAI from each image, pixels were first classified as belonging to the vegetation canopy or its background, enabling the gap fraction to be calculated. Each image was then divided into ten zenith rings of 7.5°, and each zenith ring further divided into 48 azimuth cells of 7.5°. Only zenith angles of less than 75° were considered to minimise the influence of mixed pixels at the extremes of the image [8,49]. Finally, LAI was derived as a discretisation of Miller's [50] integral, such that

$$LAI = 2 \sum_{i=1}^{10} \overline{-\ln P(\theta_i)} \cos \theta_i \sin \theta_i d\theta_i \qquad (1)$$

where $P(\theta_i)$ is the gap fraction in ring i and θ_i is its central zenith angle. The effects of foliage clumping were accounted for according to Lang and Yueqin [51] by calculating the mean of the natural logarithm of gap fraction values over all azimuth cells in each zenith ring. Please note that because the classification could not distinguish between foliage and other canopy elements, the resulting estimates also contained some contribution from other plant material such as stems and branches [7,8,52].

Canopy chlorophyll content (CCC) was calculated as the product of LAI and leaf chlorophyll concentration (LCC) in g m^{-2}. Estimates of LCC were typically obtained within several days of LAI measurements (Figure 2) using a Konica Minolta SPAD-502 optical chlorophyll meter, which determines a relative value proportional to LCC based on the ratio of incident and transmitted radiation at 650 nm and 940 nm. Relative values were converted to absolute units using species–specific calibration functions [53,54]. Three leaves were removed from the tree closest to each sampling point, and each tree was marked using coloured rope to enable re-identification during subsequent surveys. For each leaf, three replicates were performed to account for variations in LCC across its surface, yielding a total of nine measurements per sampling location, and 45 per ESU. Care was taken to avoid major veins within the leaf [9,54].

2.3. Interpolation of Field Data

When evaluating biophysical and biochemical variables derived from high spatial resolution instruments such as MSI, temporal scaling issues are of increased importance, particularly in areas of high cloud cover such as our study site. While spatial scaling issues have received considerable attention in the context of validating operational biophysical and biochemical products [55–58], approaches for dealing with the temporal mismatch between field and satellite observations are less mature [37]. Traditionally, field observations have been matched to the closest satellite acquisition, often within the bounds of an arbitrary period such as one week [59,60]. Such an approach assumes that there is little variation in vegetation condition within this period—an assumption that may be violated at the onset of greenness and senescence, during which rapid changes in biophysical and biochemical properties can occur.

In light of these issues, we adopted an interpolation-based approach that can be applied if field data are collected on multiple dates throughout the year. By interpolating multi-temporal field observations, variations in vegetation condition can better be accounted for, particularly if a parametric function is selected that encodes prior knowledge about expected vegetation dynamics. Importantly, the approach maximises the amount of cloud-free satellite data that can be used for comparison, while also enabling LAI and LCC data collected on different dates to be more easily integrated for the estimation of CCC. To facilitate interpolation, we fit double logistic functions to the LAI and LCC data collected at each ESU. The double logistic function is widely used to represent vegetation phenology [61–63], and takes the form

$$g(x) = a + \frac{b}{[1+\exp(c-dx)]\,[1+\exp(e-fx)]} \tag{2}$$

where $g(x)$ is the LAI or LCC value at a given DOY, a is the base level, b is the seasonal amplitude, c and d control the timing and rate of the onset of greenness, and e and f control the timing and rate of the onset of senescence. To assess their ability to accurately represent seasonal trajectories of the variables of interest throughout the study period, leave-one-out cross validation was carried out on double logistic functions fit to the mean values of LAI and LCC at each DOY. Each data point was sequentially removed before fitting, and its value predicted and compared to that observed. The absolute error was used to investigate the relative importance of each observation date, while overall accuracy was quantified using the root mean square error (RMSE).

2.4. MSI Data Pre-Processing

Ten L1C MSI products covering the growing season from DOY 111 to 300 were obtained over the study site and processed to L2A bottom-of-atmosphere (BOA) reflectance with Sen2Cor 2.5.5 [64], which performs atmospheric, cirrus, and terrain correction. It also provides a scene classification, in addition to estimates of aerosol optical thickness and water vapour. To restrict our analysis to high-quality, cloud-free data, pixels identified as cloud by the scene classification were discarded (three pixels from a total of 90).

2.5. Hybrid LAI and CCC Retrieval

From L2A BOA reflectance values, we first estimated LAI and CCC using the SNAP L2B retrieval algorithm as implemented in SNAP 6.0 [65]. In addition to cloud-contaminated pixels, retrievals flagged by the algorithm as having an out of range input or output were discarded (a further six pixels). Because the SNAP L2B retrieval algorithm is not optimised for forest environments, we also developed an alternative hybrid retrieval algorithm, trained using INFORM. As in the SNAP L2B retrieval algorithm, leaf reflectance spectra were simulated by PROSPECT. 50,000 simulations were carried out, in which input parameters were drawn randomly from a combination of fixed, uniform, and truncated Gaussian distributions (Table 2). Retrieval algorithms trained with 50,000 simulations were shown to provide comparable retrieval accuracies to those utilising over 100,000 simulations by Weiss et al. [66], while additional simulations may not be beneficial in the case of machine learning algorithms such as ANNs, which are prone to overfitting [40,67]. It is worth noting that a similar number of simulations (41,472) was used by Weiss and Baret [33] to train the SNAP L2B retrieval algorithm. Soil background spectra were computed by selecting randomly from a spectral library containing 25 soils [68] and applying a multiplicative soil brightness coefficient (Table 2). INFORM output spectra were convolved with the MSI spectral response functions [69] to generate simulated reflectance values in MSI's spectral bands.

To reflect uncertainties in the radiometric calibration and atmospheric correction of MSI data, simulated reflectance values were contaminated with wavelength-dependent and -independent

Gaussian white noise [23]. This consisted of both additive (0.01) and multiplicative (2%) components [33,70,71], such that

$$R_{cont}(\lambda) = R_{sim}(\lambda)\left(1 + \varepsilon[0, \sigma_{multi}(\lambda)] + \varepsilon[0, \sigma_{multi}(all)]\right) + \varepsilon[0, \sigma_{add}(\lambda)] + \varepsilon[0, \sigma_{add}(all)] \quad (3)$$

where $R_{cont}(\lambda)$ and $R_{sim}(\lambda)$ are the contaminated and simulated reflectance values in the band centered at λ, $\varepsilon[0, \sigma]$ is a Gaussian distribution with a mean of zero, while $\sigma_{add}(\lambda)$, $\sigma_{multi}(\lambda)$, $\sigma_{add}(all)$, and $\sigma_{multi}(all)$ are the additive and multiplicative components of wavelength dependent and independent Gaussian white noise respectively.

Table 2. Distributions from which Invertible Forest Reflectance Model (INFORM) input parameters were randomly drawn.

Parameter	Minimum	Maximum	Mean	Standard Deviation	Distribution	Reference
Structural parameter (N)	1.5	1.7	-	-	Uniform	[72,73]
Chlorophyll a + b (μg cm^{-2})	10	60	50	20	Gaussian	This study
Dry matter (g cm^{-2})	0.004	0.02	-	-	Uniform	[72,73]
Equivalent water thickness (g cm^{-2})	0.01	0.02	-	-	Uniform	[72,73]
Average leaf angle (°)	55	55	-	-	Fixed	[72,73]
Single tree LAI	1.0	5.0	4.0	0.5	Gaussian	This study
Understory LAI	0.5	0.5	-	-	Fixed	[72,73]
Stem density (ha^{-1})	72	536	256	106	Gaussian	[43]
Canopy height (m)	10	34	-	-	Uniform	[44]
Crown diameter (m)	6	14	8	4	Gaussian	[43]
Solar zenith angle (°)	29	64	-	-	Uniform	This study
Observer zenith angle (°)	3	11	-	-	Uniform	This study
Relative azimuth angle (°)	20	136	-	-	Uniform	This study
Soil brightness coefficient	0.5	0.5	-	-	Fixed	[72]
Fraction of diffuse radiation	0.1	0.1	-	-	Fixed	[40,72,73]

As in the SNAP L2B retrieval algorithm, an ANN-based retrieval approach was adopted. Each ANN was trained using the Levenberg-Marquardt minimisation algorithm, and comprised one hidden layer with five tangent sigmoid neurons, in addition to one output layer with a single linear neuron. Given sufficient training data, this architecture (which is also adopted by the SNAP L2B retrieval algorithm) has been found to perform as well as more complex ones [12,16,23,33]. A random subset of 50% of the simulations were used for training, while 25% were used for regularisation, and 25% were used for testing. To prevent overfitting, the regularisation subset was used to enable early stopping (i.e., training was halted when the error, as assessed using the regularisation subset, stopped continuing to decrease). The testing subset, which was not used in training or early stopping, was used to evaluate theoretical performance. Ten ANNs were trained, and the one with the best theoretical performance was selected for further analysis (Appendix A).

Because using a single ANN to estimate multiple biophysical variables can lead to comparatively poor performance [16,23], individual ANNs were trained to estimate LAI and CCC (as is also the case in the SNAP L2B retrieval algorithm). In addition to the simulated BOA reflectance in the eight MSI bands used by the SNAP L2B retrieval algorithm (Table 3), ANN input variables included the cosine of the observer zenith angle (OZA), relative azimuth angle (RAA), and solar zenith angle (SZA), enabling variations in viewing and illumination geometry to be accounted for [21,33,70]. MSI band 2 was excluded because of the likelihood of residual atmospheric contamination [33]. All bands were resampled to a common spatial resolution of 20 m, in order to take advantage of MSI's three red-edge bands (B5, B6 and B7).

Table 3. MSI bands adopted by the SNAP L2B and INFORM-based retrieval algorithms.

Band	Central Wavelength (nm)	Bandwidth (nm)	Native Spatial Resolution (m)
B3	550	35	10
B4	665	30	10
B5	705	15	20
B6	740	15	20
B7	783	20	20
B8A	865	20	20
B11	1610	90	20
B12	2190	180	20

2.6. Forward Modelling Experiments

A prerequisite to accurate biophysical and biochemical variable retrieval is that the underlying RTMs must be able to reproduce observed spectra in a satisfactory manner. To investigate the ability of SAIL and INFORM to reproduce observed MSI spectra over our study site, forward modelling experiments were carried out [74,75]. To simulate the MSI spectrum for each date and ESU, the interpolated field measurements of LAI and LCC were used in model parameterisation, as were the viewing and illumination geometries of the associated MSI scenes. Because the remaining input parameters were not measured in the field, they were instead randomly drawn from the distributions described in Section 2.5. For each date, ESU, and RTM, a database of 5,000 simulations was established. This number of simulations was considered appropriate given that nine input parameters were fixed, resulting in a substantially constrained parameter space. The reflectance mismatch was evaluated between the observed MSI spectrum and the best fitting simulation, which was itself determined as the simulation with the minimum RMSE [74,75].

2.7. Performance Metrics

To evaluate the performance of each retrieval algorithm, LAI and CCC retrievals were compared with values provided by the double logistic functions for the DOY in question. Agreement between the interpolated field data and LAI and CCC retrievals was assessed using the coefficient of determination (r^2), while retrieval accuracy was quantified using the RMSE. A normalised RMSE (NRMSE) was calculated by dividing the RMSE by the mean of observed values. Bias was determined as the mean difference, while precision was quantified as the standard deviation of differences. In addition to calculating these statistics using all available data, statistics were also computed on three subsets to evaluate phenological variations in retrieval accuracy. These represented the onset of greenness (DOY 100 to 149), peak greenness (DOY 150 to 249), and the onset of senescence (DOY 250 to 300). The same performance metrics were adopted in the forward modelling experiments.

3. Results

3.1. Field Data and Interpolation

Throughout the study period, LAI ranged from a minimum of 0.81 on DOY 101 to a maximum of 4.60 on DOY 155, while LCC ranged from a minimum of 0.09 g m^{-2} on DOY 120 to a maximum of 0.65 g m^{-2} on DOY 259. It should be noted that the minimum LAI values corresponded mainly to stems and branches, which could not be distinguished from other canopy elements using DHP. At the start of the growing season, the rate of increase in LCC was slower than that of LAI (Figure 3). Unlike LCC observations, which became more variable towards the end of the growing season, the degree of variability associated with LAI observations remained relatively consistent between measurement dates (Figure 3). The double logistic functions fit to the field data were able to successfully represent seasonal trajectories of LAI and LCC throughout the study period (Figure 3). The results of leave-one-out cross validation revealed an RMSE of 0.10 for LAI and 0.11 g m^{-2} for LCC. The largest absolute errors were observed when those observations during the onset of greenness and senescence were removed (Table 4).

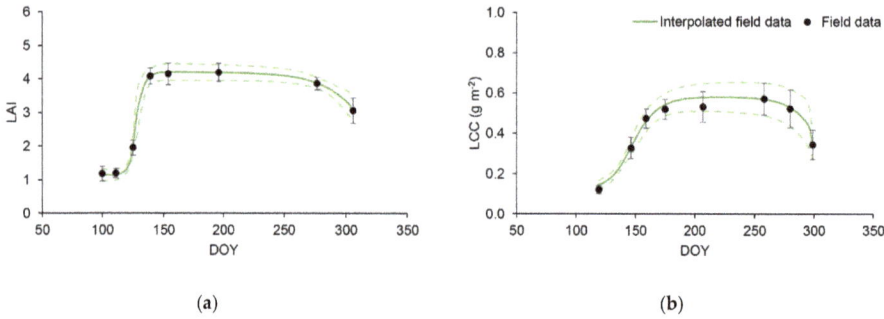

Figure 3. Time series of mean and interpolated LAI (**a**) and LCC (**b**) values calculated over all ESUs. Error bars and dashed lines represent ± 1 standard deviation.

Table 4. Leave-one-out cross-validation results indicating the absolute error associated with interpolated LAI and LCC values on each sampling date (i.e., the error that would occur if that observation were not used in function fitting).

LAI		LCC	
DOY	Absolute Error	DOY	Absolute Error (g m^{-2})
101	0.02	120	0.21
112	0.00	147	0.01
126	0.10	160	0.03
140	0.16	176	0.03
155	0.04	208	0.04
197	0.05	259	0.04
278	0.04	281	0.05
307	0.18	300	0.22

3.2. Overall Performance of the Retrieval Algorithms

Overall, the SNAP L2B retrieval algorithm was characterised by underestimation of LAI, leading to a moderate relationship and retrieval accuracy (r^2 = 0.54, RMSE = 1.55, NRMSE = 43%) (Figure 4a). Like LAI, SNAP L2B CCC retrievals were also characterised by a moderate relationship and retrieval accuracy (r^2 = 0.52, RMSE = 0.79 g m^{-2}, NRMSE = 45%) (Figure 4c). In contrast, a greater degree of consistency between the interpolated field data and INFORM-based LAI retrievals was observed, leading to a stronger relationship and increased retrieval accuracy (r^2 = 0.79, RMSE = 0.47, NRMSE = 13%) (Figure 4b). Improvements in overall performance were also evident for CCC (r^2 = 0.69, RMSE = 0.52, NRMSE = 29%) (Figure 4d).

3.3. Phenological Variations in Performance

When the phenological subsets were examined, the strongest relationships with the interpolated field data were observed for the onset of greenness (DOY 100 to 149, r^2 = 0.66 to 0.77) (Table 5). In contrast, weaker relationships occurred during peak greenness (DOY 150 to 250) and the onset of senescence (DOY 249 to 300) (r^2 = 0.04 to 0.26). In agreement with the overall performance results, the INFORM-based retrieval algorithm demonstrated the best retrieval accuracies in the majority of phenological subsets (as indicated by lower RMSE and NRMSE values when compared to the SNAP L2B retrievals). Although slightly higher r^2 values were typically achieved by the SNAP L2B retrieval algorithm when analysed by phenological stage, its retrievals were characterised by larger biases (and therefore higher RMSE and NRMSE values) than the INFORM-based retrieval algorithm (Table 5). The fact that these biases were not consistent across each phenological stage may have led to the lower r^2 values achieved when all phenological subsets were analysed together (Figure 4). Contrary to the other phenological stages, it is worth noting that the INFORM-based CCC retrievals were subject to overestimation during the onset of greenness (Figure 5b). Although the INFORM-based retrieval algorithm reduced the underestimation associated with CCC during peak greenness (Figure 5b), some degree of underestimation remained, reflected by the negative bias observed (Table 5).

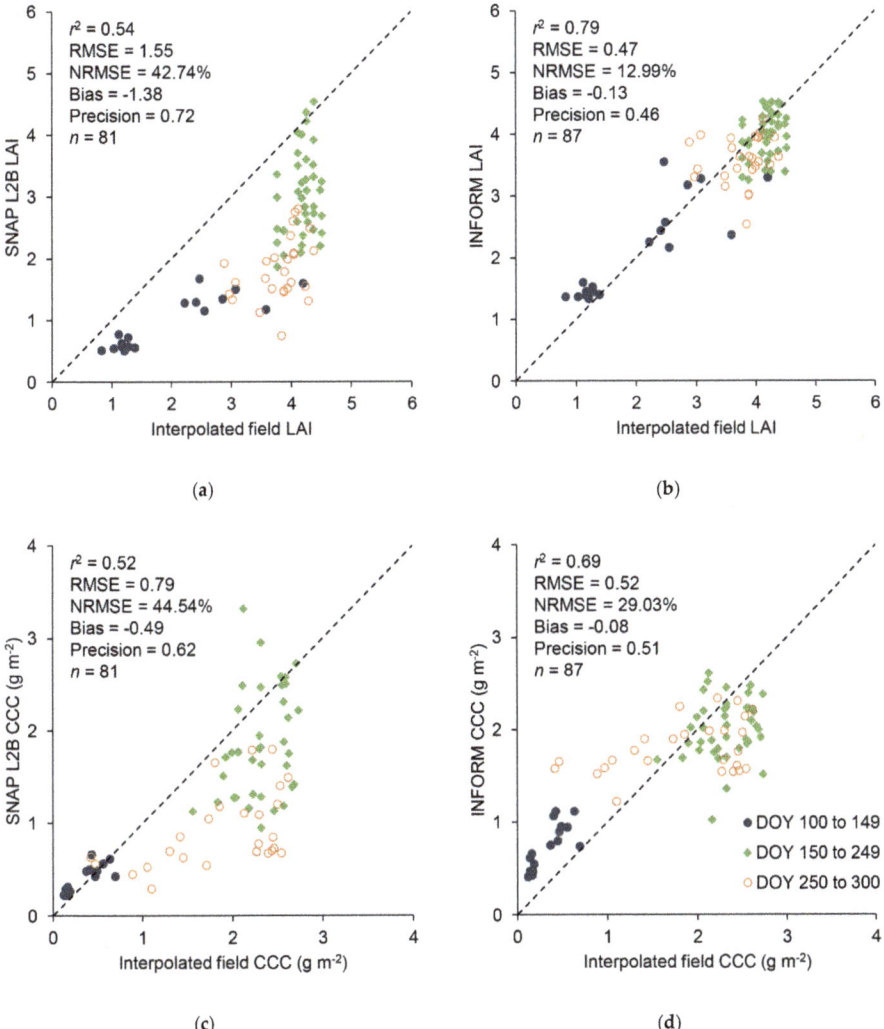

Figure 4. Comparison between interpolated field data and LAI (**a**,**b**) and CCC (**c**,**d**) retrievals from the SNAP L2B (**left**) and INFORM-based (**right**) retrieval algorithms. Points represent measurements at the ESU level, dashed lines represent a 1:1 relationship.

Table 5. Performance statistics for the SNAP L2B and INFORM-based retrieval algorithms when evaluated against interpolated field data, by phenological subset.

Variable	DOY	SNAP L2B					INFORM			
		r^2	RMSE (NRMSE)	Bias	Precision	n	r^2	RMSE (NRMSE)	Bias	Precision	n
LAI	100 to 149	0.77	1.22 (60.28%)	−1.03	0.68	17	0.74	0.51 (25.05%)	0.09	0.51	18
	150 to 249	0.07	1.34 (32.06%)	−1.17	0.66	39	0.11	0.40 (9.61%)	−0.19	0.36	43
	250 to 300	0.14	2.00 (52.76%)	−1.94	0.49	25	0.02	0.55 (14.47%)	−0.18	0.53	26
CCC	100 to 149	0.70	0.12 (36.78%)	0.06	0.11	17	0.66	0.41 (126.32%)	0.39	0.15	18
	150 to 249	0.10	0.72 (30.89%)	−0.45	0.57	39	0.04	0.49 (20.94%)	−0.30	0.39	43
	250 to 300	0.26	1.11 (59.92%)	−0.94	0.59	25	0.17	0.62 (33.56%)	−0.04	0.63	26

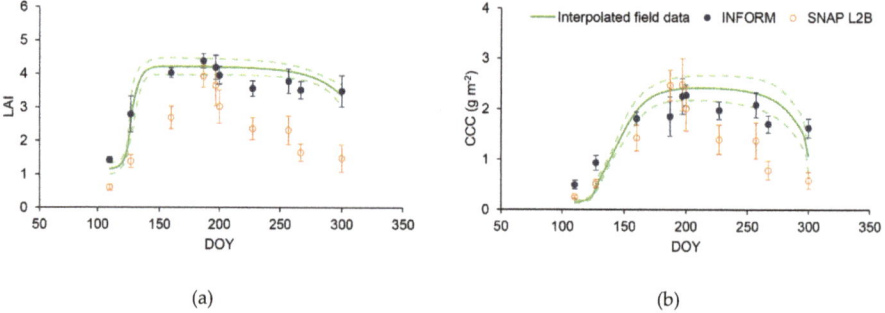

Figure 5. Time series of mean LAI (**a**) and CCC (**b**) retrievals calculated over all ESUs. Error bars and dashed lines represent ± 1 standard deviation.

3.4. Reproduction of Observed MSI Spectra by SAIL and INFORM

When evaluated against the observed MSI spectra, the SAIL reflectance simulations were characterised by greater bias and less precision than the INFORM simulations (Table 6), resulting in higher RMSE and NRMSE values for all bands except B11 (where the difference was marginal). In the case of both RTMs, the best-modelled band was B8A, and the worst-modelled band was B4. A mean absolute error of 0.02, which was used by Darvishzadeh et al. and Atzberger et al. [28,74–76] as a threshold to identify badly modelled bands, was exceeded only by B5 in the case of the INFORM simulations (Table 6). On the other hand, five out of eight bands exceeded this threshold in the case of the SAIL simulations (B4, B5, B6, B7, and B12).

Table 6. Performance statistics for modelled SAIL and INFORM reflectance values when evaluated against observed MSI spectra, by band (n = 87).

Band	SAIL				INFORM			
	r^2	RMSE (NRMSE)	Bias	Precision	r^2	RMSE (NRMSE)	Bias	Precision
B3	0.06	0.03 (56.15%)	−0.01	0.03	0.23	0.02 (39.68%)	−0.01	0.02
B4	0.15	0.03 (72.92%)	−0.02	0.02	0.12	0.02 (58.91%)	−0.01	0.02
B5	0.21	0.04 (38.68%)	−0.02	0.03	0.43	0.03 (32.60%)	−0.02	0.02
B6	0.84	0.04 (16.95%)	0.02	0.04	0.96	0.02 (6.86%)	0.00	0.02
B7	0.92	0.04 (13.60%)	0.03	0.03	0.99	0.01 (4.72%)	0.01	0.01
B8A	0.93	0.04 (10.31%)	0.00	0.04	0.99	0.01 (3.68%)	−0.01	0.01
B11	0.73	0.02 (11.36%)	−0.01	0.02	0.78	0.02 (11.68%)	−0.01	0.02
B12	0.55	0.03 (37.65%)	−0.03	0.02	0.63	0.02 (25.88%)	−0.01	0.02

4. Discussion

4.1. Utility of Interpolating Field Data

When compared to the standard approach of matching field and satellite observations, our interpolation-based approach enabled all available cloud-free MSI scenes to be utilised, as opposed to only those acquired within one week of field data collection. Thus, a substantially greater number of observations were available against which the investigated retrieval algorithms could be assessed. The results of leave-one-out cross-validation indicated that the error associated with this interpolation-based approach was small. The largest absolute errors were observed when observations during the onset of senescence were removed, although similar errors occurred when observations during the onset of greenness were left out in the case of LCC. This has important implications for the timing of field campaigns. It is during these periods, when rates of change in LAI and LCC are highest, that the collection of field data is of most importance. Thus, future efforts should place particular focus on these phenological stages.

4.2. Choice of Retrieval Algorithms for Forest-Related Applications

Previous evaluations of the SNAP L2B retrieval algorithm over crop canopies have demonstrated good performance. For example, Vuolo et al. [36] reported a strong relationship and high retrieval accuracy for SNAP L2B retrievals of LAI over an agricultural site comprised of maize, onion, potato, sugarbeet, and winter wheat (r^2 = 0.83, RMSE = 0.32, NRMSE = 12%), while Vanino et al. [39] reported comparable results over a tomato crop (r^2 = 0.69, RMSE = 0.56, NRMSE = 25%). Similarly, an overall RMSE of 0.98 was reported by Djamai et al. [37] in a recent evaluation over alfalfa, black bean, canola, corn, oat, soybean, and wheat. Nevertheless, previous work has demonstrated poor performance when retrieval algorithms based on one-dimensional RTMs such as SAIL are applied over forest environments [23]. The moderate relationships and retrieval accuracies demonstrated by the SNAP L2B retrieval algorithm in our study support these findings, as do the results of our forward modelling experiments.

When compared to our deciduous broadleaf forest site, the crop canopies investigated in previous studies better conform to the turbid medium assumption adopted by SAIL [23,77]. Due to their dense, leafy, and homogeneous nature, a one-dimensional RTM such as SAIL can provide a good approximation of these canopies, enabling high retrieval accuracies to be achieved by SAIL-based retrieval algorithms. In contrast, deciduous broadleaf forest canopies are more heterogeneous, contain a greater number of woody elements, and are subject to increased crown transmission, foliage clumping, and shadowing. As SAIL does not account for these factors, underestimation of LAI and CCC occurs when the SNAP L2B retrieval algorithm is applied over deciduous broadleaf forest. The issue may be compounded by the ANN-based design of the retrieval algorithm, as ANNs are known to perform unpredictably when inputs strongly deviate from their training data [10].

Given that five out of eight bands were considered to be badly modelled by SAIL, it is possible that improved retrieval accuracies could be obtained using a SAIL-based retrieval algorithm trained on a subset of the best modelled bands. As our objective was to evaluate the SNAP L2B retrieval algorithm available to users, which makes use of all considered bands, such an approach was not explicitly investigated in this study. However, methods to determine and select optimal subsets of bands have successfully been applied to improve retrieval accuracies in previous work [23,28,74–76,78]. For example, in a recent study using INFORM to retrieve LCC from MSI data, Darvishzadeh et al. [28] demonstrated increased retrieval accuracies when using a spectral subset of only MSI's red-edge bands as opposed to the full band set. Similarly, Inoue et al. [29] observed that models utilising a large number of spectral bands did not improve the retrieval of CCC from hyperspectral data when compared to those incorporating just two optimally selected bands. It is worth noting that, unlike Darvishzadeh et al. [28], we did not observe large discrepancies between modelled and observed MSI spectra in the near-infrared shoulder. In contrast, B8A (centred at 865 nm) was the band best modelled by both investigated RTMs in our study.

The INFORM-based retrieval algorithm was characterised by increased overall retrieval accuracies, better reflecting previous studies that have successfully applied MSI and its red-edge bands for forest biophysical and biochemical variable retrieval. For example, using a statistical approach to retrieve LAI from MSI data over a boreal forest site, Korhonen et al. [32] reported a comparable relationship and retrieval accuracy (r^2 = 0.73, RMSE = 0.60, NRMSE = 20%). Similarly, Darvishzadeh et al. [28] obtained an NRMSE of 33% when using LUT inversion of INFORM to retrieve LCC from MSI data over spruce stands. Beyond MSI, our results reflect those of Schlerf and Atzberger [40], who trained an ANN for retrieval of spruce LAI from airborne hyperspectral data using INFORM, achieving a similar relationship and retrieval accuracy ($r^2 = 0.73$, RMSE = 0.58, NRMSE = 18%). Comparable results were reported by Heiskanen et al. [79], who adopted the semi-physical forest reflectance model PARAS to retrieve LAI from High-Resolution Visible and Infrared (HRVIR) and Enhanced Thematic Mapper (ETM+) data over a boreal forest site. Using ANNs, Heiskanen et al. [79] obtained an RMSE of 0.59 (NRMSE = 25%). In a later study, Schlerf and Atzberger [72] applied LUT inversion of INFORM to retrieve LAI from multi-angular Compact High-Resolution Imaging Spectrometer (CHRIS) data, reporting similar results over beech ($r^2 = 0.57$, RMSE = 0.94, NRMSE = 26%) and spruce ($r^2 = 0.51$, RMSE = 0.74, NRMSE = 18%).

In terms of phenological variations in the relationship between MSI derived retrievals and field data, the results of our study reflect those reported by Heiskanen et al. [80], who used a statistical approach to retrieve LAI from HRVIR and Hyperion data over boreal forest throughout several seasons. As in our study, Heiskanen et al. [80] reported strong relationships with LAI during the onset of greenness (DOY 116 to 153, $r^2 = 0.69$ to 0.74). These were followed by weaker relationships during peak greenness and, in particular, the onset of senescence (DOY 273 to 279, $r^2 = 0.29$ to 0.40). An explanation for this pattern is that during the onset of greenness, a wide range of LAI values are experienced, whereas during peak greenness, this range is substantially reduced. The weaker relationships during the onset of senescence may be related to changes in leaf biochemistry, to which the MSI derived LAI retrievals are sensitive, but to which the field estimates of LAI (which do not discriminate between green and senescent leaves) are not.

Although improved overall retrieval accuracy was obtained by the INFORM-based retrieval algorithm in the case of CCC, overestimation occurred during the onset of greenness, while some degree of underestimation persisted throughout peak greenness. In terms of the apparent overestimation during the onset of greenness, a small bias was also observed in the INFORM LAI retrievals. Being the product of LAI and LCC, this may have translated into a larger bias in terms of CCC. One explanation for the apparent underestimation observed during peak greenness could be related to the field measurements of LCC themselves, which, due to the height of the forest, could only be performed on leaves from middle or bottom of the tree crown. The sunlit leaves at the top of the crown may have been characterised by reduced LCC when compared to the measured leaves, as LCC commonly decreases with increased light availability [81]. As such, the observed discrepancies may have resulted from sampling biases in the field data. Finally, while our study analysed the performance of the two retrieval algorithms throughout the growing season, it is important to note that it was restricted to a single site. Further work is required to confirm the applicability of our results over additional sites, including those characterised by different forest types.

5. Conclusions

Although it is not optimised for forest environments, because of its ease of use and integration within the image processing software, it is expected that many users will adopt the SNAP L2B retrieval algorithm for forest LAI and CCC retrieval as a first port of call. Using field data collected throughout the growing season at a deciduous broadleaf forest site in Southern England, we evaluated its performance and that of an alternative retrieval algorithm optimised for forest environments, trained using INFORM. We also developed and successfully applied an interpolation approach to address the temporal mismatch between field and satellite observations. When compared to previously investigated crop canopies, the SNAP L2B retrieval algorithm appears less well-suited to LAI and CCC retrieval over forest environments. Its moderate retrieval accuracies highlight the importance of selecting an RTM that can accurately describe the structure of the canopy of interest, as do the improvements associated with the INFORM-based retrieval algorithm. This finding is corroborated by the results of our forward modelling experiments: over forest environments such as our study site, SAIL appears less able to reproduce observed spectra than INFORM. Based on these results, for forest-related applications using MSI data, we recommend users seek retrieval algorithms optimised for forest environments.

Author Contributions: Conceptualization, L.A.B., B.O.O. and J.D.; methodology, L.A.B.; software, L.A.B.; validation, L.A.B., B.O.O. and J.D.; formal analysis, L.A.B.; investigation, L.A.B.; resources, L.A.B. and J.D.; data curation, L.A.B.; writing—original draft preparation, L.A.B.; writing—review and editing, L.A.B., B.O.O. and J.D.; visualization, L.A.B.; supervision, B.O.O. and J.D.; project administration, L.A.B.; funding acquisition, J.D.

Funding: This research was supported by the European Space Agency and a University of Southampton Vice Chancellor's Scholarship.

Acknowledgments: The authors thank the European Space Agency and the European Commission for making the Sentinel products and associated tools freely available through the Copernicus programme. The authors are also grateful to the Forestry Commission for granting permission to undertake this study, to Tracy Adole, Andrew Maclachlan, Alan Smith and Qiaoyun Xie for their assistance in the collection of field data, and to Clement Atzberger for providing a copy of INFORM.

Conflicts of Interest: The authors declare no conflict of interest.

Appendix A

Table A1. Theoretical performance of all trained ANNs (assessed using the testing subset of INFORM simulations). The best performing ANNs, indicated by the lowest RMSE, are marked with *.

ANN	LAI			CCC		
	r^2	RMSE	NRMSE (%)	r^2	RMSE	NRMSE (%)
1	0.58	0.45	11.36	0.60	0.40	20.25
2	0.59	0.45	11.41	0.61	0.40	20.09
3	0.58	0.45	11.43	0.61	0.40	20.17
4	0.58	0.45	11.35	0.61	0.40	20.11
5	0.59	0.45	11.32	0.61	0.40	20.20
6	0.58	0.45	11.47	0.59	0.41	20.60
7	0.59	0.44*	11.27	0.61	0.40	20.22
8	0.58	0.45	11.42	0.62	0.40	20.13
9	0.59	0.45	11.37	0.60	0.40	20.35
10	0.58	0.45	11.45	0.61	0.39*	20.00

References

1. Running, S.W.; Nemani, R.R.; Heinsch, F.A.; Zhao, M.; Reeves, M.; Hashimoto, H. A Continuous Satellite-Derived Measure of Global Terrestrial Primary Production. *Bioscience* **2004**, *54*, 547. [CrossRef]
2. Sellers, P.J.; Tucker, C.J.; Collatz, G.J.; Los, S.O.; Justice, C.O.; Dazlich, D.A.; Randall, D.A. A Revised Land Surface Parameterization (SiB2) for Atmospheric GCMS. Part II: The Generation of Global Fields of Terrestrial Biophysical Parameters from Satellite Data. *J. Clim.* **1996**, *9*, 706–737. [CrossRef]
3. Sellers, P.J.; Dickinson, R.E.; Randall, D.A.; Betts, A.K.; Hall, F.G.; Berry, J.A.; Collatz, G.J.; Denning, A.S.; Mooney, H.A.; Nobre, C.A.; et al. Modeling the Exchanges of Energy, Water, and Carbon Between Continents and the Atmosphere. *Science* **1997**, *275*, 502–509. [CrossRef] [PubMed]
4. Beer, C.; Reichstein, M.; Tomelleri, E.; Ciais, P.; Jung, M.; Carvalhais, N.; Rodenbeck, C.; Arain, M.A.; Baldocchi, D.; Bonan, G.B.; et al. Terrestrial Gross Carbon Dioxide Uptake: Global Distribution and Covariation with Climate. *Science.* **2010**, *329*, 834–838. [CrossRef] [PubMed]
5. *FAO Global Forest Resources Assessment 2015*; Food and Agriculture Organizaion of the United Nations: Rome, Italy, 2015; ISBN 978-92-5-108826-5.
6. Barredo, J.I.; Bastrup-Birk, A.; Teller, A.; Onaindia, M.; Fernández de Manuel, B.; Madariaga, I.; Rodríguez-Loinaz, G.; Pinho, P.; Nunes, A.; Ramos, A.; et al. *Mapping and Assessment of Forest Ecosystems and Their Services—Applications and Guidance for Decision Making in the Framework of MAES*; European Commission Joint Research Centre: Ispra, Italy, 2015; ISBN 978-92-79-55332-5.
7. Bréda, N.J.J. Ground-based measurements of leaf area index: A review of methods, instruments and current controversies. *J. Exp. Bot.* **2003**, *54*, 2403–2417. [CrossRef]
8. Jonckheere, I.; Fleck, S.; Nackaerts, K.; Muys, B.; Coppin, P.; Weiss, M.; Baret, F. Review of methods for in situ leaf area index determination. *Agric. For. Meteorol.* **2004**, *121*, 19–35. [CrossRef]
9. Markwell, J.; Osterman, J.C.; Mitchell, J.L. Calibration of the Minolta SPAD-502 leaf chlorophyll meter. *Photosynth. Res.* **1995**, *46*, 467–472. [CrossRef]

10. Verrelst, J.; Camps-Valls, G.; Muñoz-Marí, J.; Rivera, J.P.; Veroustraete, F.; Clevers, J.G.P.W.; Moreno, J. Optical remote sensing and the retrieval of terrestrial vegetation bio-geophysical properties—A review. *ISPRS J. Photogramm. Remote Sens.* **2015**, *108*, 273–290. [CrossRef]
11. Liang, S. Recent developments in estimating land surface biogeophysical variables from optical remote sensing. *Prog. Phys. Geogr.* **2007**, *31*, 501–516. [CrossRef]
12. Richter, K.; Hank, T.B.; Vuolo, F.; Mauser, W.; D'Urso, G. Optimal Exploitation of the Sentinel-2 Spectral Capabilities for Crop Leaf Area Index Mapping. *Remote Sens.* **2012**, *4*, 561–582. [CrossRef]
13. Knyazikhin, Y.; Martonchik, J.V.; Myneni, R.B.; Diner, D.J.; Running, S.W. Synergistic algorithm for estimating vegetation canopy leaf area index and fraction of absorbed photosynthetically active radiation from MODIS and MISR data. *J. Geophys. Res. Atmos.* **1998**, *103*, 32257–32275. [CrossRef]
14. Myneni, R.B.; Hoffman, S.; Knyazikhin, Y.; Privette, J.L.; Glassy, J.; Tian, Y.; Wang, Y.; Song, X.; Zhang, Y.; Smith, G.R.; et al. Global products of vegetation leaf area and fraction absorbed PAR from year one of MODIS data. *Remote Sens. Environ.* **2002**, *83*, 214–231. [CrossRef]
15. Baret, F.; Weiss, M.; Lacaze, R.; Camacho, F.; Makhmara, H.; Pacholcyzk, P.; Smets, B. GEOV1: LAI and FAPAR essential climate variables and FCOVER global time series capitalizing over existing products. Part 1: Principles of development and production. *Remote Sens. Environ.* **2013**, *137*, 299–309. [CrossRef]
16. Baret, F.; Hagolle, O.; Geiger, B.; Bicheron, P.; Miras, B.; Huc, M.; Berthelot, B.; Niño, F.; Weiss, M.; Samain, O.; et al. LAI, fAPAR and fCover CYCLOPES global products derived from VEGETATION. *Remote Sens. Environ.* **2007**, *110*, 275–286. [CrossRef]
17. Lacaze, R.; Smets, B.; Baret, F.; Weiss, M.; Ramon, D.; Montersleet, B.; Wandrebeck, L.; Calvet, J.-C.; Roujean, J.-L.; Camacho, F. OPERATIONAL 333m BIOPHYSICAL PRODUCTS OF THE COPERNICUS GLOBAL LAND SERVICE FOR AGRICULTURE MONITORING. *ISPRS - Int. Arch. Photogramm. Remote Sens. Spat. Inf. Sci.* **2015**, *XL-7/W3*, 53–56. [CrossRef]
18. Vuolo, F.; Atzberger, C.; Richter, K.; Dash, J. Retrieval of Biophysical Vegetation Products From Rapideye Imagery. In *Proceedings of the ISPRS TC VII Symposium, Vienna, Austria, 5–7 July 2010*; Wagner, W., Székely, B., Eds.; International Society for Photogrammetry and Remote Sensing: Hannover, Germany; pp. 281–286.
19. Verrelst, J.; Rivera, J.P.; Veroustraete, F.; Muñoz-Marí, J.; Clevers, J.G.P.W.; Camps-Valls, G.; Moreno, J. Experimental Sentinel-2 LAI estimation using parametric, non-parametric and physical retrieval methods – A comparison. *ISPRS J. Photogramm. Remote Sens.* **2015**, *108*, 260–272. [CrossRef]
20. Verrelst, J.; Muñoz, J.; Alonso, L.; Delegido, J.; Rivera, J.P.; Camps-Valls, G.; Moreno, J. Machine learning regression algorithms for biophysical parameter retrieval: Opportunities for Sentinel-2 and -3. *Remote Sens. Environ.* **2012**, *118*, 127–139. [CrossRef]
21. Bacour, C.; Baret, F.; Béal, D.; Weiss, M.; Pavageau, K. Neural network estimation of LAI, fAPAR, fCover and LAIxCab, from top of canopy MERIS reflectance data: Principles and validation. *Remote Sens. Environ.* **2006**, *105*, 313–325. [CrossRef]
22. Kimes, D.S.; Knyazikhin, Y.; Privette, J.L.; Abuelgasim, A.A.; Gao, F. Inversion methods for physically-based models. *Remote Sens. Rev.* **2000**, *18*, 381–439. [CrossRef]
23. Verger, A.; Baret, F.; Camacho, F. Optimal modalities for radiative transfer-neural network estimation of canopy biophysical characteristics: Evaluation over an agricultural area with CHRIS/PROBA observations. *Remote Sens. Environ.* **2011**, *115*, 415–426. [CrossRef]
24. Drusch, M.; Del Bello, U.; Carlier, S.; Colin, O.; Fernandez, V.; Gascon, F.; Hoersch, B.; Isola, C.; Laberinti, P.; Martimort, P.; et al. Sentinel-2: ESA's Optical High-Resolution Mission for GMES Operational Services. *Remote Sens. Environ.* **2012**, *120*, 25–36. [CrossRef]
25. Clevers, J.G.P.W.; Gitelson, A.A. Remote estimation of crop and grass chlorophyll and nitrogen content using red-edge bands on Sentinel-2 and -3. *Int. J. Appl. Earth Obs. Geoinf.* **2013**, *23*, 344–351. [CrossRef]
26. Clevers, J.; Kooistra, L.; van den Brande, M. Using Sentinel-2 Data for Retrieving LAI and Leaf and Canopy Chlorophyll Content of a Potato Crop. *Remote Sens.* **2017**, *9*, 405. [CrossRef]
27. Frampton, W.J.; Dash, J.; Watmough, G.; Milton, E.J. Evaluating the capabilities of Sentinel-2 for quantitative estimation of biophysical variables in vegetation. *ISPRS J. Photogramm. Remote Sens.* **2013**, *82*, 83–92. [CrossRef]
28. Darvishzadeh, R.; Skidmore, A.; Abdullah, H.; Cherenet, E.; Ali, A.; Wang, T.; Nieuwenhuis, W.; Heurich, M.; Vrieling, A.; O'Connor, B.; et al. Mapping leaf chlorophyll content from Sentinel-2 and RapidEye data in spruce stands using the invertible forest reflectance model. *Int. J. Appl. Earth Obs. Geoinf.* **2019**, *79*, 58–70. [CrossRef]

29. Inoue, Y.; Guérif, M.; Baret, F.; Skidmore, A.; Gitelson, A.; Schlerf, M.; Darvishzadeh, R.; Olioso, A. Simple and robust methods for remote sensing of canopy chlorophyll content: A comparative analysis of hyperspectral data for different types of vegetation. *Plant. Cell Environ.* **2016**, *39*, 2609–2623. [CrossRef] [PubMed]
30. Gholizadeh, A.; Mišurec, J.; Kopačková, V.; Mielke, C.; Rogass, C. Assessment of Red-Edge Position Extraction Techniques: A Case Study for Norway Spruce Forests Using HyMap and Simulated Sentinel-2 Data. *Forests* **2016**, *7*, 226. [CrossRef]
31. Majasalmi, T.; Rautiainen, M. The potential of Sentinel-2 data for estimating biophysical variables in a boreal forest: A simulation study. *Remote Sens. Lett.* **2016**, *7*, 427–436. [CrossRef]
32. Korhonen, L.; Hadi; Packalen, P.; Rautiainen, M. Comparison of Sentinel-2 and Landsat 8 in the estimation of boreal forest canopy cover and leaf area index. *Remote Sens. Environ.* **2017**, *195*, 259–274. [CrossRef]
33. Weiss, M.; Baret, F. *S2ToolBox Level 2 Products: LAI, FAPAR, FCOVER*; 1.1.; Institut National de la Recherche Agronomique: Avignon, France, 2016.
34. Verhoef, W.; Jia, L.; Xiao, Q.; Su, Z. Unified Optical-Thermal Four-Stream Radiative Transfer Theory for Homogeneous Vegetation Canopies. *IEEE Trans. Geosci. Remote Sens.* **2007**, *45*, 1808–1822. [CrossRef]
35. Feret, J.-B.; François, C.; Asner, G.P.; Gitelson, A.A.; Martin, R.E.; Bidel, L.P.R.; Ustin, S.L.; le Maire, G.; Jacquemoud, S. PROSPECT-4 and 5: Advances in the leaf optical properties model separating photosynthetic pigments. *Remote Sens. Environ.* **2008**, *112*, 3030–3043. [CrossRef]
36. Vuolo, F.; Żółtak, M.; Pipitone, C.; Zappa, L.; Wenng, H.; Immitzer, M.; Weiss, M.; Baret, F.; Atzberger, C. Data Service Platform for Sentinel-2 Surface Reflectance and Value-Added Products: System Use and Examples. *Remote Sens.* **2016**, *8*, 938. [CrossRef]
37. Djamai, N.; Fernandes, R.; Weiss, M.; McNairn, H.; Goïta, K. Validation of the Sentinel Simplified Level 2 Product Prototype Processor (SL2P) for mapping cropland biophysical variables using Sentinel-2/MSI and Landsat-8/OLI data. *Remote Sens. Environ.* **2019**, *225*, 416–430. [CrossRef]
38. Xie, Q.; Dash, J.; Huete, A.; Jiang, A.; Yin, G.; Ding, Y.; Peng, D.; Hall, C.C.; Brown, L.; Shi, Y.; et al. Retrieval of crop biophysical parameters from Sentinel-2 remote sensing imagery. *Int. J. Appl. Earth Obs. Geoinf.* **2019**, *80*, 187–195. [CrossRef]
39. Vanino, S.; Nino, P.; De Michele, C.; Falanga Bolognesi, S.; D'Urso, G.; Di Bene, C.; Pennelli, B.; Vuolo, F.; Farina, R.; Pulighe, G.; et al. Capability of Sentinel-2 data for estimating maximum evapotranspiration and irrigation requirements for tomato crop in Central Italy. *Remote Sens. Environ.* **2018**, *215*, 452–470. [CrossRef]
40. Schlerf, M.; Atzberger, C. Inversion of a forest reflectance model to estimate structural canopy variables from hyperspectral remote sensing data. *Remote Sens. Environ.* **2006**, *100*, 281–294. [CrossRef]
41. Dash, J.; Almond, S.F.; Boyd, D.; Curran, P.J. Multi-scale analysis and validation of the Envisat MERIS Terrestrial Chlorophyll Index (MTCI) in woodland. In *Proceedings of the 2nd MERIS/(A)ATSR User Workshop, Frascati, Italy, 22–26 September 2008*; European Space Agency: Noordwijk, The Netherlands.
42. Ogutu, B.; Dash, J.; Dawson, T.P. Evaluation of leaf area index estimated from medium spatial resolution remote sensing data in a broadleaf deciduous forest in southern England, UK. *Can. J. Remote Sens.* **2012**, *37*, 333–347. [CrossRef]
43. Cantarello, E.; Newton, A.C. Identifying cost-effective indicators to assess the conservation status of forested habitats in Natura 2000 sites. *For. Ecol. Manage.* **2008**, *256*, 815–826. [CrossRef]
44. Mountford, E.P.; Peterken, G.F.; Edwards, P.J.; Manners, J.G. Long-term change in growth, mortality and regeneration of trees in Denny Wood, an old-growth wood-pasture in the New Forest (UK). *Perspect. Plant Ecol. Evol. Syst.* **1999**, *2*, 223–272. [CrossRef]
45. Justice, C.O.; Townshend, J.R.G. Integrating ground data with remote sensing. In *Terrain Analysis and Remote Sensing*; Townshend, J.R.G., Ed.; Allen and Unwin: London, UK, 1981; pp. 35–58.
46. Gascon, F.; Bouzinac, C.; Thépaut, O.; Jung, M.; Francesconi, B.; Louis, J.; Lonjou, V.; Lafrance, B.; Massera, S.; Gaudel-Vacaresse, A.; et al. Copernicus Sentinel-2A Calibration and Products Validation Status. *Remote Sens.* **2017**, *9*, 584. [CrossRef]
47. Garmin. *Garmin eTrex H Owner's Manual*; Garmin: Olathe, KS, USA, 2007; ISBN 0808238000.
48. Campbell, J.L.; Burrows, S.; Gower, S.T.; Cohen, W.B. *BigFoot: Characterizing Land Cover, LAI and NPP at the Landscape Scale for EOS/MODIS Validation - Field Manual*; 2.1.; Oak Ridge National Laboratory: Oak Ridge, TN, USA, 1999; ISBN 0071601201.
49. Weiss, M.; Baret, F. *CAN-EYE V6.4.91 User Manual*; Institut National de la Recherche Agronomique: Avignon, France, 2017.
50. Miller, J. A formula for average foliage density. *Aust. J. Bot.* **1967**, *15*, 141–144. [CrossRef]

51. Lang, A.R.G.; Yueqin, X. Estimation of leaf area index from transmission of direct sunlight in discontinuous canopies. *Agric. For. Meteorol.* **1986**, *37*, 229–243. [CrossRef]
52. Weiss, M.; Baret, F.; Smith, G.J.; Jonckheere, I.; Coppin, P. Review of methods for in situ leaf area index (LAI) determination Part II: Estimation of LAI, errors and sampling. *Agric. For. Meteorol.* **2004**, *121*, 37–53. [CrossRef]
53. Demarez, V. Seasonal variation of leaf chlorophyll content of a temperate forest. Inversion of the PROSPECT model. *Int. J. Remote Sens.* **1999**, *20*, 879–894. [CrossRef]
54. Uddling, J.; Gelang-Alfredsson, J.; Piikki, K.; Pleijel, H. Evaluating the relationship between leaf chlorophyll concentration and SPAD-502 chlorophyll meter readings. *Photosynth. Res.* **2007**, *91*, 37–46. [CrossRef]
55. Brown, L.A.; Dash, J.; Lidón, A.L.; Lopez-Baeza, E.; Dransfeld, S. Synergetic Exploitation of the Sentinel-2 Missions for Validating the Sentinel-3 Ocean and Land Colour Instrument Terrestrial Chlorophyll Index over a Vineyard Dominated Mediterranean Environment. *IEEE J. Sel. Top. Appl. Earth Obs. Remote Sens.* **2019**, *12*. [CrossRef]
56. Fernandes, R.; Plummer, S.; Nightingale, J.; Baret, F.; Camacho, F.; Fang, H.; Garrigues, S.; Gobron, N.; Lang, M.; Lacaze, R.; et al. Global Leaf Area Index Product Validation Good Practices. In *Best Practice for Satellite-Derived Land Product Validation*; Fernandes, R., Plummer, S., Nightingale, J., Eds.; Land Product Validation Subgroup (Committee on Earth Observation Satellites Working Group on Calibration and Validation): Greenbelt, MD, USA, 2014.
57. Morisette, J.T.; Baret, F.; Privette, J.L.; Myneni, R.B.; Nickeson, J.E.; Garrigues, S.; Shabanov, N.V.; Weiss, M.; Fernandes, R.A.; Leblanc, S.G.; et al. Validation of global moderate-resolution LAI products: A framework proposed within the CEOS land product validation subgroup. *IEEE Trans. Geosci. Remote Sens.* **2006**, *44*, 1804–1817. [CrossRef]
58. Vuolo, F.; Dash, J.; Curran, P.J.; Lajas, D.; Kwiatkowska, E. Methodologies and Uncertainties in the Use of the Terrestrial Chlorophyll Index for the Sentinel-3 Mission. *Remote Sens.* **2012**, *4*, 1112–1133. [CrossRef]
59. Baret, F.; Weiss, M.; Allard, D.; Garrigues, S.; Leroy, M.; Jeanjean, H.; Fernandes, R.; Myneni, R.; Privette, J.; Morisette, J.; et al. *VALERI: A Network of Sites and a Methodology for the Validation of Medium Spatial Resolution Land Satellite Products*; Institut National de la Recherche Agronomique: Avignon, France, 2005.
60. De Kauwe, M.G.; Disney, M.I.; Quaife, T.; Lewis, P.; Williams, M. An assessment of the MODIS collection 5 leaf area index product for a region of mixed coniferous forest. *Remote Sens. Environ.* **2011**, *115*, 767–780. [CrossRef]
61. Atkinson, P.M.; Jeganathan, C.; Dash, J.; Atzberger, C. Inter-comparison of four models for smoothing satellite sensor time-series data to estimate vegetation phenology. *Remote Sens. Environ.* **2012**, *123*, 400–417. [CrossRef]
62. Beck, P.S.A.; Atzberger, C.; Høgda, K.A.; Johansen, B.; Skidmore, A.K. Improved monitoring of vegetation dynamics at very high latitudes: A new method using MODIS NDVI. *Remote Sens. Environ.* **2006**, *100*, 321–334. [CrossRef]
63. Zhang, X.; Friedl, M.A.; Schaaf, C.B.; Strahler, A.H.; Hodges, J.C.F.; Gao, F.; Reed, B.C.; Huete, A. Monitoring vegetation phenology using MODIS. *Remote Sens. Environ.* **2003**, *84*, 471–475. [CrossRef]
64. Müller-Wilm, U. *Sentinel-2 MSI – Level-2A Prototype Processor Installation and User Manual*; Telespazio VEGA: Darmstadt, Germany, 2016.
65. ESA SNAP. Available online: http://step.esa.int/main/toolboxes/snap/ (accessed on 24 August 2018).
66. Weiss, M.; Baret, F.; Myneni, R.B.; Pragnère, A.; Knyazikhin, Y. Investigation of a model inversion technique to estimate canopy biophysical variables from spectral and directional reflectance data. *Agronomie* **2000**, *20*, 3–22. [CrossRef]
67. Combal, B.; Baret, F.; Weiss, M.; Trubuil, A.; Macé, D.; Pragnère, A.; Myneni, R.; Knyazikhin, Y.; Wang, L. Retrieval of canopy biophysical variables from bidirectional reflectance. *Remote Sens. Environ.* **2003**, *84*, 1–15. [CrossRef]
68. Baldridge, A.M.; Hook, S.J.; Grove, C.I.; Rivera, G. The ASTER spectral library version 2.0. *Remote Sens. Environ.* **2009**, *113*, 711–715. [CrossRef]
69. ESA Sentinel-2 Spectral Response Functions (S2-SRF). Available online: https://earth.esa.int/documents/247904/685211/Sentinel-2+MSI+Spectral+Responses/ (accessed on 16 May 2017).
70. Li, W.; Weiss, M.; Waldner, F.; Defourny, P.; Demarez, V.; Morin, D.; Hagolle, O.; Baret, F. A Generic Algorithm to Estimate LAI, FAPAR and FCOVER Variables from SPOT4_HRVIR and Landsat Sensors: Evaluation of the Consistency and Comparison with Ground Measurements. *Remote Sens.* **2015**, *7*, 15494–15516. [CrossRef]

71. Upreti, D.; Huang, W.; Kong, W.; Pascucci, S.; Pignatti, S.; Zhou, X.; Ye, H.; Casa, R. A Comparison of Hybrid Machine Learning Algorithms for the Retrieval of Wheat Biophysical Variables from Sentinel-2. *Remote Sens.* **2019**, *11*, 481. [CrossRef]
72. Schlerf, M.; Atzberger, C. Vegetation Structure Retrieval in Beech and Spruce Forests Using Spectrodirectional Satellite Data. *IEEE J. Sel. Top. Appl. Earth Obs. Remote Sens.* **2012**, *5*, 8–17. [CrossRef]
73. Yuan, H.; Ma, R.; Atzberger, C.; Li, F.; Loiselle, S.; Luo, J. Estimating Forest fAPAR from Multispectral Landsat-8 Data Using the Invertible Forest Reflectance Model INFORM. *Remote Sens.* **2015**, *7*, 7425–7446. [CrossRef]
74. Atzberger, C.; Darvishzadeh, R.; Schlerf, M.; Le Maire, G. Suitability and adaptation of PROSAIL radiative transfer model for hyperspectral grassland studies. *Remote Sens. Lett.* **2013**, *4*, 55–64. [CrossRef]
75. Darvishzadeh, R.; Atzberger, C.; Skidmore, A.; Schlerf, M. Mapping grassland leaf area index with airborne hyperspectral imagery: A comparison study of statistical approaches and inversion of radiative transfer models. *ISPRS J. Photogramm. Remote Sens.* **2011**, *66*, 894–906. [CrossRef]
76. Darvishzadeh, R.; Skidmore, A.; Schlerf, M.; Atzberger, C. Inversion of a radiative transfer model for estimating vegetation LAI and chlorophyll in a heterogeneous grassland. *Remote Sens. Environ.* **2008**, *112*, 2592–2604. [CrossRef]
77. Richter, K.; Atzberger, C.; Vuolo, F.; Weihs, P.; D'Urso, G. Experimental assessment of the Sentinel-2 band setting for RTM-based LAI retrieval of sugar beet and maize. *Can. J. Remote Sens.* **2009**, *35*, 230–247. [CrossRef]
78. Verrelst, J.; Rivera, J.P.; Gitelson, A.; Delegido, J.; Moreno, J.; Camps-Valls, G. Spectral band selection for vegetation properties retrieval using Gaussian processes regression. *Int. J. Appl. Earth Obs. Geoinf.* **2016**, *52*, 554–567. [CrossRef]
79. Heiskanen, J.; Rautiainen, M.; Korhonen, L.; Mõttus, M.; Stenberg, P. Retrieval of boreal forest LAI using a forest reflectance model and empirical regressions. *Int. J. Appl. Earth Obs. Geoinf.* **2011**, *13*, 595–606. [CrossRef]
80. Heiskanen, J.; Rautiainen, M.; Stenberg, P.; Mõttus, M.; Vesanto, V.-H.; Korhonen, L.; Majasalmi, T. Seasonal variation in MODIS LAI for a boreal forest area in Finland. *Remote Sens. Environ.* **2012**, *126*, 104–115. [CrossRef]
81. Baltzer, J.L.; Thomas, S.C. Leaf optical responses to light and soil nutrient availability in temperate deciduous trees. *Am. J. Bot.* **2005**, *92*, 214–223. [CrossRef]

© 2019 by the authors. Licensee MDPI, Basel, Switzerland. This article is an open access article distributed under the terms and conditions of the Creative Commons Attribution (CC BY) license (http://creativecommons.org/licenses/by/4.0/).

Article

A Simplified and Robust Surface Reflectance Estimation Method (SREM) for Use over Diverse Land Surfaces Using Multi-Sensor Data

Muhammad Bilal [1,†], Majid Nazeer [2,3,†], Janet E. Nichol [4], Max P. Bleiweiss [5], Zhongfeng Qiu [1,*], Evelyn Jäkel [6], James R. Campbell [7], Luqman Atique [8], Xiaolan Huang [1] and Simone Lolli [9,10]

1. School of Marine Sciences, Nanjing University of Information Science and Technology, Nanjing 210044, China; muhammad.bilal@connect.polyu.hk (M.B.); 20158301035@nuist.edu.cn (X.H.)
2. Key Laboratory of Digital Land and Resources, East China University of Technology, Nanchang 330013, China; majid.nazeer@comsats.edu.pk
3. Earth and Atmospheric Remote Sensing Lab (EARL), Department of Meteorology, COMSATS University Islamabad, Islamabad 45550, Pakistan
4. Department of Geography, School of Global Studies, University of Sussex, Brighton BN19RH, UK; janet.nichol@connect.polyu.hk
5. Department of Entomology, Plant Pathology and Weed Science, New Mexico State University (NMSU), Las Cruces, NM 88003, USA; maxb@nmsu.edu
6. Leipzig Institute for Meteorology (LIM), University of Leipzig, Stephanstr. 3, 04103 Leipzig, Germany; e.jaekel@uni-leipzig.de
7. Naval Research Laboratory, Monterey, CA 93943, USA; james.campbell@nrlmry.navy.mil
8. School of Earth Sciences, Zhejiang University, Hangzhou 310027, China; lagondal@zju.edu.cn
9. Institute of Methodologies for Environmental Analysis, National Research Council (CNR), 85050 Tito Scalo (PZ), Italy; simone.lolli@imaa.cnr.it
10. SSAI-NASA, Science Systems and Applications Inc, Lanham, MD 20706, USA
* Correspondence: zhongfeng.qiu@nuist.edu.cn; Tel.: +86-025-5869-5696
† Authors with equal contributions.

Received: 8 May 2019; Accepted: 2 June 2019; Published: 4 June 2019

Abstract: Surface reflectance (SR) estimation is the most critical preprocessing step for deriving geophysical parameters in multi-sensor remote sensing. Most state-of-the-art SR estimation methods, such as the vector version of the Second Simulation of the Satellite Signal in the Solar Spectrum (6SV) radiative transfer (RT) model, depend on accurate information on aerosol and atmospheric gases. In this study, a Simplified and Robust Surface Reflectance Estimation Method (SREM) based on the equations from 6SV RT model, without integrating information of aerosol particles and atmospheric gasses, is proposed and tested using Landsat 5 Thematic Mapper (TM), Landsat 7 Enhanced Thematic Mapper plus (ETM+), and Landsat 8 Operational Land Imager (OLI) data from 2000 to 2018. For evaluation purposes, (i) the SREM SR retrievals are validated against in situ SR measurements collected by Analytical Spectral Devices (ASD) from the South Dakota State University (SDSU) site, USA; (ii) cross-comparison between the SREM and Landsat spectral SR products, i.e., Landsat Ecosystem Disturbance Adaptive Processing System (LEDAPS) and Landsat 8 Surface Reflectance Code (LaSRC), are conducted over 11 urban (2013–2018), 13 vegetated (2013–2018), and 11 desert/arid (2000 to 2018) sites located over different climatic zones at a global scale; (iii) the performance of the SREM spectral SR retrievals for low to high aerosol loadings is evaluated; (iv) spatio-temporal cross-comparison is conducted for six Landsat paths/rows located in Asia, Africa, Europe, and the United States of America from 2013 to 2018 to consider a large variety of land surfaces and atmospheric conditions; (v) cross-comparison is also performed for the Normalized Difference Vegetation Index (NDVI), the Enhanced Vegetation Index (EVI), and the Soil Adjusted Vegetation Index (SAVI) calculated from both the SREM and Landsat SR data; (vi) the SREM is also applied to the Sentinel-2A and Moderate Resolution Imaging Spectrometer (MODIS) data to explore its applicability; and (vii) errors in the SR retrievals are reported using the mean bias error (MBE), root mean squared deviation (RMSD), and mean systematic error (MSE). Results depict significant and strong positive

Pearson's correlation (r), small MBE, RMSD, and MSE for each spectral band against in situ ASD data and Landsat (LEDAPS and LaSRC) SR products. Consistency in SREM performance against Sentinel-2A (r = 0.994, MBE = −0.009, and RMSD = 0.014) and MODIS (r = 0.925, MBE = 0.007, and RMSD = 0.014) data suggests that SREM can be applied to other multispectral satellites data. Overall, the findings demonstrate the potential and promise of SREM for use over diverse surfaces and under varying atmospheric conditions using multi-sensor data on a global scale.

Keywords: Landsat 8; surface reflectance; LEDAPS; LaSRC; 6SV; SREM; NDVI

1. Introduction

Due to the cost effectiveness and ready availability of data, satellite remote sensing is now extensively used for deriving various geophysical parameters at a global scale; but, it mostly depends on accurate retrievals of the surface reflectance (SR), i.e., "the fraction of incoming sunlight that the surface reflects", from the remotely sensed data. SR is thus the most basic remotely sensed parameter in the solar reflective spectral bands (visible and infrared) that is used as an essential input parameter to obtain many parameters including vegetation indices [1], leaf area index [2], burned area identification [3], land cover classification [4], aerosol optical depth [5–7], and water quality parameters [8]. It is estimated from the reflectance received by satellites at top of the atmosphere (TOA). However, TOA reflectance is affected by atmospheric constituents that introduce nonnegligible offset and uncertainty in the satellite data. Therefore, before performing any qualitative or quantitative analysis, it is critical to eliminate atmospheric contributions and accurately estimate the SR of a ground target.

Normally, the elimination of the atmospheric contributions and estimation of SR are performed using image-based and physical methods. Image-based methods such as, dark object subtraction (DOS) [9], the empirical line method (ELM) [10], and histogram matching [11] do not employ any physical parameters, e.g., atmospheric direct and diffuse transmissions, water vapor, and ozone, etc., to estimate SR, as they obtain the required ancillary information (solar and sensor viewing geometry) from the image metadata [12]. The most common physical methods are the Atmospheric Correction (ATCOR) [11], the Fast Line-of-sight Atmospheric Analysis of Spectral Hypercubes (FLAASH) [13], the Image Correction (iCOR, previously known as OPERA) [14], the Framework for Operational Radiometric Correction for Environmental monitoring (FORCE) [15], the Landsat Ecosystem Disturbance Adaptive Processing System (LEDAPS) [16], and the Landsat Surface Reflectance Code (LaSRC) [17]. All of these methods estimate surface reflectance based on physical parameters and precalculated comprehensive lookup tables (LUT) which are constructed by radiative transfer (RT) models [18–21]. The ATCOR, FLASH, and iCOR use the Moderate-resolution Atmospheric Transmission (MODTRAN) RT model [18], FORCE is based on the Simulation of Satellite Signal in the Solar Spectrum (5S) [19], and the LEDAPS and LaSRC applies the vector version of the Second Simulation of the Satellite Signal in the Solar Spectrum (6SV) [20,21] to simulate atmospheric conditions. Such methods are complex in nature compared to image-based methods due to the requirement of several ancillary parameters to simulate atmospheric conditions and correct the data degraded by atmospheric constituents. For example, the 6SV, the most commonly used method in the remote sensing community to eliminate atmospheric contributions and estimate accurate surface reflectance, requires information on vertical profiles of air pressure, water vapor concentration, air temperature, ozone concentration, a digital elevation model (DEM), aerosol optical depth, and an aerosol model as well as the solar and sensor viewing geometry to simulate atmospheric conditions and construct the LUT. The attributed accuracy of 6S, of within 1% [22] is nevertheless subject to inherited errors from the ancillary input parameters [12,23], which are obtained from different sources.

Previous studies have evaluated the various atmospheric correction methods over different types of land covers, atmospheric conditions, and geographical locations [12,24–27]. For example, Nazeer et al. (2014) [12] validated the SR from both image-based (DOS and ELM) and physical methods

(ATCOR, FLAASH, and 6S) over sand, artificial turf, grass, and water surfaces using in situ measured SR, and found the 6S to be robust and more accurate for SR estimation compared to the other methods. Nguyen et al. (2015) [24] tested the adequacy of the DOS, FLAASH, and 6S for above-ground biomass (AGB) estimations of the Gongju and Sejong regions of South Korea and found that 6S outperforms the other methods. López-Serrano et al. (2016) [25] compared the ATCOR, COST (cosine of the Sun zenith angle), FLAASH, and 6S for estimating AGB in the temperate forest area of northeast Durango, Mexico, and concluded that the 6S method is more efficient and reliable than other methods. These validation exercises suggested that (i) the physical methods performed much better than the image-based methods, and (ii) the 6S is the most reliable physical method.

To date, operational SR satellite products are available from the Moderate Resolution Imaging Spectroradiometer (MODIS) [28], the Visible Infrared Imaging Radiometer Suite (VIIRS) [29], the Landsat 4–7 Thematic Mappers (TM) and Enhanced Thematic Mapper Plus (ETM+) (LEDAPS) [16], the Landsat 8 Operational Land Imager (OLI, LaSRC) [17], and the Sentinel-2 A/B Multispectral Instruments (MSI) [30]. In addition to the operational SR products from the Landsat 8 OLI and Sentinel-2 MSI sensors, a Harmonized Landsat and Sentinel (HLS) SR product is generated to improve the global coverage for every 2 to 3 days at a spatial resolution <30 m [31]. A number of efforts have also been made to validate the LEDAPS, LaSRC, and S2 MSI SR products under different conditions and reference data sets [12,17,32–38].

As evident from all these publications, an accurate estimation of SR based on RT models requires precise retrieval of AOD and vice versa. In other words, these two parameters, i.e., SR and AOD, complement and depend on each other for their inversions. The available state-of-the-art SR methods based on the RT algorithms do not provide a meaningful solution without incorporating information on aerosol particles and atmospheric gases which is a daunting task given the inherent errors in satellite AOD retrievals, which vary spatially and seasonally across the globe [39,40]. Therefore, the prime intent of this study is to provide a user-friendly SR method, which can easily be applied. This study proposes a new Simplified and Robust Surface Reflectance Estimation Method (SREM) based on the equations of the 6SV RT model that can perform SR inversion without using precalculated comprehensive LUT, and information on aerosol particles and atmospheric gases and furnishes results similar to well-known state-of-the-art methods. The outline of the manuscript is as follow: Section 2 is related to the datasets and selection of the validation sites, Section 3 is based on the research methodology for SREM and evaluation process, results and discussions are described in Section 4, and conclusions of this study are summarized in Section 5.

2. Datasets

In this study, archived datasets are used from the satellite sensors, i.e., Landsat 5 (L5) TM, Landsat 7 (L7) ETM+, and Landsat 8 (L8) OLI, for the development of SREM. For validation purposes, in situ SR data from South Dakota State University (SDSU: grassland site) in South Dakota (Figure 1) are taken by Maiersperger et al. (2013) [35]. For comparison purposes, LEDAPS and LaSRC SR products are obtained for 11 urban, 13 vegetated, and 11 desert (arid) sites (Figure 1) from January 2000 to October 2018.

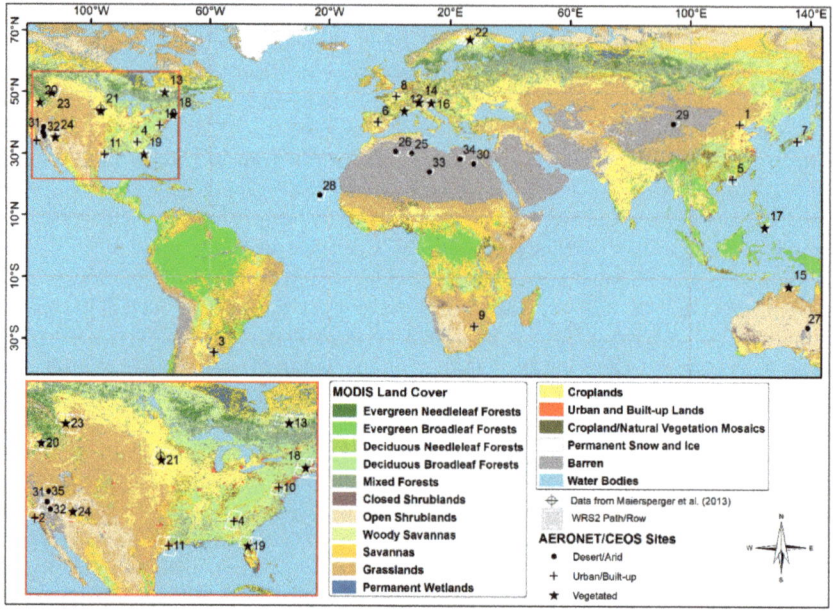

Figure 1. Sites used for validation and comparison of SREM derived surface reflectance (SR) products for Landsat and Sentinel-2A. Please refer to the "S/N" in Table 2 for labeled values, and WRS-2 (Worldwide Reference System) path/row numbers.

2.1. Satellite Data

Landsat TM/ETM+/OLI

The Landsat TM, ETM+, and OLI sensors were launched in March 1984, April 1999, and February 2013, respectively. Landsat ETM+ scenes acquired since May 30, 2003, have data gaps due to the Scan Line Corrector (SLC) failure and the scenes acquired before this date are defined as non-SLC. All sensors have a spatial resolution of 30 m for the multispectral bands and a revisit time of 16 days. In this study, Landsat TM, ETM+, and OLI data for the visible to near-infrared bands (Table 1) are used in the SR retrieval by applying SREM. The new SR data set is compared against the readily available SR products (i.e., LEDAPS and LaSRC) from the respective sensors.

Table 1. Spectral bands, band numbers, and central wavelengths (nm) of the L8, L7, and L5 data used in this study.

Spectral Band	Band Numbers		
	L8 OLI	L7 ETM+	L4/5 TM
Coastal Aerosol	B1 [443.0]	–	–
Blue	B2 [482.0]	B1 [485.0]	B1 [485.0]
Green	B3 [561.5]	B2 [560.0]	B2 [560.0]
RED	B4 [654.5]	B3 [660.0]	B3 [660.0]
NIR	B5 [865.0]	B4 [835.0]	B4 [830.0]
SWIR1	B6 [1608.5]	B5 [1650.0]	B5 [1650.0]
SWIR2	B7 [2200.5]	B7 [2220.0]	B7 [2215.0]

The LEDAPS algorithm is applied to process the L5 TM and L7 ETM+ Level-1 products to SR (Level-2), where the SR is derived automatically from the calibrated TOA reflectance using the 6S atmospheric correction method and atmospheric parameters [16], similar to the MODIS SR products (i.e., MOD09 for Terra and MYD09 for Aqua sensors). In contrast to the LEDAPS algorithm, the LaSRC

SR product [17] for the L8 OLI sensor includes improved estimation of the atmospheric parameters (pressure, water vapor, air temperature, ozone, and AOD) essential as input to the RT-model-based SR estimation. The atmospheric parameters for the LaSRC and LEDAPS are obtained from the MODIS Climate Modeling Gridded (CMG) data products and National Centers for Environmental Prediction (NCEP) gridded products, respectively, both based on the 6SV RT model.

For this study, the Collection-1 Landsat data are taken from the United States Geological Survey (USGS), Earth Resources Observation and Science (EROS), Center's Science Processing Architecture (ESPA) on-demand interface (https://espa.cr.usgs.gov) as Level-1 and Level-2 products for the period of January 2000 to October 2018.

2.2. In Situ Surface Reflectance Data

The ground-truth SR data were collected by Maiersperger et al. [35] from the South Dakota State University's (SDSU) grassland site in South Dakota (Figure 1) using the Analytical Spectral Devices (ASD) FieldSpec spectrometer. Generally, this site is used by the Committee on Earth Observation Satellites (CEOS) as a reference site for vicarious calibrations. For this site, a total of 10 L5 and L7 scenes were found coincident with the ASD data (Appendix A) for validation.

2.3. Site Selection for Comparison Purpose

In order to compare the SREM with the standard algorithms, e.g. LEDAPS and LaSRC, three broad land cover types (urban, vegetation, and arid) are selected based on the global mosaics of the standard MODIS land cover type data product (MCD12Q1). This product describes land cover properties derived from yearly MODIS observations. The primary land cover scheme identifies 17 land cover classes defined by the International Geosphere Biosphere Programme (IGBP), which includes 11 natural vegetation classes, 3 developed and mosaicked land classes, and 3 nonvegetated land classes [4]. The spatially aggregated MODIS Collection 6 land cover product at a spatial resolution of 500 m [41] is obtained from the NASA Earthdata Search (https://search.earthdata.nasa.gov/) for the year 2017. A total of 35 AERONET (Aerosol Robotic Network) and CEOS pseudo-invariant and instrumented sites located in urban/built-up (11), vegetated (13) and arid/desert (11) regions are selected based on the MODIS land cover classification (Table 2). For the AERONET sites, only the L8 OLI sensor data were used (from 1 April, 2013 to 15 October, 2018), while for the CEOS sites, the L5 TM, L7 ETM+, and L8 OLI sensors data are used from January 2000 to May 2012, July 1999 to May 2003 (non-SLC affected data), and April 2013 to 15 October, 2018, respectively. For each selected site, the Landsat scenes are ordered through the ESPA on-demand interface (https://espa.cr.usgs.gov).

Table 2. Aerosol Robotic Network (AERONET) and Committee on Earth Observation Satellites (CEOS) sites involved in a comparison between the Landsat surface reflectance (Landsat Ecosystem Disturbance Adaptive Processing System (LEDAPS) and Landsat 8 Surface Reflectance Code (LaSRC)) and SREM.

S/N	Site Name	Longitude (dd)	Latitude (dd)	Land Cover	Subtype	Path/Row
1	Beijing [a]	116.38	39.98	Urban	Urban	123/32
2	CalTech [a]	−118.13	34.14	Urban	Near Coast	41/36
3	CEILAP-BA [a]	−58.51	−34.56	Urban	Urban	225/84
4	Georgia_Tech [a]	−84.40	33.78	Urban	Near Vegetation	19/36, 19/37
5	Hong_Kong_PolyU [a]	114.18	22.30	Urban	Urban	121/45, 122/44, 122/45
6	Madrid [a]	−3.72	40.45	Urban	Urban	201/32
7	Osaka [a]	135.59	34.65	Urban	Urban	109/36, 110/36
8	Paris [a]	2.36	48.85	Urban	Urban	199/26
9	Pretoria_CSIR-DPSS [a]	28.28	−25.76	Urban	Urban	170/78
10	UMBC [a]	−76.71	39.25	Urban	Urban	15/33
11	Univ_of_Houston [a]	−95.34	29.72	Urban	Urban	25/39, 25/40, 26/39
12	Carpentras [a]	5.06	44.08	Vegetation	Cropland	196/29
13	Chapais [a]	−74.98	49.82	Vegetation	Forest	16/25, 17/25
14	Davos [a]	9.84	46.81	Vegetation	Grassland	193/27, 193/28, 194/27
15	Jabiru [a]	132.89	−12.66	Vegetation	Savanna	104/69, 105/69
16	Kanzelhohe_Obs [a]	13.90	46.68	Vegetation	Forest	191/27, 191/28
17	ND_Marbel_Univ [a]	124.84	6.50	Vegetation	Cropland	112/55, 112/56
18	NEON_Harvard [a]	−72.17	42.54	Vegetation	Forest	13/30, 13/31, 12/30, 12/31
19	NEON_OSBS [a]	−81.99	29.69	Vegetation	Savanna	16/39, 16/40, 17/39
20	Rimrock [a]	−116.99	46.49	Vegetation	Savanna	42/28, 43/28
21	Sioux_Falls [a]	−96.63	43.74	Vegetation	Cropland	29/29, 29/30
22	Sodankyla [a]	26.63	67.37	Vegetation	Savanna	191/13, 190/13, 192/12, 192/13
23	Univ_of_Lethbridge [a]	−112.87	49.68	Vegetation	Grassland	40/25, 40/26, 41/25
24	USGS_Flagstaff_ROLO [a]	−111.63	35.21	Vegetation	Savanna	37/35, 37/36
25	Algeria 3 [b]	7.66	30.32	Desert	Arid	192/39
26	Algeria 5 [b]	2.23	31.02	Desert	Arid	195/39
27	Birdsville [a]	139.35	−25.90	Desert	Arid	98/78
28	Capo_Verde [a]	−22.94	16.73	Desert	Shrubland	209/48, 209/49
29	Dunhuang [b]	94.34	40.13	Desert	Arid	137/32
30	El_Farafra [a]	27.99	27.06	Desert	Barren	178/41
31	Frenchman_Flat [a]	−115.93	36.81	Desert	Barren	40/34, 40/35
32	Ivanpah Playa [b]	−115.40	35.57	Desert	Arid	39/35
33	Libya 1 [b]	13.35	24.42	Desert	Arid	187/43
34	Libya 4 [b]	23.39	28.55	Desert	Arid	181/40
35	Railroad Valley Playa [b]	−115.69	38.50	Desert	Arid	40/33

[a] AERONET sites (L8 OLI sensor's data was used from 1 April, 2013 to 15 October, 2018). [b] CEOS sites (L5 TM, L7ETM+ and L8 OLI sensors data were used from 1 January, 2000 to 15, October 2018).

3. Methodology

3.1. Surface Reflectance Inversion

In general, SR retrievals are derived from the TOA reflectance that can be simulated based on the following 6SV RT model equation (Equation (1)) for the Lambertian uniform target [17,20,42]:

$$\begin{aligned}\rho_{TOA}&(\lambda, \theta_s, \theta_v, \varphi, \tau_a, \omega_a, P_A, U_{H_2O}, U_{O_3}) \\ &= Tg_{OG}Tg_{O_3}[\rho_{atm}(\lambda, \theta_s, \theta_v, \varphi, \tau_a, \omega_a, P_A, U_{H_2O}) \\ &+ T_s(\lambda, \theta_s, \tau_a, \omega_a, P_A)T_v(\lambda, \theta_v, \tau_a, \omega_a, P_A)\frac{\rho_s(\lambda)}{1-S_{atm}(\lambda, \tau_a, \omega_a, P_A)\rho_s(\lambda)}Tg_{H_2O}]\end{aligned} \quad (1)$$

where

ρ_{TOA} = reflectance received by satellite at the top of the atmosphere,
ρ_{atm} = atmospheric intrinsic path reflectance,
λ = wavelength
T_s = atmospheric transmittance of sun-surface path (downward),
T_v = atmospheric transmittance of surface-sensor path (upward),
ρ_s = surface reflectance to be estimated,
S_{atm} = atmospheric backscattering ratio to count multiple reflections between the surface and atmosphere,
θ_s = solar zenith angle,
θ_v = sensor zenith angle,
φ = relative azimuth angle,
U_{H_2O}, = the integrated water vapor content,
U_{O_3}, = the integrated ozone content,
τ_a, ω_a, P_A = aerosol optical depth, aerosol single scatter albedo, and aerosol phase function, respectively, and
$Tg_{H_2O}, Tg_{O_3}, Tg_{OG}$ = gaseous transmission by water vapor, ozone, and other gases, respectively.

The atmospheric intrinsic reflectance can be approximated using Equation (2) [17]:

$$\rho_{atm}(\lambda, \theta_s, \theta_v, \varphi, \tau_a, \omega_a, P_A, U_{H_2O}) = \rho_R(\lambda, \theta_s, \theta_v, \varphi) + (\rho_{A+R}(\lambda, \theta_s, \theta_v, \varphi) - \rho_R(\lambda, \theta_s, \theta_v, \varphi))Tg_{H_2O} \quad (2)$$

where

ρ_R = atmospheric reflectance due to Rayleigh scattering and
ρ_{A+R} = combined atmospheric reflectance due to Rayleigh and aerosols.

The objective of this study is to perform an SR inversion using an equation based on the 6SV RT model without using aerosol information such as $\tau_a, \omega_a,$ and P_A (i.e., $\rho_A = 0$) and other atmospheric parameters such as, U_{H_2O}, U_{O_3}, and OG (i.e., $Tg_{OG}, Tg_{O_3},$ and $Tg_{H_2O} = 1$). Therefore, these parameters are neglected on the right-hand sides of Equations (1) and (2), such that the TOA reflectance can be approximated as Equation (3):

$$\rho_{TOA}(\lambda, \theta_s, \theta_v, \varphi, \tau_a, \omega_a, P_A) = \rho_R(\lambda, \theta_s, \theta_v, \varphi) + T_s(\lambda)T_v(\lambda)\frac{\rho_s(\lambda)}{1-S_{atm}(\lambda)\rho_s(\lambda)} \quad (3)$$

From Equation (3), ρ_s for the SREM method is approximated as Equation (4), and for simplicity, Equation (4) is expressed as Equation (5):

$$\rho_s(\lambda) = \frac{\rho_{TOA}(\lambda, \theta_s, \theta_v, \varphi, \tau_a, \omega_a, P_A) - \rho_R(\lambda, \theta_s, \theta_v, \varphi)}{(\rho_{TOA}(\lambda, \theta_s, \theta_v, \varphi, \tau_a, \omega_a, P_A) - \rho_R(\lambda, \theta_s, \theta_v, \varphi))S_{atm}(\lambda) + T_s(\lambda)T_v(\lambda)} \quad (4)$$

$$\rho_s = \frac{\rho_{TOA} - \rho_R}{(\rho_{TOA} - \rho_R)S_{atm} + T_sT_v} \quad (5)$$

where ρ_s = SREM estimated surface reflectance.

It should be noted that SREM SR is different than the Rayleigh corrected TOA reflectance, which can be obtained by simple subtraction of Rayleigh reflectance from the TOA reflectance.

In Equation (4), the TOA reflectance and Rayleigh reflectance is computed using Equations (6) and (7) [43], respectively:

$$\rho_{TOA}(\lambda, \theta_s, \theta_v, \varphi, \tau_a, \omega_a, P_A) = \frac{\pi L_{TOA}(\lambda, \theta_s, \theta_v, \varphi)d^2}{ESUN_\lambda \mu_s} \quad (6)$$

where

L_{TOA} = radiance received by satellite at the top of the atmosphere,
d = distance between the Earth and Sun in the astronomical unit,
$ESUN$ = mean solar exoatmospheric radiation,
μ_s = cosine of solar zenith angle, and
λ = wavelength.

$$\rho_R(\lambda, \theta_s, \theta_v, \varphi) = P_R(\theta_s, \theta_v, \varphi) \frac{\left(1 - e^{-M\tau_r}\right)}{4(\mu_s + \mu_v)} \quad (7)$$

where

M = air mass calculated using Equation (8) [43],
τ_r = Rayleigh optical depth calculated using Equation (9) [44],
P_R = Rayleigh phase function calculated using Equation (10) [43], and
μ_v = cosine of sensor zenith angle.

$$M = \frac{1}{\mu_s} + \frac{1}{\mu_v} \quad (8)$$

$$\tau_r = 0.008569(\lambda)^{-4}\left(1 + 0.0113(\lambda)^{-2} + 0.0013(\lambda)^{-4}\right) \quad (9)$$

$$P_R = \frac{3A}{4+B}\left(1 + \cos^2\Theta\right); A = 0.9587256, B = 1 - A \quad (10)$$

where

Θ = scattering angle, and
A and B are coefficients that account for the molecular asymmetry.

In Equation (4), the atmospheric backscattering ratio and total atmospheric transmission, without integrating aerosol information, is expressed as Equations (11)–(13) [7,45,46], respectively:

$$S_{atm}(\lambda) = (0.92\tau_r)e^{-\tau_r} \quad (11)$$

$$T_s(\lambda) = e^{(-\tau_r/\mu_s)} + e^{(-\tau_r/\mu_s)}\left\{e^{(0.52\tau_r/\mu_s)} - 1\right\} \quad (12)$$

$$T_v(\lambda) = e^{(-\tau_r/\mu_v)} + e^{(-\tau_r/\mu_v)}\left\{e^{(0.52\tau_r/\mu_v)} - 1\right\} \quad (13)$$

The SREM SR retrievals were estimated for each Landsat band (TM and ETM+: B1-B5 and B7, and OLI: B1-B7) using Equation (4), and for a clear understanding of the SREM, the step-by-step methodology is described in Figure 2.

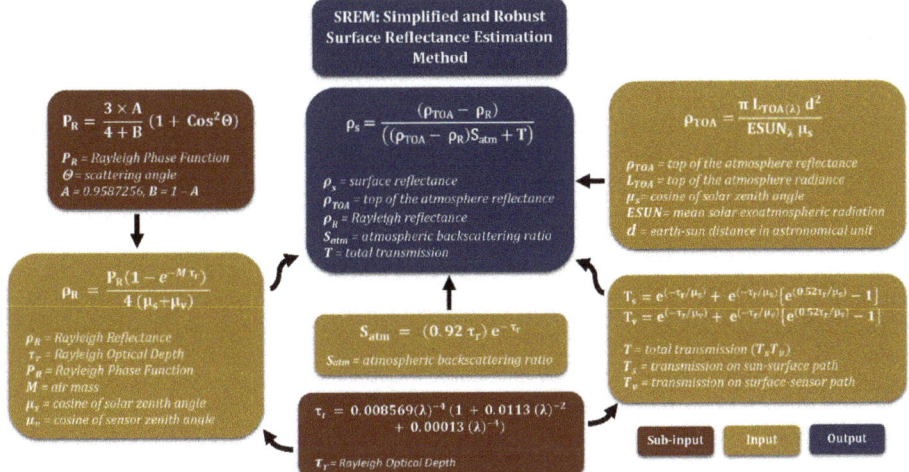

Figure 2. Systematic methodology of the SREM (Simplified and Robust Surface Reflectance Estimation Method).

3.2. Evaluation Process

The SREM is validated against a range of criteria and features, extensive in itself, in order to test its robustness and explore its potential application. The evaluation process comprises eight steps: (1) The SREM estimated SR and LEDAPS (TM5 and ETM+) SR observations are compared with in situ SR measurements collected by Maiersperger, Scaramuzza, Leigh, Shrestha, Gallo, Jenkerson, and Dwyer [35]. The SREM and LEDAPS SR retrievals are averaged from the spatial window of 3 × 3 pixels if at least 2 out of 9 pixels are available centered on the measurement site. (2) The SREM and Landsat (LEDAPS and LaSRC) SR retrievals are compared for 35 sites located over urban (2013–2018), vegetated (2013–2018), and desert surfaces (2000 to 2018) (Figure 1 and Table 2). To obtain the collocated SREM and Landsat data, (i) retrievals are filtered for the quality flag "66" (clear and low-confidence cloud) for LEDAPS and "322" (clear and low-confidence cloud) for LaSRC, and (ii) matched for the same time and location. (3) In order to evaluate the performance of the SR inversion methods during different aerosol loadings, SR retrievals for each channel were filtered based on the AOD at 550 nm obtained from AERONET sites. (4) A spatio-temporal cross-comparison is conducted for six Landsat paths/rows (122/44, 199/26, 201/32, 170/78, 15/33, and 25/39) located over different regions and climatic zones. For a comprehensive comparison for diverse land surfaces and varying atmospheric conditions, 3000 data points are randomly selected from each image-pair of SREM and LaSRC for each path/row. For this comparison, those Landsat retrievals, that may have values outside the theoretical limits, i.e., 0 < SR < 1, are removed. These unusual retrievals are available due to over-correction for atmosphere and Landsat calibration errors [47–50] or retrievals with SR > 1 might be available for those surfaces that reflect more strongly than Lambertian surfaces [51]. (5) The Normalized Difference Vegetation Index (NDVI, Equation (14)) [1,52], Enhanced Vegetation Index (EVI, Equation (15)) [1], and Soil Adjusted Vegetation Index (SAVI, Equation (16)) [1] are calculated using SREM and Landsat SR data and compared with each other to demonstrate the ability of the SREM to monitor vegetation and crops.

$$NDVI = \frac{(NIR - Red)}{(NIR + Red)} \quad (14)$$

$$EVI = 2.5 \times \left\{ \frac{(NIR - Red)}{(NIR + 6 \times Red - 7.5 \times Blue + 1)} \right\} \quad (15)$$

$$SAVI = 1.5 \times \left\{ \frac{(NIR - Red)}{(NIR + Red + 0.5)} \right\} \quad (16)$$

(6) To further explore its applicability, the SREM is applied to the Sentinel-2A and MODIS datasets and compared with the Sentinel-2A SR observations estimated by the latest version (2.5.5) of the Sen2Cor and MOD09 level 2 surface reflectance products, respectively. For this purpose, Beijing, a city with mixed bright urban surfaces that mostly remains under frequent haze and dust pollution effects, is selected as a test site. (7) In order to calculate the slope and intercept between the SREM and Landsat retrievals, the reduced major axis (RMA) is used, which can simultaneously account for errors in both dependent and independent variables [53,54]. In RMA, slope (β) and intercept (α) are determined using Equations (17) and (18):

$$\beta = \frac{\sigma_y}{\sigma_x} \tag{17}$$

$$\alpha = \overline{Y} - \left(\frac{\sigma_y}{\sigma_x}\right)\overline{X} \tag{18}$$

where

$\overline{X}$ and $\overline{Y}$ = means of X and Y, respectively, and
σ_x and σ_y = standard deviations of X and Y, respectively.

(8) To report the consistency and errors in the SREM SR product, the Pearson's correlation coefficient (r), mean bias error (MBE, Equation (19)), root-mean-squared difference (RMSD, Equation (20)), and mean systematic error (MSE, Equation (21)) are computed. The MSE is useful to report the difference between the trend of X and Y data; small MSE indicates a good trend.

$$MBE = \frac{1}{n}\sum_{i=1}^{n}(Y_i - X_i) \tag{19}$$

$$RMSD = \sqrt{\frac{1}{n}\sum_{i=1}^{n}(Y_i - X_i)^2} \tag{20}$$

$$MSE = \frac{1}{n}\sum_{i=1}^{n}(\hat{Y}_i - X_i)^2 \tag{21}$$

where $\hat{Y}$ = predicted value based on RMA relationship (Y = βX + α) between X and Y.

4. Results and Discussion

4.1. Cross-Comparison of ASD, LEDAPS, and SREM SR Data

The SR data collected by ASD for the SDSU site are available for only 10 days [35] (Appendix A) and are compared with the LEDAPS and SREM SR retrievals (Table 3). Table 3 shows comparable values of Pearson's correlation coefficient (r) for LEDAPS and SREM with ASD data. Overestimation is observed in LEDAPS from B1 to B3, and underestimation in SREM is from B4 to B7. The maximum positive MBE for LEDAPS is 0.006 for B2 and for SREM it is 0.018 for B1. Similarly, the maximum negative MBE is for B5, with −0.009 and −0.035 for LEDAPS and SREM, respectively. The results for SREM are satisfactory with a high value of r and the reason for large negative values of MBE are investigated in the following analysis.

Table 3. Validation summary of LEDAPS and SREM SR retrievals against Analytical Spectral Devices (ASD) FieldSpec spectrometer data.

Date	Sensor	Band 1			Band 2			Band 3		
		ASD	LEDAPS	SREM	ASD	LEDAPS	SREM	ASD	LEDAPS	SREM
20030826	ETM+	0.045	0.053	0.067	0.075	0.080	0.076	0.086	0.090	0.085
20060615	ETM+	0.054	0.063	0.078	0.092	0.099	0.094	0.106	0.108	0.102
20070720	ETM+	0.051	0.057	0.070	0.085	0.091	0.087	0.110	0.116	0.110
20080612	TM5	0.072	0.063	0.073	0.114	0.105	0.096	0.123	0.108	0.087
20080714	TM5	0.056	0.058	0.068	0.086	0.095	0.088	0.108	0.111	0.101
20080823	ETM+	0.051	0.052	0.064	0.080	0.080	0.076	0.093	0.092	0.104
20080916	TM5	0.037	0.054	0.063	0.059	0.083	0.076	0.068	0.100	0.093
20090530	TM5	0.052	0.055	0.065	0.087	0.090	0.084	0.084	0.087	0.057
20100805	TM5	0.030	0.040	0.056	0.057	0.067	0.066	0.052	0.057	0.418
20100821	TM5	0.030	0.037	0.052	0.059	0.068	0.066	0.054	0.058	0.359
Average		0.048	0.053	0.066	0.079	0.086	0.081	0.088	0.093	0.088
[1] StDev		0.012	0.008	0.007	0.017	0.012	0.010	0.023	0.020	0.018
[2] CV		0.255	0.154	0.108	0.212	0.138	0.126	0.261	0.214	0.201
MBE			0.005	0.018		0.006	0.002		0.004	0.000
r			0.869	0.809		0.905	0.914		0.883	0.888

Date	Sensor	Band 4			Band 5			Band 7		
		ASD	LEDAPS	SREM	ASD	LEDAPS	SREM	ASD	LEDAPS	SREM
20030826	ETM+	0.277	0.276	0.253	0.319	0.310	0.289	0.172	0.174	0.148
20060615	ETM+	0.312	0.299	0.268	0.317	0.296	0.271	0.170	0.163	0.135
20070720	ETM+	0.259	0.256	0.239	0.344	0.338	0.318	0.197	0.206	0.179
20080612	TM5	0.328	0.301	0.281	0.317	0.289	0.262	0.174	0.153	0.136
20080714	TM5	0.246	0.277	0.254	0.335	0.323	0.289	0.203	0.186	0.163
20080823	ETM+	0.280	0.264	0.248	0.335	0.324	0.305	0.183	0.183	0.159
20080916	TM5	0.236	0.244	0.225	0.277	0.300	0.269	0.148	0.175	0.153
20090530	TM5	0.307	0.280	0.263	0.295	0.282	0.258	0.159	0.156	0.140
20100805	TM5	0.315	0.317	0.284	0.233	0.226	0.199	0.109	0.105	0.090
20100821	TM5	0.339	0.334	0.299	0.236	0.227	0.200	0.106	0.095	0.082
Average		0.290	0.285	0.261	0.301	0.292	0.266	0.162	0.160	0.138
StDev		0.034	0.026	0.021	0.038	0.036	0.038	0.031	0.033	0.029
CV		0.116	0.930	0.815	0.125	0.125	0.142	0.193	0.208	0.210
MBE			−0.005	−0.028		−0.009	−0.035		−0.002	−0.024
r			0.878	0.919		0.944	0.949		0.921	0.922

[1] StDev = Standard deviation. [2] CV = Coefficient of variations (StDev/average).

For this purpose, LEDAPS and SREM SR retrievals are compared with the TOA reflectance observations obtained for the same dates (Table 4). The hypothesis of this analysis is that the method, LEDAPS or SREM, with the larger negative value of MBE would be considered as superior, as it represents the greater removal of atmospheric effects. On the other hand, a large positive value represents "under-correction"; hence, the respective method is unable to remove atmospheric effects significantly. The results (Table 4) show that LEDAPS has a larger negative MBE for B1 compared to SREM, whereas, SREM has a larger negative MBE for B2 and B3 than LEDAPS, which indicate the better performances of LEDAPS for B1 and SREM for B2 and B3. For B4 to B7, LEDAPS has larger positive values of MBE than SREM, which might be due to its sensitivity to the absorption by atmospheric gases in the infrared spectral region and shows "under correction (lack of atmospheric correction)" of LEDAPS. These results (Tables 3 and 4) suggest that SREM is less sensitive to the absorption by atmospheric gases and performs within the expected range, as the average values of SR retrievals are less than the TOA reflectance, whereas LEDAPS values are even greater than the TOA reflectance; especially, for B4 to B7. Therefore, it can be concluded that SREM, without integrating aerosol information and a comprehensive precalculated LUT in the inversion, can provide SR retrievals that are comparable with both the ASD and LEDAPS observations.

Table 4. Summary of cross-comparison between LEDAPS and SREM SR retrievals and top of atmosphere (TOA) reflectance observations.

Bands	Average			MBE		R	
	TOA	LEDAPS	SREM	LEDAPS	SREM	LEDAPS	SREM
B1	0.107	0.053	0.066	−0.054	−0.041	0.963	0.997
B2	0.104	0.086	0.081	−0.018	−0.023	0.994	0.999
B3	0.100	0.093	0.088	−0.008	−0.013	1.000	1.000
B4	0.265	0.285	0.261	0.020	−0.003	0.984	1.000
B5	0.266	0.292	0.266	0.025	0.000	0.993	1.000
B7	0.139	0.160	0.138	0.021	−0.001	0.995	1.000

4.2. Cross-Comparison between SREM and Landsat SR Retrievals

A cross-comparison between SREM and Landsat (LEDAPS and LaSRC) SR retrievals is conducted over urban as well as vegetated surfaces from 2013 to 2018, and desert (arid) surfaces from 2000 to 2018. Figure 3 shows the scatter (dashed line = 1:1 line) and line plots (black line = Landsat, gray line = SREM) between SREM and Landsat retrievals. The results are summarized in Table 5 which shows that the SREM SR retrievals for coastal aerosol (Figure 3a–c) and blue (Figure 3d–f) spectral bands over desert (arid) sites (Figure 3d,f) are well correlated with the LaSRC and LEDAPS SR retrievals with r ~ 0.990–0.991 (Table 5), and small values of MBE ~ 0.004–0.022, and RMSD ~ 0.009–0.024, compared with urban (Figure 3a,d) and vegetated sites (Figure 3b,e). The values of MSE ~ 0.000–0.002 indicate only minor differences between the trend of SREM and Landsat (LaSRC and LEDAPS) retrievals independent from the surface type. The performance of the SREM for the blue band is better than for the coastal aerosol band with significant small values of MBE and RMSD. The SREM retrievals appear overestimated as indicated by the positive values of MBE, and these values over urban and vegetated sites are high compared to the results from desert sites and are acceptable according to previous studies [31,32,48]. Overestimation in coastal aerosol and blue bands may be due to the enhanced aerosol extinction and Rayleigh contribution in these wavelengths.

The performance of the SREM for green, red, NIR, SWIR1, and SWIR2 bands over the urban and vegetated sites is robust with r ~ 0.951–1.00, MBE ~ −0.01–0.011, RMSD ~ 0.001–0.012, and MSE ~ 0.00. The low values of MBE, RMSD, and MSE represent (i) very small differences between the SREM and Landsat retrievals, (ii) scatter points close to the 1:1 line, and (iii) minimal differences between the trend of SREM and Landsat SR retrievals. The comparison of all these bands over the desert sites was relatively less reliable with larger differences (MBE and RMSD) due to underestimation in the SREM retrievals, especially for SR > 0.50. The underestimation in the SREM retrievals in comparison with Landsat, especially for the desert surfaces, might be due to the "under-correction" of the Landsat atmospheric correction algorithm as observed in Section 4.1, which describes the cross-comparison between SR retrievals and TOA reflectance observations. Therefore, overall performance of the SREM appears robust, as results show a high consistency in the SREM with very high values of r, and small values of MBE and RMSD which suggest that the SREM renders consistent spatial (i.e., from site to site) and temporal (i.e., from 2000 to 2018) variations in SR as generated by LaSRC and LEDAPS products over heterogeneous surfaces.

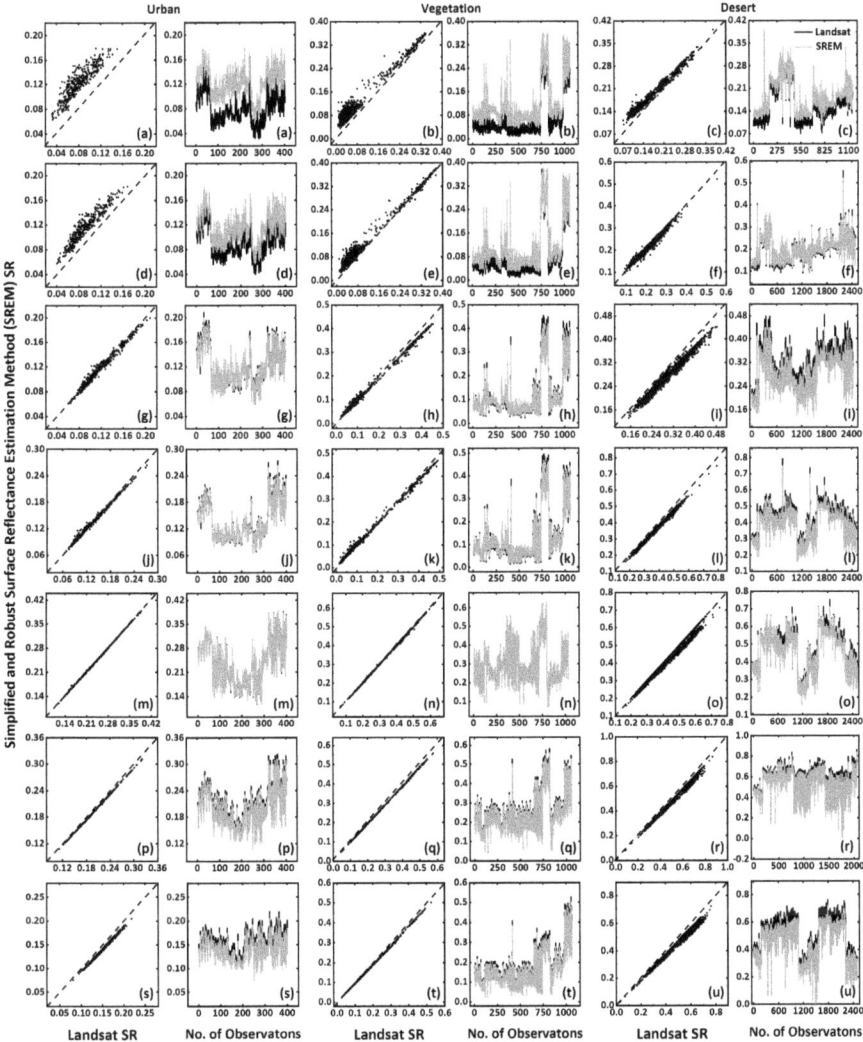

Figure 3. Cross-comparison between coincident SREM and Landsat (LEDAPS and LaSRC) SR retrievals over urban and vegetated sites from 2013–2018, and desert (arid) sites from 2013–2018. Where, coastal aerosol band = (**a**) urban sites, (**b**) vegetated sites, and (**c**) desert sites; blue band = (**d**) urban sites, (**e**) vegetated sites, and (**f**) desert sites; green band = (**g**) urban sites, (**h**) vegetated sites, and (**i**) desert sites; red band = (**j**) urban sites, (**k**) vegetated sites, and (**l**) desert sites; NIR = (**m**) urban sites, (**n**) vegetated sites, and (**o**) desert sites; SWIR1 = (**p**) urban sites, (**q**) vegetated sites, and (**r**) desert sites; SWIR2 = (**s**) urban sites, (**t**) vegetated sites, and (**u**) desert sites; the black line = Landsat retrievals; the grey line = SREM retrievals; and the dashed line = 1:1 line.

Table 5. Summary of cross-comparison between coincident SREM and Landsat (LEDAPS and LaSRC) SR retrievals over urban and vegetated sites from 2013–2018, and desert (arid) sites from 2000–2018.

[1] LC	[2] TP	Sensor	Band	[3] n	[4] β	[5] α	[6] r	MBE	RMSD	MSE
Urban	2013–2018	OLI	Coastal Aerosol	402	1.057	0.037	0.891	0.042	0.044	0.002
			Blue	402	1.018	0.022	0.951	0.024	0.025	0.001
			Green	402	0.943	0.006	0.990	−0.001	0.005	0.000
			Red	402	0.939	0.007	0.997	−0.002	0.005	0.000
			NIR	402	0.989	0.003	1.000	0.000	0.001	0.000
			SWIR1	402	0.972	−0.002	1.000	−0.007	0.008	0.000
			SWIR2	402	0.949	−0.003	0.997	−0.01	0.011	0.000
			All	2814	0.874	0.025	0.963	0.006	0.020	0.000
Vegetation	2013–2018	OLI	CA	1062	0.928	0.043	0.983	0.038	0.041	0.001
			B	1056	0.931	0.027	0.991	0.021	0.025	0.000
			G	1056	0.904	0.008	0.997	−0.003	0.012	0.000
			R	1056	0.929	0.007	0.998	−0.002	0.010	0.000
			NIR	1032	0.989	0.002	1.000	−0.001	0.003	0.000
			SWIR1	1056	0.966	−0.001	1.000	−0.009	0.010	0.000
			SWIR2	1056	0.944	−0.002	1.000	−0.011	0.012	0.000
			All	7374	0.919	0.018	0.990	0.005	0.020	0.000
Desert	2013–2018 2000–2018	OLI TM ETM+ OLI	CA	1148	0.914	0.036	0.991	0.022	0.024	0.001
			B	2482	0.927	0.018	0.990	0.004	0.009	0.000
			G	2440	0.929	−0.002	0.991	−0.024	0.026	0.001
			R	2516	0.954	−0.007	0.997	−0.026	0.027	0.001
			NIR	2520	0.975	−0.006	0.990	−0.018	0.024	0.000
			SWIR1	2065	0.967	−0.011	0.995	−0.029	0.032	0.001
			SWIR2	2499	0.900	0.002	0.994	−0.048	0.052	0.003
			All	15789	0.907	0.016	0.994	−0.020	0.031	0.001

[1] LC = Land cover; [2] TP = Time Period; [3] n = Total number of observations; [4] β = Slope; [5] α = Intercept; [6] r = Pearson's correlation.

To investigate the underestimation in the SREM retrievals over the desert sites for green to SWIR2 bands, cross-comparisons between the Landsat (LEDAPS and LaSRC) and SREM SR retrievals and TOA reflectance observations are conducted, similar to that in Section 4.1. The results (Table 6) show that the Landsat retrievals have positive MBE of 0.014 for green, 0.017 for red, 0.025 for SWIR1, and 0.036 for SWIR2 bands, compared with the SREM retrievals which are within the expected range of below or equal to TOA. These results represent "under-correction" of data by the Landsat atmospheric correction algorithms which might be due to their sensitivity to the atmospheric scattering and absorption in the visible and infrared spectral regions, respectively. Therefore, these results suggest that the apparent "underestimation" in the SREM retrievals over desert sites when compared to Landsat, is mainly due to the under-correction (positive bias) of the Landsat retrievals.

Table 6. Summary of cross-comparison between Landsat (LEDAPS and LaSRC) and SREM SR retrievals and TOA reflectance observations over desert sites.

Bands	Average			MBE		r	
	TOA	Landsat	SREM	Landsat	SREM	Landsat	SREM
Coastal Aerosol	0.268	0.217	0.238	−0.051	−0.030	0.997	0.998
Blue	0.284	0.256	0.261	−0.028	−0.023	0.998	0.998
Green	0.350	0.364	0.338	0.014	−0.012	0.997	0.998
Red	0.433	0.451	0.427	0.017	−0.006	0.997	0.998
NIR	0.517	0.520	0.517	0.003	0.000	0.994	0.995
SWIR1	0.582	0.607	0.585	0.025	0.003	0.977	0.982
SWIR2	0.499	0.525	0.498	0.036	−0.001	0.977	0.990

4.3. Impact of Aerosol Particles on SR Retrievals

In order to evaluate the performance of the SREM method during low to high aerosol loadings, the SR retrievals for each band are filtered based on five levels of AOD at 550 nm obtained from the AERONET sites, i.e., (i) $0.0 < AOD < 0.1$, (ii) $0.1 < AOD < 0.2$, (iii) $0.2 < AOD < 0.3$, (iv) $0.3 < AOD <$

0.4, and (v) 0.4 < AOD < 1.1. The results are presented in Figure 4, where different colors represent different levels of AOD. The cross-comparison is summarized in Table 7, showing that the number of coincident retrievals decreases with the increase in AOD levels. Figure 4 shows that most of the scatter points for each band are close to the 1:1 line and hence well correlated with each other with a value of r from 0.881 to 1.00. According to the statistical summary (Table 7), the values of MBE increase with the increase in aerosol loadings for the coastal aerosol and blue bands, which suggests that the accuracy of SR retrievals for these bands are affected by the aerosol loadings. However, no direct or linear relationship between MBE and aerosol loadings was observed for the other bands (green to SWIR2). This suggests that the performance of the SREM improves for longer wavelength bands (green to SWIR), independent of the aerosol load. Interestingly, it is observed for these bands that the MBE and RMSD for high aerosol loading (0.4 < AOD < 1.1) is smaller than for low aerosol loadings (0.0 < AOD < 0.1), which suggests that SREM retrievals are less sensitive to the high aerosol load. Overall, results are significant and robust, showing consistency between the SREM and LaSRC retrievals during low to high aerosol loadings, and these justify the application of SREM, without integrating information of aerosol particles and atmospheric gases, to estimate SR similar to the LaSRC product.

Table 7. Statistical summary of cross-comparison between coincident Landsat and SREM SR retrievals for low to high AOD levels.

Band	[1] AOD	[2] n	[3] β	[4] α	[5] r	MBE	RMSD
Coastal Aerosol	0.0 < AOD < 0.1	319	0.920	0.041	0.987	0.035	0.038
	0.1 < AOD < 0.2	125	0.893	0.049	0.985	0.039	0.042
	0.2 < AOD < 0.3	56	0.835	0.060	0.963	0.045	0.049
	0.3 < AOD < 0.4	13	0.864	0.055	0.988	0.039	0.042
	0.4 < AOD < 1.1	12	0.903	0.060	0.881	0.052	0.055
Blue	0.0 < AOD < 0.1	319	0.933	0.025	0.994	0.019	0.022
	0.1 < AOD < 0.2	125	0.906	0.032	0.995	0.021	0.025
	0.2 < AOD < 0.3	56	0.871	0.040	0.985	0.026	0.030
	0.3 < AOD < 0.4	13	0.894	0.036	0.997	0.021	0.024
	0.4 < AOD < 1.1	12	0.914	0.040	0.955	0.031	0.034
Green	0.0 < AOD < 0.1	319	0.913	0.007	0.999	−0.005	0.013
	0.1 < AOD < 0.2	125	0.899	0.011	0.999	−0.005	0.014
	0.2 < AOD < 0.3	56	0.890	0.015	0.998	−0.002	0.013
	0.3 < AOD < 0.4	13	0.905	0.013	1.000	−0.006	0.014
	0.4 < AOD < 1.1	12	0.906	0.016	0.995	0.003	0.010
Red	0.0 < AOD < 0.1	319	0.938	0.005	1.000	−0.005	0.011
	0.1 < AOD < 0.2	125	0.926	0.009	1.000	−0.005	0.013
	0.2 < AOD < 0.3	56	0.923	0.011	0.999	−0.003	0.012
	0.3 < AOD < 0.4	13	0.932	0.009	1.000	−0.007	0.013
	0.4 < AOD < 1.1	12	0.925	0.013	0.998	0.001	0.009
NIR	0.0 < AOD < 0.1	319	0.991	0.001	1.000	−0.002	0.002
	0.1 < AOD < 0.2	125	0.986	0.003	1.000	−0.002	0.003
	0.2 < AOD < 0.3	56	0.985	0.004	1.000	−0.001	0.003
	0.3 < AOD < 0.4	13	0.987	0.004	1.000	0.000	0.003
	0.4 < AOD < 1.1	12	0.981	0.006	1.000	0.001	0.004
SWIR1	0.0 < AOD < 0.1	319	0.964	−0.001	1.000	−0.011	0.012
	0.1 < AOD < 0.2	125	0.960	0.001	1.000	−0.012	0.014
	0.2 < AOD < 0.3	56	0.961	0.001	1.000	−0.011	0.013
	0.3 < AOD < 0.4	13	0.963	0.000	1.000	−0.013	0.015
	0.4 < AOD < 1.1	12	0.964	0.000	1.000	−0.008	0.009
SWIR2	0.0 < AOD < 0.1	319	0.935	−0.001	1.000	−0.014	0.018
	0.1 < AOD < 0.2	125	0.927	0.000	1.000	−0.017	0.022
	0.2 < AOD < 0.3	56	0.930	−0.001	1.000	−0.016	0.020
	0.3 < AOD < 0.4	13	0.925	0.000	1.000	−0.021	0.026
	0.4 < AOD < 1.1	12	0.925	0.000	1.000	−0.013	0.015

[1] AOD = AERONET AOD at 550 nm; [2] n = Total number of observations; [3] β = Slope; [4] α = Intercept; [5] r = Pearson's correlation.

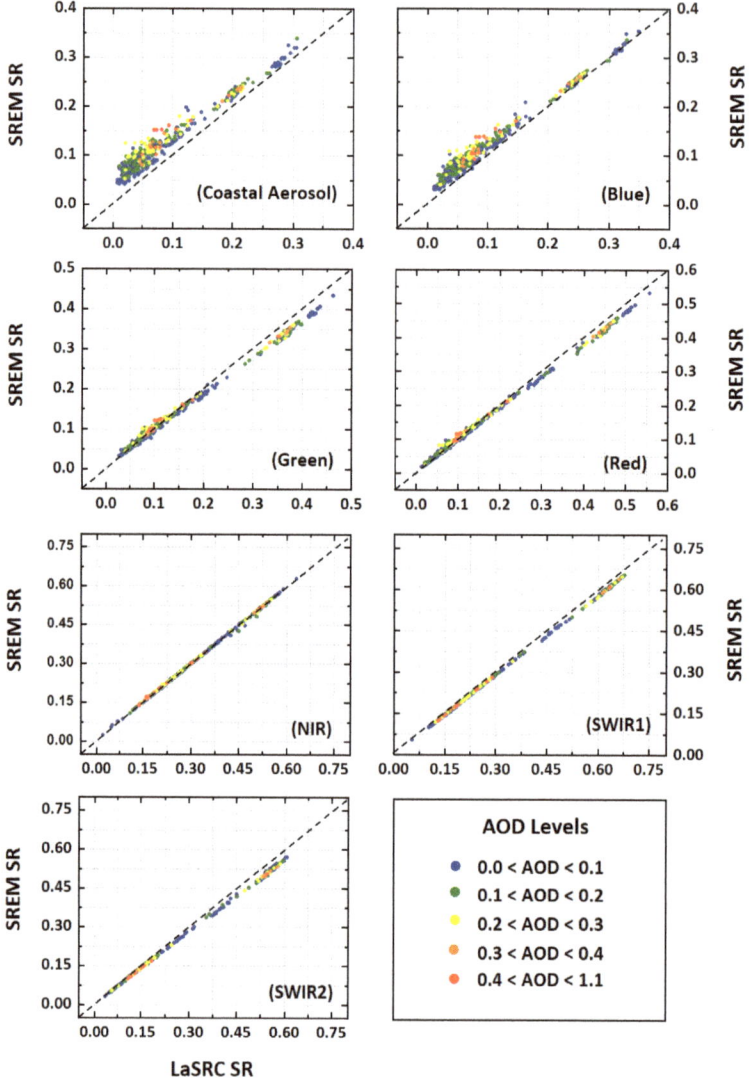

Figure 4. Cross-comparison between coincident Landsat (LaSRC) and SREM SR retrievals over AERONET sites for low to high AOD levels from 2013 to 2018. The dashed line is the 1:1 line.

4.4. Spatio-Temporal Cross-Comparison between SREM and LaSRC Data

For performing spatial cross-comparison between SREM and Landsat SR products, six paths/rows located in Asia (122/44), Africa (170/78), Europe (119/26 and 201/32), and the United States of America (15/33 and 25/39) are selected to represent the diversity of land cover types, and climatic as well as air quality conditions. Figure 5 shows the LaSRC (Landsat 8) and SREM SR displayed as RGB false composites of bands 6, 5, 4, and the results indicate that the SREM yields SR images, which are spatially comparable with the LaSRC SR images. A spatial differences map between SREM and Sentinel-2A SR retrievals for the mentioned paths/rows is added as Figure S1 (supplementary data). A careful visual comparison of any land cover feature, from any panel in Figure 5, exhibits strong alikeness and agreement. These results show that the SREM has the ability to remove atmospheric effects without incorporating atmospheric parameters and a precalculated comprehensive LUT based on the RTM.

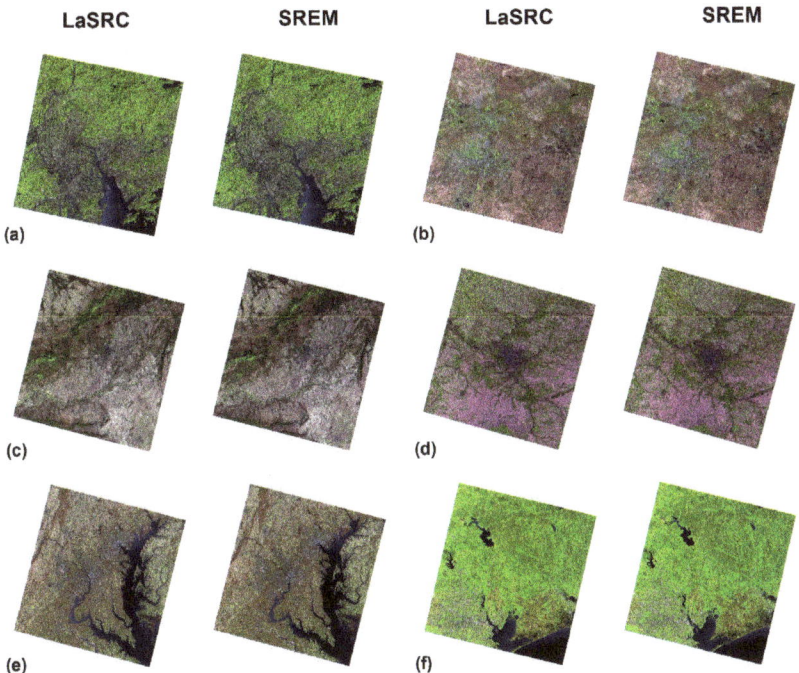

Figure 5. Spatial comparison between LaSRC (left image in each panel) and SREM (right image in each panel) corrected images for different path/rows including (**a**) 122/44, (**b**) 170/78, (**c**) 201/32, (**d**) 199/26, (**e**) 15/33 and (**f**) 25/39. All images are composed using the "natural looking" false color composite of 654 as RGB. All images are North up. No stretch/contrast is applied to the images.

Temporal analysis is also conducted to consider diverse surface types and atmospheric conditions by selecting 3000 random points from each image of each path/row. This approach found 13 (122/44), 66 (170/78), 21 (199/26), 52 (201/32), 22 (15/33), and 26 (25/39) image-pairs, from 2013 to 2018. Total numbers of coincident points for analysis are 531,462 for blue, 532,179 for green, 533,102 for red, 533,140 for NIR, 534230 for SWIR1, and 533,717 for SWIR2 bands (Figure 6). Overall, results reveal a very good correlation between SREM and LaSRC SR products, with r close to unity (r = 0.993 to 1.00) and small values of MBE ≤ −0.002 for green, red, and NIR bands. A large value of MBE (0.020) is observed for the blue band, which may be due to the enhanced aerosol extinction and Rayleigh contribution. The SREM retrievals for the SWIR1 and SWIR2 are also correlated well with LaSRC retrievals but a slight underestimation is found as indicated by the negative value of MBE (−0.009 to −0.011). This underestimation may be due to under-correction by the Landsat atmospheric correction algorithm as discussed in the previous sections on cross-comparison with TOA reflectance. This investigation considers Landsat data as a "true and standard" data for cross-comparison which in fact has its own uncertainties due to aerosol retrieval algorithm, cloud contamination, and under or over atmospheric correction [47–49,55]. Overall, all these findings demonstrate the robust promise of SREM to retrieve SR for diverse surfaces and under varying atmospheric conditions, without incorporating aerosol and atmospheric parameters, in good agreement with the Landsat SR product.

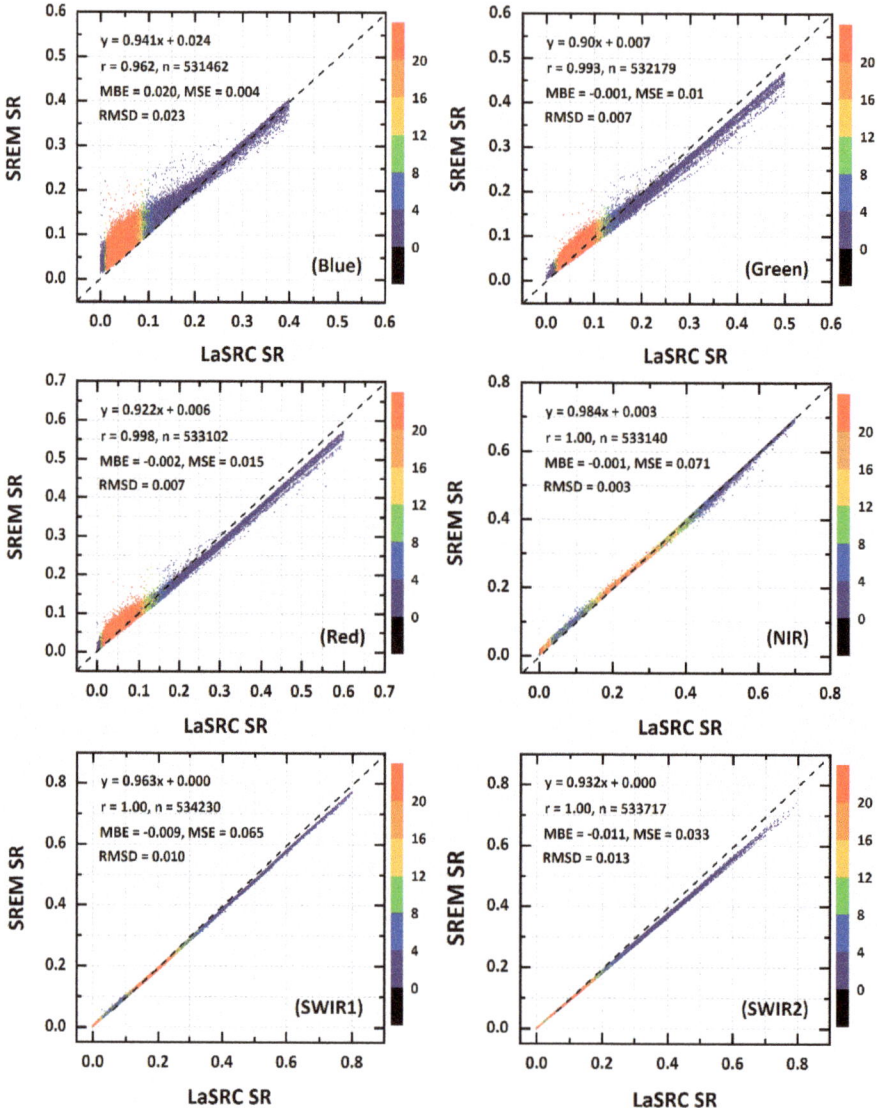

Figure 6. Cross-comparison between SREM and LaSRC data for six paths/rows located in Asia (122/44), Africa (170/78), Europe (119/26 and 201/32), and the United States of America (15/33 and 25/39) from 2013 to 2018 for a large variety of surface types under varying atmospheric conditions. The dashed line is the 1:1 line and the color bar represents the relative frequency of the coincident points.

4.5. Application of SREM to Derive Vegetation Indices

Vegetation indices such as NDVI, EVI, and SAVI data are computed using SREM and compared with Landsat vegetation indices (Figure 7) for the urban (2013–2018), vegetated (2013–2018), and desert sites (2000–2018), in order to test the suitability of the SREM data for vegetation and crop monitoring. Results reveal high consistency in the SREM computed vegetation indices NDVI (Figure 7a), EVI (Figure 7b), and SAVI (Figure 7c) compared to Landsat, as most of the observations are found close to the 1:1 line (dotted line) with slope from 0.951 to 1.086, intercept from 0.013 and 0.017, Pearson's correlation from 0.995 to 0.997, and MBE from 0.007 to 0.024. This comparison is worthy, as an error in the surface reflectance can introduce error in the indices and their potential applications. For example, the SREM SR slightly overestimates in the blue band compared to Landsat, which leads to a larger MBE (0.024) and slopes in the SREM EVI (which uses blue, red, and NIR bands) compared to the NDVI and SAVI, which do not incorporate the blue band. These results show that the SREM SR product is faithful and reliable and can be used for vegetation mapping and monitoring on a global scale.

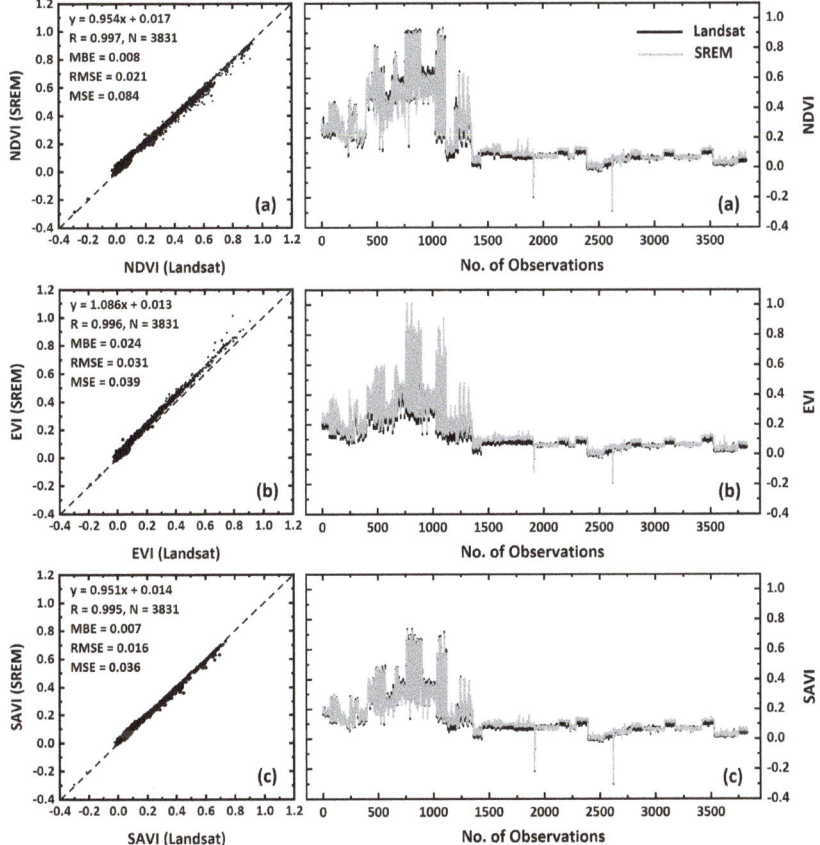

Figure 7. Cross-comparison of NDVI (**a**), EVI (**b**), and SAVI (**c**) data based on the SREM and Landsat (LEDAPS and LaSRC) SR products from 2000 to 2018 for the 35 selected urban, vegetated, and desert sites. The dashed line is the 1:1 line; the grey line = SREM retrievals; the black line = Landsat retrievals which are partially hidden by SREM.

4.6. SREM Implementation in Sentinel-2A and MODIS Data

To further substantiate the applicability of the SREM, preliminary analyses are conducted by SREM using Sentinel-2A and Aqua-MODIS data for Beijing, a city with mixed bright urban surfaces and under effects of severe air pollution episodes. The SREM is applied to cloud-free green band images of Sentinel-2A at 10 m spatial resolution from 8 January to 18 May, 2017, and MODIS at 500 m spatial resolution for the year 2014 (Figure 8). For comparison purposes, the Sentinel-2A SR images are processed using the latest version 2.5.5 of the Sen2Cor atmospheric correction processor, and the Aqua-MODIS Level 2 surface reflectance swath product (MYD09) is used. Figure 8 indicates that most of the scatter points are on or close to the 1:1 line with a high value of r from 0.925 (MODIS) to 0.994 (Sentinel-2A) and small values of MBE from −0.009 (Sentinel-2A) to 0.007 (MODIS) and RMSD of 0.014 for both. The slope between SREM and Sentinel-2A is less than the slope observed for SREM vs.

MODIS due to "under-correction "of the Sentinel-2A SR data by the atmospheric correction algorithm, i.e., Sentinel-2A SR values are greater than TOA reflectance over bright surfaces, whereas, SREM SR values are less than TOA reflectance over these surfaces (Figure S2). Overall, these preliminary results suggest that the SREM has the potential to estimate SR also for other multispectral satellite data.

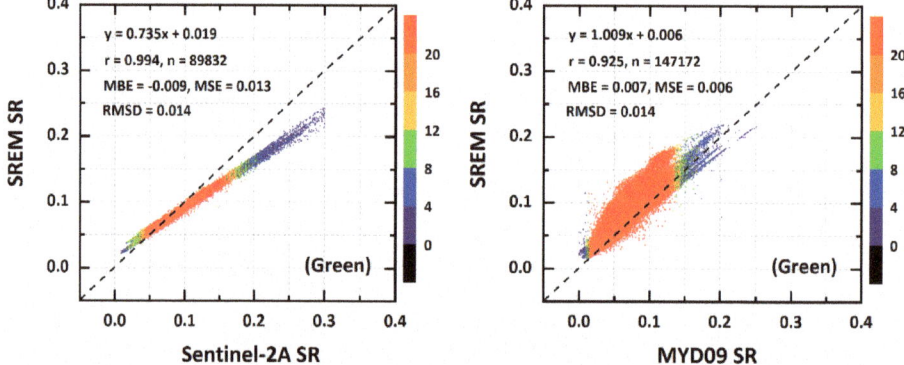

Figure 8. Cross-comparison of SREM vs. Sen2Cor 2.5.5, and SREM vs. MYD09 for the Beijing site. Cloud-free images of Sentinel-2A at 10 m spatial resolution from 8 January to 18 May, 2017, and MODIS at 500 m spatial resolution for the year 2014 are used for cross-comparison. The dashed line represents the 1:1 (y = x) line and the color bar represent the relative frequency of the coincident points.

5. Conclusions

The prime objective of this study was to develop a new Simplified and Robust Surface Reflectance Estimation Method (SREM) based on the Satellite Signal in the Solar Spectrum (6SV) radiative transfer (RT) model equations, without integrating information on aerosol particles and atmospheric gases. The SREM surface reflectance (SR) retrievals were validated against in situ measurements collected by an Analytical Spectral Devices (ASD) spectrometer, and cross-compared with Landsat (LEDAPS and LaSRC) SR products for diverse land surfaces and varying atmospheric conditions, as well as tested on Sentinel2A and MODIS data products. This study concluded that the SREM is capable of accurately estimating spectral surface reflectance (SR) without incorporating information on aerosol particles and atmospheric parameters, and the SR retrievals are comparable with the SR data collected by the ASD spectrometer as well as those provided by Landsat SR products (LEDAPS and LaSRC) which use the 6SV model. Larger positive values of MBE were observed for coastal aerosol band compared to longer wavelengths, which may be related to increase scattering effects at lower wavelengths. Large negative values of MBE were observed in SREM from green to SWIR2 bands when compared to Landsat, which were mainly due to "under-correction (lack of atmospheric correction)" of data by the Landsat atmospheric correction algorithms when compared to TOA reflectance. The preliminary analysis implies that SREM has a strong potential for augmenting vegetation and crop monitoring and it can be implemented with Sentinel-2A and MODIS data or other multispectral satellite data sets.

Supplementary Materials: The following are available online at http://www.mdpi.com/2072-4292/11/11/1344/s1, Figure S1: A spatial difference map between SREM and LaSRC corrected images for different path/rows including (a) 122/44, (b) 170/78, (c) 201/32, (d) 199/26, (e) 15/33, and (f) 25/39. The color bar represents the spatial differences between −0.02 and +0.02; Figure S2: Map shows the "under-correction" of Sentinel-2A SR data over bright surfaces compared to the SREM SR data, i.e., Sentinel-2A SR values are greater than the TOA reflectance. No stretch/contrast is applied to the images.

Author Contributions: Conceptualization, M.B. and M.N.; methodology, M.B. and M.N.; validation, M.B., M.N., J.E.N., M.P.B., Z.Q., X.H. and E.J.; formal analysis, M.B. and M.N.; investigation, M.B., M.N., J.E.N., M.P.B., and E.J.; resources, M.B., M.N., J.E.N., M.P.B., Z.Q., E.J., J.R.C., S.L., X.H. and L.A.; data curation, M.B. and M.N.; writing—original draft preparation, M.B. and M.N.; writing—review and editing, M.B., M.N., J.E.N., M.P.B., Z.Q., E.J., J.R.C., S.L., and L.A.; funding acquisition, M.B. and Z.Q.

Funding: This research is supported by the Special Project of Jiangsu Distinguished Professor (1421061801003), the Startup Foundation for Introduction Talent of NUIST (2017r107), the National Key Research and Development Program of China by Jiangsu Chair Professorship (2016YFC1400901), and Jiangsu Provincial Programs for Marine Science and Technology Innovation (HY2017-5). Additional support came from the New Mexico State University College of Agriculture Consumer and Environmental Sciences' Agricultural Experiment Station.

Acknowledgments: The authors would like to acknowledge USGS for Landsat data, and principal investigators of AERONET sites for aerosol data. The authors also would like to thank Devin White (Oak Ridge National Laboratory) for MODIS Conversion Tool Kit (MCTK), and Vitor S. Martins (Iowa State University, USA) for data sampling procedure, and Yingjie Li (Jiangsu Normal University, China) for Sentinel-2A data processing.

Conflicts of Interest: The authors declare no conflict of interest. The funders had no role in the design of the study; in the collection, analyses, or interpretation of data; in the writing of the manuscript, or in the decision to publish the results.

Appendix A

Table A1. List of the Landsat 5 and 7 images used coincident with the ASD spectrometer data for SDSU site obtained from Maiersperger et al. (2013) [35].

Date	Image ID
2003-08-26	LE07_L1TP_029029_20030826_20160927_01_T1
2006-06-15	LE07_L1TP_029029_20060615_20160925_01_T1
2007-07-20	LE07_L1TP_029029_20070720_20160922_01_T1
2008-06-12	LT05_L1TP_029029_20080612_20160906_01_T1
2008-07-14	LT05_L1TP_029029_20080714_20160906_01_T1
2008-08-23	LE07_L1TP_029029_20080823_20160922_01_T1
2008-09-16	LT05_L1TP_029029_20080916_20160905_01_T1
2009-05-30	LT05_L1TP_029029_20090530_20160905_01_T1
2010-08-05	LT05_L1TP_029029_20100805_20160831_01_T1
2010-08-21	LT05_L1TP_029029_20100821_20160901_01_T1

References

1. Huete, A.; Didan, K.; Miura, T.; Rodriguez, E.; Gao, X.; Ferreira, L. Overview of the radiometric and biophysical performance of the MODIS vegetation indices. *Remote Sens. Environ.* **2002**, *83*, 195–213. [CrossRef]
2. Peng, G.; Ruiliang, P.; Biging, G.S.; Larrieu, M.R. Estimation of forest leaf area index using vegetation indices derived from hyperion hyperspectral data. *IEEE Trans. Geosci. Remote. Sens.* **2003**, *41*, 1355–1362. [CrossRef]
3. Zhang, R.; Qu, J.J.; Liu, Y.; Hao, X.; Huang, C.; Zhan, X. Detection of burned areas from mega-fires using daily and historical MODIS surface reflectance. *Int. J. Remote Sens.* **2015**, *36*, 1167–1187.
4. Friedl, M.A.; Sulla-Menashe, D.; Tan, B.; Schneider, A.; Ramankutty, N.; Sibley, A.; Huang, X. MODIS Collection 5 global land cover: Algorithm refinements and characterization of new datasets. *Remote Sens. Environ.* **2010**, *114*, 168–182. [CrossRef]
5. Bilal, M.; Nichol, J.E. Evaluation of MODIS aerosol retrieval algorithms over the Beijing-Tianjin-Hebei region during low to very high pollution events. *J. Geophys. Res. Atmos.* **2015**, *120*, 7941–7957. [CrossRef]
6. Bilal, M.; Nichol, J.E.; Chan, P.W. Validation and accuracy assessment of a Simplified Aerosol Retrieval Algorithm (SARA) over Beijing under low and high aerosol loadings and dust storms. *Remote Sens. Environ.* **2014**, *153*, 50–60. [CrossRef]
7. Bilal, M.; Nichol, J.E.; Bleiweiss, M.P.; Dubois, D. A Simplified high resolution MODIS Aerosol Retrieval Algorithm (SARA) for use over mixed surfaces. *Remote Sens. Environ.* **2013**, *136*, 135–145. [CrossRef]
8. Nazeer, M.; Wong, M.S.; Nichol, J.E. A new approach for the estimation of phytoplankton cell counts associated with algal blooms. *Sci. Total Environ.* **2017**, *590*, 125–138. [CrossRef] [PubMed]
9. Chavez, P.S. An improved dark-object subtraction technique for atmospheric scattering correction of multispectral data. *Remote Sens. Environ.* **1988**, *24*, 459–479. [CrossRef]
10. Smith, G.M.; Milton, E.J. The use of the empirical line method to calibrate remotely sensed data to reflectance. *Int. J. Remote Sens.* **1999**, *20*, 2653–2662. [CrossRef]
11. Richter, R. A spatially adaptive fast atmospheric correction algorithm. *Int. J. Remote Sens.* **1996**, *17*, 1201–1214. [CrossRef]

12. Nazeer, M.; Nichol, J.E.; Yung, Y.K. Evaluation of atmospheric correction models and Landsat surface reflectance product in an urban coastal environment. *Int. J. Remote Sens.* **2014**, *35*, 6271–6291. [CrossRef]
13. Matthew, M.W.; Adler-Golden, S.M.; Berk, A.; Richtsmeier, S.C.; Levine, R.Y.; Bernstein, L.S.; Acharya, P.K.; Anderson, G.P.; Felde, G.W.; Hoke, M.L.; et al. Status of atmospheric correction using a MODTRAN4-based algorithm. In *Algorithms for Multispectral, Hyperspectral, and Ultraspectral Imagery VI*; International Society for Optics and Photonics: Bellingham, WA, USA, 2000; Volume 4049, p. 199.
14. Sterckx, S.; Knaeps, E.; Adriaensen, S.; Reusen, I.; De Keukelaere, L.; Hunter, P.; Giardino, C.; Odermatt, D. OPERA: An atmospheric correction for land and water. In Proceedings of the Sentinel-3 for Science Workshop, Venice, Italy, 2–5 June 2015.
15. Frantz, D.; Roder, A.; Stellmes, M.; Hill, J. An operational radiometric landsat preprocessing framework for large-area time series applications. *IEEE Trans. Geosci. Remote. Sens.* **2016**, *54*, 3928–3943. [CrossRef]
16. Masek, J.G.; Vermote, E.F.; Saleous, N.E.; Wolfe, R.; Hall, F.G.; Huemmrich, K.F.; Gao, F.; Kutler, J.; Lim, T.-K. A Landsat surface reflectance dataset for North America, 1990-2000—IEEE Xplore Document. *IEEE Geosci. Remote Sens. Lett.* **2006**, *3*, 68–72. [CrossRef]
17. Vermote, E.; Justice, C.; Claverie, M.; Franch, B. Preliminary analysis of the performance of the Landsat 8/OLI land surface reflectance product. *Remote Sens. Environ.* **2016**, *185*, 46–56. [CrossRef]
18. Berk, A.; Bernstein, L.S.; Robertson, D.C. *MODTRAN: A Moderate Resolution Model for LOWTRAN 7*; Spectral Sciences Inc.: Burlington, MA, USA, 1989.
19. Tanré, D.; Deroo, C.; Duhaut, P.; Herman, M.; Morcrette, J.J.; Perbos, J.; Deschamps, P.Y. Description of a computer code to simulate the satellite signal in the solar spectrum: The 5S code. *Int. J. Remote Sens.* **1990**, *11*, 659–668. [CrossRef]
20. Vermote, E.F.; Tanre, D.; Deuze, J.L.; Herman, M.; Morcette, J.-J. Second simulation of the satellite signal in the solar spectrum, 6S: An overview. *IEEE Trans. Geosci. Remote. Sens.* **1997**, *35*, 675–686. [CrossRef]
21. Wilson, R.T. Py6S: A Python interface to the 6S radiative transfer model. *Comput. Geosci.* **2013**, *51*, 166. [CrossRef]
22. Kotchenova, S.Y.; Vermote, E.F.; Levy, R.; Lyapustin, A. Radiative transfer codes for atmospheric correction and aerosol retrieval: Intercomparison study. *Appl. Opt.* **2008**, *47*, 2215. [CrossRef]
23. Wilson, R.T.; Milton, E.J.; Nield, J.M. Are visibility-derived AOT estimates suitable for parameterizing satellite data atmospheric correction algorithms? *Int. J. Remote Sens.* **2015**, *36*, 1675–1688. [CrossRef]
24. Nguyen, H.; Jung, J.; Lee, J.; Choi, S.-U.; Hong, S.-Y.; Heo, J.; Nguyen, H.C.; Jung, J.; Lee, J.; Choi, S.-U.; et al. Optimal atmospheric correction for above-ground forest biomass estimation with the ETM+ remote sensor. *Sensors* **2015**, *15*, 18865–18886. [CrossRef]
25. López-Serrano, P.; Corral-Rivas, J.; Díaz-Varela, R.; Álvarez-González, J.; López-Sánchez, C. Evaluation of radiometric and atmospheric correction algorithms for aboveground forest biomass estimation using Landsat 5 TM data. *Remote Sens.* **2016**, *8*, 369. [CrossRef]
26. Lolli, S.; Alparone, L.; Garzelli, A.; Vivone, G. Haze correction for contrast-based multispectral pansharpening. *IEEE Geosci. Remote. Sens. Lett.* **2017**, *14*, 2255–2259. [CrossRef]
27. Doxani, G.; Vermote, E.; Roger, J.-C.; Gascon, F.; Adriaensen, S.; Frantz, D.; Hagolle, O.; Hollstein, A.; Kirches, G.; Li, F.; et al. Atmospheric correction inter-comparison exercise. *Remote Sens.* **2018**, *10*, 352. [CrossRef]
28. Justice, C.; Townshend, J.R.; Vermote, E.; Masuoka, E.; Wolfe, R.; Saleous, N.; Roy, D.; Morisette, J. An overview of MODIS Land data processing and product status. *Remote Sens. Environ.* **2002**, *83*, 3–15. [CrossRef]
29. Vermote, E.; Justice, C.; Csiszar, I. Early evaluation of the VIIRS calibration, cloud mask and surface reflectance Earth data records. *Remote Sens. Environ.* **2014**, *148*, 134–145. [CrossRef]
30. Muller-Wilm, U.; Louis, J.; Richter, R.; Gascon, F.; Niezette, M. Sentinel-2 Level 2A prototype processor: Architecture, algorithms and first results. In Proceedings of the ESA Living Planet Symposium, Edinburgh, UK, 9–13 September 2013.
31. Claverie, M.; Ju, J.; Masek, J.G.; Dungan, J.L.; Vermote, E.F.; Roger, J.C.; Skakun, S.V.; Justice, C. The Harmonized Landsat and Sentinel-2 surface reflectance data set. *Remote Sens. Environ.* **2018**, *219*, 145–161. [CrossRef]
32. Claverie, M.; Vermote, E.F.; Franch, B.; Masek, J.G. Evaluation of the Landsat-5 TM and Landsat-7 ETM + surface reflectance products. *Remote Sens. Environ.* **2015**, *169*, 390–403. [CrossRef]

33. Gascon, F.; Bouzinac, C.; Thépaut, O.; Jung, M.; Francesconi, B.; Louis, J.; Lonjou, V.; Lafrance, B.; Massera, S.; Gaudel-Vacaresse, A.; et al. Copernicus Sentinel-2A Calibration and products validation status. *Remote Sens.* **2017**, *9*, 584. [CrossRef]
34. Li, Y.; Chen, J.; Ma, Q.; Zhang, H.K.; Liu, J. Evaluation of Sentinel-2A surface reflectance derived using Sen2Cor in North America. *IEEE J. Sel. Top. Appl. Earth Obs. Remote. Sens.* **2018**, *11*, 1997–2021. [CrossRef]
35. Maiersperger, T.K.; Scaramuzza, P.L.; Leigh, L.; Shrestha, S.; Gallo, K.P.; Jenkerson, C.B.; Dwyer, J.L. Characterizing LEDAPS surface reflectance products by comparisons with AERONET, field spectrometer, and MODIS data. *Remote Sens. Environ.* **2013**, *136*, 1–13. [CrossRef]
36. Vuolo, F.; Żółtak, M.; Pipitone, C.; Zappa, L.; Wenng, H.; Immitzer, M.; Weiss, M.; Baret, F.; Atzberger, C. Data service platform for Sentinel-2 surface reflectance and value-added products: System use and examples. *Remote Sens.* **2016**, *8*, 938. [CrossRef]
37. Vuolo, F.; Mattiuzzi, M.; Atzberger, C. Comparison of the Landsat Surface Reflectance Climate Data Record (CDR) and manually atmospherically corrected data in a semi-arid European study area. *Int. J. Appl. Earth Obs. Geoinf.* **2015**, *41*, 1–10. [CrossRef]
38. Choi, M.; Kim, J.; Lee, J.; Kim, M.; Park, Y.-J.; Holben, B.; Eck, T.F.; Li, Z.; Song, H.H. GOCI Yonsei aerosol retrieval version 2 products: An improved algorithm and error analysis with uncertainty estimation from 5-year validation over East Asia. *Atmos. Meas. Tech.* **2018**, *11*, 385–408. [CrossRef]
39. Bilal, M.; Nichol, J.; Wang, L. New customized methods for improvement of the MODIS C6 Dark Target and Deep Blue merged aerosol product. *Remote Sens. Environ.* **2017**, *197*, 115–124. [CrossRef]
40. Bilal, M.; Nichol, J. Evaluation of the NDVI-based pixel selection criteria of the MODIS C6 Dark Target and Deep Blue combined aerosol product. *IEEE J. Sel. Top. Appl. Earth Obs. Remote Sens.* **2017**, *10*, 3448–3453. [CrossRef]
41. Sulla-Menashe, D.; Gray, J.M.; Abercrombie, S.P.; Friedl, M.A. Hierarchical mapping of annual global land cover 2001 to present: The MODIS Collection 6 Land Cover product. *Remote Sens. Environ.* **2019**, *222*, 183–194. [CrossRef]
42. Kotchenova, S.Y.; Vermote, E.F.; Matarrese, R.; Frank, J.; Klemm, J. Validation of a vector version of the 6S radiative transfer code for atmospheric correction of satellite data. Part I: Path radiance. *Appl. Opt.* **2006**, *45*, 6762–6774. [CrossRef]
43. LISE. OLCI Level 2: Rayleigh Correction Over Land (S3-L2-SD-03-C15-LISE-ATBD). Available online: https://sentinels.copernicus.eu/documents/247904/349589/OLCI_L2_Rayleigh_Correction_Land.pdf (accessed on 17 October 2018).
44. Hansen, J.E.; Travis, L.D. Light scattering in planetary atmospheres. *Space Sci. Rev.* **1974**, *16*, 527–610. [CrossRef]
45. Tanre, D.; Herman, M.; Deschamps, P.Y.; de Leffe, A. Atmospheric modeling for space measurements of ground reflectances, including bidirectional properties. *Appl. Opt.* **1979**, *18*, 3587–3594. [CrossRef]
46. Liu, C.-H.; Liu, G.-R. Aerosol optical depth retrieval for spot HRV images. *J. Mar. Sci. Technol.* **2009**, *17*, 300–305.
47. Roy, D.P.; Qin, Y.; Kovalskyy, V.; Vermote, E.F.; Ju, J.; Egorov, A.; Hansen, M.C.; Kommareddy, I.; Yan, L. Conterminous United States demonstration and characterization of MODIS-based Landsat ETM+ atmospheric correction. *Remote Sens. Environ.* **2014**, *140*, 433–449. [CrossRef]
48. Ju, J.; Roy, D.P.; Vermote, E.; Masek, J.; Kovalskyy, V. Continental-scale validation of MODIS-based and LEDAPS Landsat ETM+ atmospheric correction methods. *Remote Sens. Environ.* **2012**, *122*, 175–184. [CrossRef]
49. Roy, D.P.; Ju, J.; Kline, K.; Scaramuzza, P.L.; Kovalskyy, V.; Hansen, M.; Loveland, T.R.; Vermote, E.; Zhang, C. Web-enabled Landsat Data (WELD): Landsat ETM+ composited mosaics of the conterminous United States. *Remote Sens. Environ.* **2010**, *114*, 35–49. [CrossRef]
50. Markham, B.L.; Helder, D.L. Forty-year calibrated record of earth-reflected radiance from Landsat: A review. *Remote Sens. Environ.* **2012**, *122*, 30–40. [CrossRef]
51. Schaepman-Strub, G.; Schaepman, M.E.; Painter, T.H.; Dangel, S.; Martonchik, J.V. Reflectance quantities in optical remote sensing—Definitions and case studies. *Remote Sens. Environ.* **2006**, *103*, 27–42. [CrossRef]
52. Rouse, J.; Haas, R.; Schell, J.; Deering, D. Monitoring vegetation systems in the great plains with ERTS. In Proceedings of the Third ERTS Symposium, NASA, Washington, DC, USA, 10–14 December 1973; pp. 309–317.

53. Berterretche, M.; Hudak, A.T.; Cohen, W.B.; Maiersperger, T.K.; Gower, S.T.; Dungan, J. Comparison of regression and geostatistical methods for mapping Leaf Area Index (LAI) with Landsat ETM+ data over a boreal forest. *Remote Sens. Environ.* **2005**, *96*, 49–61. [CrossRef]
54. Curran, P.; Hay, A. The importance of measurement error for certain procedures in remote sensing at optical wavelengths. *Photogramm. Eng. Remote Sens.* **1986**, *52*, 229–241.
55. Martins, V.S.; Soares, J.V.; Novo, E.M.L.M.; Barbosa, C.C.F.; Pinto, C.T.; Arcanjo, J.S.; Kaleita, A. Continental-scale surface reflectance product from CBERS-4 MUX data: Assessment of atmospheric correction method using coincident Landsat observations. *Remote Sens. Environ.* **2018**, *218*, 55–68. [CrossRef]

© 2019 by the authors. Licensee MDPI, Basel, Switzerland. This article is an open access article distributed under the terms and conditions of the Creative Commons Attribution (CC BY) license (http://creativecommons.org/licenses/by/4.0/).

Article

A Comparison of Hybrid Machine Learning Algorithms for the Retrieval of Wheat Biophysical Variables from Sentinel-2

Deepak Upreti [1], Wenjiang Huang [2], Weiping Kong [2,3], Simone Pascucci [4], Stefano Pignatti [4], Xianfeng Zhou [5], Huichun Ye [2] and Raffaele Casa [1,*]

1. DAFNE, Università della Tuscia, Via San Camillo de Lellis, 01100 Viterbo, Italy; dupreti@unitus.it
2. Key laboratory of Digital Earth Science, Institute of Remote Sensing and Digital Earth, Chinese Academy of Sciences, Beijing 100094, China; huangwj@radi.ac.cn (W.H.); kongwp@radi.ac.cn (W.K.); yehc@radi.ac.cn (H.Y.)
3. Key Laboratory of Quantitative Remote Sensing Information Technology, Academy of Opto-Electronics, Chinese Academy of Sciences, Beijing 100094, China
4. Consiglio Nazionale delle Ricerche, Institute of Methodologies for Environmental Analysis (CNR, IMAA), Via del Fosso del Cavaliere, 100, 00133 Rome, Italy; stefano.pignatti@cnr.it (S.P.); simone.pascucci@cnr.it (S.P.)
5. College of Life Information Science and Instrument Engineering, Hangzhou Dianzi University, Hangzhou 310018, China; zhouxf@radi.ac.cn
* Correspondence: rcasa@unitus.it; Tel.: +39-0761-357555

Received: 8 February 2019; Accepted: 21 February 2019; Published: 26 February 2019

Abstract: This study focuses on the comparison of hybrid methods of estimation of biophysical variables such as leaf area index (LAI), leaf chlorophyll content (LCC), fraction of absorbed photosynthetically active radiation (FAPAR), fraction of vegetation cover (FVC), and canopy chlorophyll content (CCC) from Sentinel-2 satellite data. Different machine learning algorithms were trained with simulated spectra generated by the physically-based radiative transfer model PROSAIL and subsequently applied to Sentinel-2 reflectance spectra. The algorithms were assessed against a standard operational approach, i.e., the European Space Agency (ESA) Sentinel Application Platform (SNAP) toolbox, based on neural networks. Since kernel-based algorithms have a heavy computational cost when trained with large datasets, an active learning (AL) strategy was explored to try to alleviate this issue. Validation was carried out using ground data from two study sites: one in Shunyi (China) and the other in Maccarese (Italy). In general, the performance of the algorithms was consistent for the two study sites, though a different level of accuracy was found between the two sites, possibly due to slightly different ground sampling protocols and the range and variability of the values of the biophysical variables in the two ground datasets. For LAI estimation, the best ground validation results were obtained for both sites using least squares linear regression (LSLR) and partial least squares regression, with the best performances values of R^2 of 0.78, rott mean squared error (RMSE) of 0.68 m^2 m^{-2} and a relative RMSE (RRMSE) of 19.48% obtained in the Maccarese site with LSLR. The best results for LCC were obtained using Random Forest Tree Bagger (RFTB) and Bagging Trees (BagT) with the best performances obtained in Maccarese using RFTB (R^2 = 0.26, RMSE = 8.88 µg cm^{-2}, RRMSE = 17.43%). Gaussian Process Regression (GPR) was the best algorithm for all variables only in the cross-validation phase, but not in the ground validation, where it ranked as the best only for FVC in Maccarese (R^2 = 0.90, RMSE = 0.08, RRMSE = 9.86%). It was found that the AL strategy was more efficient than the random selection of samples for training the GPR algorithm.

Keywords: LAI; LCC; FAPAR; FVC; CCC; PROSAIL; GPR; machine learning; active learning

1. Introduction

Accurate quantitative estimation of biophysical variables is of crucial importance for different agricultural and ecological applications. Such variables include leaf area index (LAI), leaf chlorophyll content (LCC), fraction of absorbed photosynthetically active radiation (FAPAR), fractional vegetation cover (FVC), and canopy chlorophyll content (CCC). Their knowledge is valuable, for example, for precision agriculture [1], crop traits monitoring [2], and improved yield prediction and reduction of fertilizer usage [3,4]. The retrieval of vegetation biophysical variables from multispectral optical satellite data has been performed by many studies in the last decades [2,5]. In time, there have been significant developments in the algorithms used for such tasks, with a shift from empirical to more physically-based approaches. Starting from simple parametric regressions between vegetation indices and biophysical variables such as LAI and LCC [6–8], the current state of the art relies on hybrid methods exploiting at the same time radiative transfer models (RTM) and non-linear non-parametric regression algorithms [5,9,10]. The methods based on the relationship of, for example, vegetation indices and biophysical variables by the means of fitting functions, typically use the information provided by two or a few spectral bands. This limits the strength of such methods in today's scenarios where tens or even hundreds of spectral bands are available, respectively, in current super-spectral [11] or forthcoming hyperspectral [12,13] spaceborne sensors.

On the other hand, many of the approaches that have been more recently developed for the retrieval of biophysical variables exploit machine learning regression algorithms (MLRAs). These algorithms have gained wide popularity since they allow the use of more complex models, by requiring only one time-consuming training phase, but allowing their fast application any time thereafter for the retrieval [14,15]. Another advantage of these algorithms is the possibility of training them with full spectral information, thus overcoming issues of band selection or transformation, as in the case of parametric regression methods [16]. These algorithms are adaptive and have the ability to cope with the strong non-linearity inherent in remote sensing data [17,18]. Despite their advantages, these algorithms are computationally demanding, so for some MLRAs it is difficult to carry out a training phase with a large number of samples, though there are some that perform well also when trained with small datasets [19]. Some of the MLRAs are also considered black boxes, e.g., neural networks (NNs) [11,20], in which no insight is given about the physical processes linking the spectral reflectance with the biophysical variables. Others, such as partial least squares regression (PLSR) or Gaussian processes regression (GPR), can instead provide some information on what spectral bands are more relevant. In hybrid methods, MLRA can be trained using physically-based modelling approaches that describe the transfer and interaction of radiation inside the canopy based on physical laws, thus providing explicit relations between the biophysical variables and canopy spectral reflectance. When used in the direct mode, biophysical variables are used as input to the physical based RTM, which in turn simulates the top-of-canopy (TOC) spectral reflectance [9]. Using turbid-medium RTMs it is easy to generate during a short time a large simulated database, with realistic ranges of variation of biophysical variables and their corresponding simulated spectra, within the limits of the assumptions of the RTM on canopy properties. These pairs of biophysical variables and reflectance spectra can be used either in numerical optimization or in look up table (LUT) retrieval methods, by minimizing a cost function expressing the differences between observed (e.g., from satellite data) and RTM model's simulated reflectance and looking up the corresponding values of the biophysical variables [21]. For operational applications, due to the pixel-by-pixel calculations, these algorithms are computationally very demanding. These methods are also strongly affected by noise and measurement uncertainties [11]. As inversion methods of physically-based RTMs, they suffer from the limitation of the ill-posed nature of model inversion [22,23], by which different combinations of canopy variables lead to local minima having similar reflectance spectra [24].

In the hybrid methods of retrieval, the database obtained from RTM simulations can be used to train MLRA which are able to establish complex non-linear, non-parametric models linking the biophysical variables and the spectral reflectance [9,11]. Hybrid methods have found widespread application and have reached the operational stage, in particular with the application of NNs [25–27] trained with the RTM PROSAIL [28,29], such as the algorithm implemented in Sentinel Application Platform (SNAP) biophysical processor tool [29], developed by the European Space Agency (ESA). Leaf area index retrieval using NNs trained with PROSAIL is also carried out operationally for Cyclopes and Moderate Resolution Imaging Spectrometer (MODIS) products [30]. However, NN training is a delicate phase and requires the tuning of multiple parameters, which greatly impacts the robustness of the approach [18]. Thus, in recent years, increasing attention has been paid to alternatives to NNs that are simpler to train and have better potential for retrieval accuracy, such as kernel-based methods [31]. These methods solve non-linear regression problems by transferring the data to a higher dimensional space by the means of a kernel function [16]. Kernel-based algorithms have been suggested to offer some advantages in comparison to NNs. Kernel ridge regression (KRR) has proven to be simple for training and to yield competitive accuracy [18]. Gaussian process regression (GPR) is partially transparent compared to the black box nature of NNs. It allows the use of different kernel functions, ranging from simple to very complex, and it also provides uncertainty estimates with the mean value of prediction [10,32]. Gaussian process regression and KRR have been used to estimate successfully LAI and LCC [10,33]. Unfortunately, some kernel-based algorithms, such as GPR and KRR, are computationally very expensive if trained on large sets of simulations [9,10]. In order to have a general-purpose database, including a wide range of vegetation types, generally a huge number of simulations are performed [10]. However, not all pairs of the reflectance and corresponding biophysical variables will be relevant. Few attempts have been made by the researchers to optimize the simulations that are generated by the physically-based models [10,34]. Active learning (AL) has been proposed as a useful strategy to reduce the size of the RTM-generated database, to make the training of the kernel-based algorithms such as GPR more feasible [10]. Active learning is a sub-field of machine learning, also called optimal experimental design in statistics [35]. It initially starts with a small subset of the samples, and then, based on query strategies using either uncertainty or diversity measurement criteria, adds iteratively new samples to the initial training set of samples. In this way, the most informative samples in a dataset are selected, avoiding redundancy [10]. Active learning techniques have a great potential to optimally sample sets generated by RTM.

Very few studies [9,10] have reported the potential of kernel-based methods using AL techniques in relation to the retrieval of biophysical variables such as LAI, LCC, FVC, FAPAR, and CCC. In particular, although the estimation of these variables from hyperspectral data has been demonstrated [28], the suitability of multispectral satellites has yet to be fully proven, especially for biochemical variables such as LCC. The European Space Agency Sentinel-2 mission has been shown to provide data of a high radiometric quality [36] and has a higher revisit frequency than what is planned for hyperspectral satellites in the near future. Because of the availability of a larger number of spectral bands than other multispectral satellites, and of the inclusion of the red edge region of the spectrum, Sentinel-2 has a good potential for the estimation of these variables [9].

This work explores the application of kernel-based GPR using AL for biophysical variables retrieval from Sentinel-2, comparing the potential of this algorithm with other MLRAs and in particular of the version implemented operationally in SNAP.

The main objectives of this study are thus: (1) to compare the performances of different MLRAs, in particular with respect to the algorithm based on NNs implemented in the biophysical processor tool of the ESA SNAP toolbox, by using the same database of PROSAIL simulations [29]; (2) to assess the accuracy of estimation of the biophysical variables LAI, LCC, FAPAR, CCC, and FVC in the wheat crop, for kernel-based (GPR) and non-kernel-based MLRA hybrid methods of retrieval; (3) to explore the feasibility and potential of the use of AL strategies to optimally sample redundant PROSAIL simulations, to minimize computational time and complexity and allow the use of computationally demanding MLRAs (such as GPR) in hybrid methods.

2. Materials and Methods

The methodology adopted in this work (Figure 1), for the comparison of hybrid methods of biophysical variables retrieval, employed simulations carried out with the model PROSAIL [37] for training MLRA and finally applying them to Sentinel-2 data for assessment with ground measurements collected in two study areas, respectively in Italy and China. The procedure involved the following steps, illustrated in detail in the following sections: (1) generation of simulations with the model PROSAIL; (2) training MLRAs with model simulations and cross-validation; (3) Sentinel-2 data pre-processing; and (4) validation using ground data.

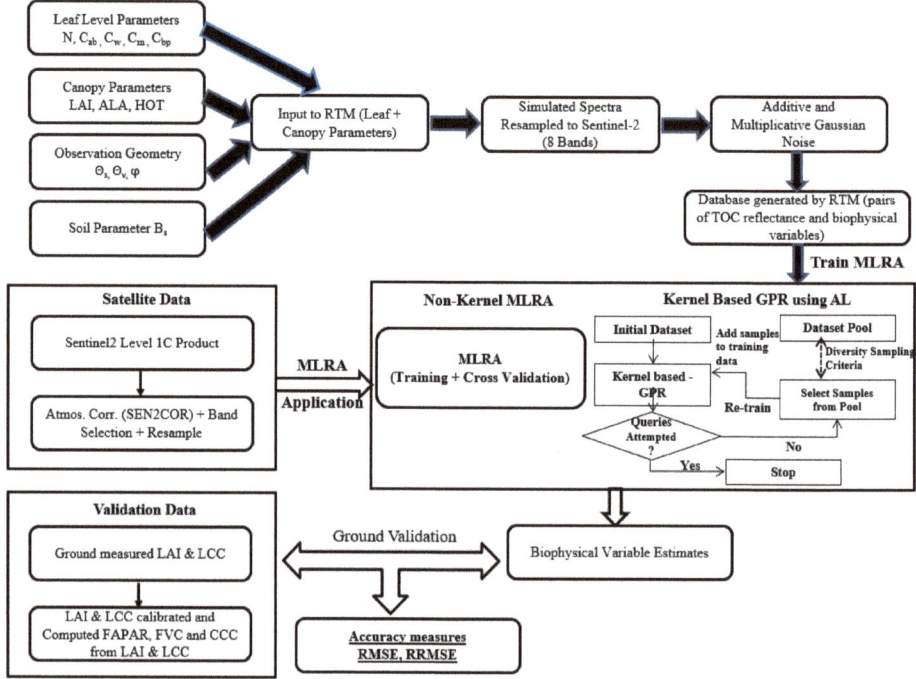

Figure 1. Flow-chart of the methodology used for the present study. Retrieval of biophysical variables: Leaf area index (LAI), Leaf chlorophyll content (LCC), Fraction of vegetation cover (FVC), Fraction of absorbed photosynthetically active radiation (FAPAR) and Canopy chlorophyll content (CCC). Radiative Transfer Model (RTM) parameters abbreviations are reported in Table 1. The RTM was used to train Machine Learning Algorithms (MLRA) including Gaussian Process Regression (GPR) using Active Learning (AL). Accuracy was assessed using the root mean squared error (RMSE) and relative RMSE (RRMSE).

2.1. Generation of PROSAIL Simulations

The model PROSAIL [37] is a widely used RTM for generating realizations of TOC canopy reflectances by considering measured biophysical variables of the crop of interest. It is a combination of the leaf level model PROSPECT [38] and canopy level model SAIL [39]. It has been inverted over a large range of vegetation canopies [40–46] with varying degrees of success. Reflectance and transmittance of a leaf from the PROSPECT model was used as an input for the SAIL model with the soil optical properties and illumination and observation geometry.

The benchmark against which the different hybrid methods were assessed in this study was the NN algorithm implemented in the ESA SNAP Sentinel-2 Toolbox, as biophysical processor. For this reason, PROSAIL simulations were generated using the same input parameter values and configuration as used for SNAP (M.Weiss personal communication), by following the Sentinel-2 Toolbox level 2 products Algorithm Theoretical Basis Document (ATBD) [29], with the exception of the geometry of observation and illumination which was set only for the conditions of our validation dataset (Section 2.3). The PROSAIL parameters were sampled using a fully orthogonal experimental plan [47], using a component of the BV-NNET (Biophysical Variables Neural Network) tool developed by Reference [48]. This procedure consists of identifying classes of values for each variable. Then all the combinations of classes are sampled once. Finally, the actual values of each variable are randomly drawn within the range of variation defined by the corresponding class, according to the distribution law specified for the variable considered. This process allows to take into account all the interactions among parameters, while having the range of variation for each variable densely and near randomly populated [29].

Table 1 reports the parameter ranges, statistical distribution, and number of classes of the PROSAIL model that were used in the present work, which were the same as those employed for training NNs in SNAP (M. Weiss personal communication). A total of 41,472 simulations were generated, representing an extensive range of vegetation properties.

The PROSAIL model top of the canopy (TOC) full spectra (at 1 nm resolution) were then resampled, using Sentinel-2 spectral response functions, to the same eight bands as used in the Sentinel-2 level 2 products ATBD [29], i.e., bands 3 to 7, 8a, 11, and 12, excluding the bands most affected by atmospheric attenuation. The spectral resampling was carried out according to Equation (1):

$$\rho = \sum_i^N \left[\frac{\beta(\lambda)}{\sum_i^N \beta(\lambda)} \right] * \rho(\lambda) \quad (1)$$

where ρ and $\rho(\lambda)$ are, respectively, the resampled Sentinel-2 reflectance and the PROSAIL reflectance at wavelength λ. $\beta(\lambda)$ represents the weight of the band's spectral response function of the Sentinel-2 MSI sensor, whereas i and N are, respectively, the minimum and maximum wavelengths of the Sentinel-2 band. Similarly, a Gaussian noise model was used to better describe actual Sentinel-2 characteristics as well as the ability of the RTM used to represent actual reflectance. The noise was computed as follows [29]:

$$R^*(\lambda) = R(\lambda)(1 + MD(\lambda) + MI)/100) + AD(\lambda) + AI \quad (2)$$

where $R(\lambda)$ and $R^*(\lambda)$ are the raw simulated reflectance and reflectance contaminated with noise, respectively. Multiplicative wavelength dependent noise and multiplicative wavelength independent noise are represented as MD and MI. Similarly, AD and AI are the additive wavelength dependent noise and independent noise, respectively. A value of 0.01 was used for AD and AI and a value of 2% was used for MD and MI for all the bands.

Table 1. Input parameter value ranges, number of classes, and statistical distributions used for generating the training database with the PROSAIL model.

	Variable	Units	Min	Max	Mode	SD	Class	Law
Canopy	Leaf area index (LAI)	-	0	15	2	2	6	log_normal
	Avearge leaf angle (ALA)	-	30	80	60	20	3	gaussian
	Hot spot parameter (HsD)	-	0.1	0.5	0.2	0.5	1	gaussian
	Leaf structure index (N)	-	1.2	1.8	1.5	0.3	3	gaussian
Leaf	Leaf chlorophyll content (LCC)	$\mu g\ cm^{-2}$	20	90	45	30	4	gaussian
	Leaf dry matter content (Cdm)	$g\ cm^{-2}$	0.003	0.011	0.005	0.005	4	gaussian
	Leaf water content (Cw_Rel)	-	0.6	0.85	0.75	0.08	4	uniform
	Brown pigments (Cbp)	-	0	2	0	0.3	3	gaussian
Soil	Soil brightness (Bs)	-	0.5	3.5	1.2	2	4	gaussian

2.2. Training Machine Learning Algorithms

Some of the most widely-used parametric and non-parametric regression algorithms were trained (or calibrated) with the simulations generated by the PROSAIL model, including linear regression, non-linear regression, hierarchical tree-based approaches, and kernel-based algorithms (Table 2).

The non-kernel-based regression methods evaluated in this study were: least squares linear regression (LSLR), partial least square regression (PLSR), NNs, regression trees (RegT), bagging trees (BagT), boosting trees (BooT), random forest fit ensemble (RFFE), and random forest tree bagger (RFTB). We also assessed the kernel-based MLRA GPR, by combining it with an AL strategy. A brief description and the references where these methods are described in detail are reported in Table 2.

Table 2. Summary of the algorithms compared in this study.

Algorithm	Brief Description	References
LSLR	Least Square Linear Regression relies on the linear relationship between explanatory variables, parameters, and the output variable.	[49]
PLSR	Partial Least Square Regression performs the regression on the projections generated using partial least squares approach.	[50]
NN	Neural Networks is an architecture composed of multiple layers of artificial neurons. For this study, hyperbolic tangent function was used and NN architecture was optimized with the Levenberg–Marquardt learning algorithm with a loss function.	[5,18,51]
RegT	Regression Trees was initially a classical approach of decision trees, in which sorting and grouping methods were added to model non-linear relationships.	[52]
BooT	Boosting Trees works by sequentially applying a classification algorithm to reweighted versions of the training data and then taking a weighted majority vote of the sequence of the classifiers thus produced.	[5,53]
BagT	Bagging Trees is a method for generating multiple versions of a predictor and using them to obtain an aggregated predictor.	[54]
RFFE	Random Forest Fit Ensemble is an ensemble to decision tree-based approach for improving the prediction accuracy, such that each tree depends on the values of a random vector sampled independently and with the same distribution for all trees in the forest.	[55]
RFTB	Random Forest Tree Bagger combines the bagging approach with the generalized approach of random forest of aggregating multiple decision trees.	[54,55]
GPR	Gaussian Process Regression are non-parametric kernel-based probabilistic models, based on Gaussian processes. They are described as a collection of random variables, any finite number of which have a joint Gaussian distribution. GPR also provides uncertainty estimates with the mean value of prediction.	[56,57]

The simulation dataset of 41,472 spectra was used for training the different algorithms. Because of the long processing time required for the training, and of the necessity of having replicates in order to carry out a statistical analysis of the results, 10 subsets of 2500 samples were extracted randomly (with replacement) from the PROSAIL simulated spectra dataset. All the MLRA were trained (and subsequently validated) ten times with the subsets of 2500 simulations each, except for GPR. Since the latter was more computationally demanding, we adopted an AL procedure for selectively further reducing the size of the training set.

Active learning (AL) is a technique which starts from a small set of initial training data and optimally selects samples from a larger data pool, based on different criteria, such as uncertainty or diversity indicators. Samples from the data pool are picked up and added to the initial training dataset, based on diversity criteria. Selecting samples by diversity ensures that added samples are dissimilar from those already accounted for. To identify the best method for selecting the samples from the data pool, we have used three different diversity-based criteria which performed as the best in a previous study [10]. These are angle-based diversity (ABD), Euclidean distance-based diversity (EBD), and cluster-based diversity (CBD). In the implementation used in this work, these criteria were applied also to the label information (i.e., the biophysical variables), since this was available in the data pool (Verreslt J., personal communication). Each process in AL was started with 50 samples as initial training data and the remaining 2450 samples in a data pool. Subsequently, a maximum of 50 samples were added per iteration, with stopping criteria of 100 iterations or a RMSE decrease lower than 50%. Only when performance was improved were the samples added. The whole AL process was repeated 10 times starting from the different subsets of 2500 records each.

In the comparison of the different diversity-based criteria, the full training set of 2500 samples was also used as a reference. Following this assessment, since EBD was the best performing method (see Results section), we only used EBD to further train and validate GPR with AL.

For all the algorithms described above, a cross-validation was performed with a k-fold strategy, where k is the number of folds with a value of 10.

The implementation of all the MLRAs was performed with the Matlab ARTMO Toolbox version 3.24 [58].

2.3. Validation with Sentinel-2 Data and Ground Measurements

Once the regression algorithms described in the previous section were trained using the PROSAIL simulations, they were applied to Sentinel-2 data acquired over two sites, Maccarese (Central Italy) and Shunyi (China). For both sites, cloud-free Sentinel-2 images were selected for the closest date to ground measurement collection (Table 3). Only the same 8 Sentinel-2 bands that had been used to resample PROSAIL simulations, i.e., bands 3 to 7, 8a, 11, and 12, were used. This band selection was established as optimal for biophysical variables retrieval, by reducing the noise introduced by atmospheric effects [29,59]. The images were initially processed for atmospheric correction with the Sen2cor tool implemented in SNAP (version 6.0) to obtain level 2A TOC reflectance and then resampled to a 10-m pixel size.

Maccarese (41.833° lat. N, 12.217° long. E, alt. 8 m a.s.l.) is located on the west coast of Central Italy, near Rome. It is a private farm of 3200 ha in a flat area with large fields. Field campaigns to measure biophysical variables on the durum wheat (*Triticum durum* Desf.) crop were carried out for this location from January 2018 to April 2018, for different growing conditions of the crop of interest, at dates close to Sentinel-2 acquisitions. The variables were sampled according to an elementary sampling unit (ESU) scheme, to capture the variability within and among different fields. Each ESU consisted of a quadrat of 20 m by 20 m size, to easily accommodate the Sentinel-2 10-m pixel resolution. A total of 15 ESUs, placed at different locations were employed at different sampling dates. Each of the ESUs contained nine points, where LAI and LCC measurements were collected. Leaf area idex was measured using the LAI-2000 or the LAI-2200C Plant Canopy Analyzer (LI-COR, Lincoln, NE, USA). Since LAI 2000/2200C measurements were not always acquired under diffuse light conditions, data were pre-processed by applying the recommended scattering correction procedure [60]. Four readings were taken above the canopy, with and without diffused cap and six readings below the canopy with 90° cap. While processing for scattering corrections five angles of directions 7°, 23°, 38°, 53°, and 68° were considered, i.e., LAI measurements including the extreme rings.

Table 3. Sentinel-2 acquisitions and ground measurements dates.

Site	Date-Ground Measurements	Date-S2 Acquisition	Difference (Days)
Maccarese, Italy	31 January 2018	29 January 2018	2
	16 February 2018	13 February 2018	3
	6 April 2018	6 April 2018	0
	20 April 2018	19 April 18	1
Shunyi, China	7–8 April 2016	10 April 2016	2
	20–21 April 2016	23 April 2016	2
	3–5 May 2016	3 May 2016	0
	18–19 May 2016	14 May 2016	4

Chlorophyll measurements were obtained using the Force-A Dualex leaf clip reader on the top-most leaves. The calibration of Dualex on durum wheat was previously performed [61] with an estimation accuracy with root mean square error (RMSE) values ranging between 7 and 11 µg cm^{-2}. The following equation [61] was used to recalibrate and compute LCC from the Dualex measured readings

$$LCC = -3.12 + 1.55 \times (Dualex) \tag{3}$$

LAI 2000/2200C measurements were also used to derive other biophysical variables, i.e., FVC and FAPAR. FVC was obtained from the LAI values measured on the ground by using the following equation [62]:

$$FVC = \left(1 - e^{-0.43 \times LAI}\right)^{0.52} \tag{4}$$

Previous studies have shown that the instantaneous FAPAR value at 10:00 (or 14:00) solar time is very close to the daily integrated value under clear sky conditions [29,48]. A Matlab script was written to compute FAPAR from the raw LAI-2200 data measured on the ground. The script extracts time dimensions, angles, and gaps fraction data from raw data files, computes day of the year and calculates FAPAR at 10:00 hour solar time. The CCC was estimated by multiplying LAI with chlorophyll content at leaf level.

A dataset of ground measurements corresponding to Sentinel-2 data was also collected in the Shunyi district (40.130 lat. N, 116.654 long. E), Beijing, in China. Ground campaigns were carried out in the period between April 2016 and May 2016 on the winter wheat (*Triticum aestivum* L.) crop. Leaf area index was measured using Licor LAI 2000/2200C and chlorophyll was measured using Force-A Dualex readings and calibrated using laboratory analysis. From April 2016 to May 2016, at four different dates, with 24 observations per date, LAI and LCC were collected in different wheat fields. The FVC was computed as mentioned using Equation (4). Due to the unavailability of raw LAI data, FAPAR was not computed for this test site.

The ground validation dataset was thus composed of 45 samples for Maccarese (each one the average of an ESU) and 96 samples for Shunyi.

2.4. Comparative Assessment of MLRAs

We used r^2, root mean square error (RMSE), and relative RMSE (RRMSE %) to assess the accuracy of the retrievals for both the k-fold cross validation and the ground validation. The RRMSE was used to compare the performances across different MLRAs and variables [18]. In general, the performances of the models were estimated by comparing the differences among RRMSE of the estimated and measured values of different biophysical variables. Lower RMSE values corresponded to higher accuracy of the algorithms for the retrieval of biophysical variables from Sentinel-2.

To assess the presence of statistically significant differences among the performances of the models, we applied a Friedman test to the computed metrics [63]. Whenever the Friedman test revealed significant differences ($p < 0.001$) among the metric means, which indicated the presence of significant differences among the models, a Friedman's aligned ranks post-hoc test, followed by Benjamini/Hochberg (non-negative) adjusted for *p*-values, was performed for multiple pairwise comparisons.

3. Results

The full simulation dataset generated with PROSAIL was used as a source of subsets for training the different MLAs, as described in the methods section, except for GPR for which a data reduction procedure based on AL was employed. Preliminarily, for GPR we compared three different diversity-based criteria used to perform AL, with respect to the full training with 2500 samples. Figure 2 shows the comparison of the different methods of sample selection in terms of R^2 and RMSE with respect to the number of iterations, for all the variables, applied to the Maccarese dataset. Only when the performance was improved were the samples added. This explains why fewer samples were added than potentially possible, i.e., less than 50 samples were added for each iteration (e.g., in Figure 2a the final sample size for CBD was of 1347 at 75 iterations).

The Euclidean distance-based diversity metric EBD surpassed ABD and CBD, achieving higher r^2 and lower RMSE values, i.e., higher accuracy and lower error, with a lower number of samples and iterations. The best performing EBD was thus subsequently used in all further applications of AL in the present work. It should be noted that the training of GPR with the full set of 2500 samples, i.e., without AL, provided worse performance than AL in all cases except for FAPAR, despite using a larger dataset for training, revealing some redundancy in the full set.

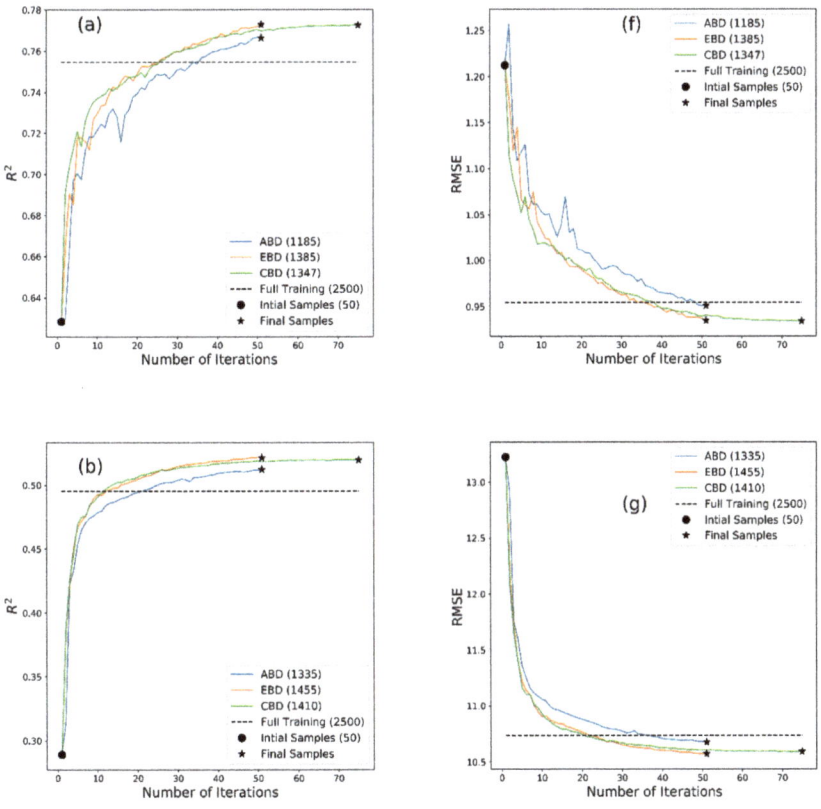

Figure 2. *Cont.*

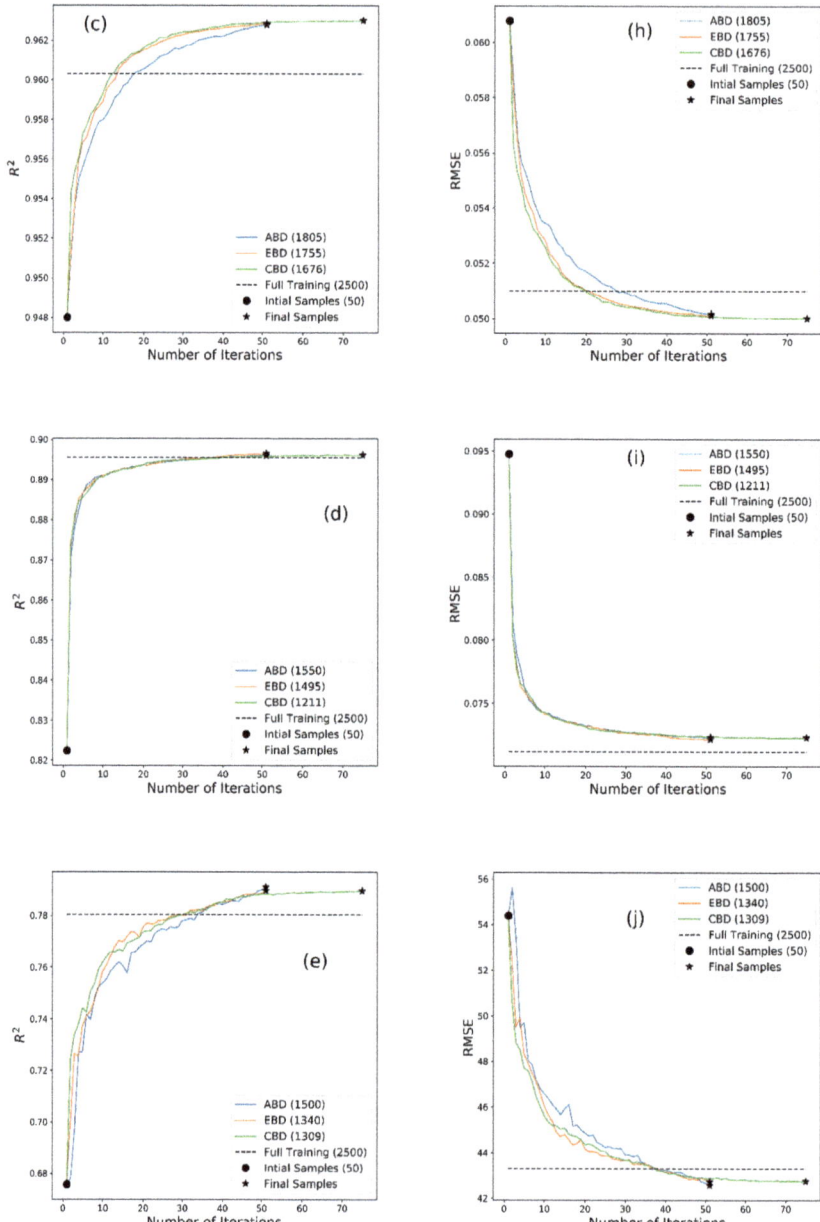

Figure 2. Comparison of three diversity criteria of data pool samples selection in AL, i.e. angle-based diversity (ABD), Euclidean distance-based diversity (EBD), and cluster-based diversity (CBD) for the different biophysical variables for the Maccarese dataset. (**a,f**) LAI; (**b,g**) LCC; (**c,h**) FVC (**d,i**) FAPAR; (**e,j**) CCC. The values in parentheses in the legend indicate the final size of the training dataset after the Active learning process. The star symbol indicates the final samples after reaching the AL stopping criteria (see text). The horizontal dashed line indicates the performance of the GPR training with the full dataset of 2500 samples without AL.

In the comparison to all the algorithms tested, the cross-validation metrics give an indication of their performance, though the actual accuracy can be more realistically assessed using ground validation results. Tables 4 and 5 report the summary of RMSE estimates for both cross-validation and ground validation for the Maccarese and Shunyi datasets, respectively, for all the algorithms tested.

Table 4. Mean RMSE values of the 10 replicates for the estimation of LAI, LCC, CCC, FVC, and FAPAR using different machine learning algorithms (MLRAs) for cross-validation (CV) and ground validation (GV), for the Maccarese test site. Abbreviations for the MLRAs are reported in Table 2. Values sharing the same letters are not significantly different according to Friedman's aligned rank post-hoc test. The best results for each variable are highlighted in bold.

MLRA	LAI		LCC		CCC		FVC		FAPAR	
	CV	GV	CV	GV	CV	GV	CV	GV	CV	GV
BagT	0.99 [b]	1.01 [d]	11.4 [c]	8.90 [a]	45.83 [a,b]	46.12 [b]	0.05 [c]	**0.08** [a]	0.08 [c]	0.11 [a]
NN (ARTMO)	1.04 [c]	0.90 [b]	11.02 [b]	10.68 [b,c]	46.53 [b]	52.1 [c]	0.05 [b]	0.09 [b,c]	0.07 [b]	0.11 [a]
RFTB	1.00 [b]	0.93b [c]	11.55 [c,d]	**8.88** [a]	46.47 [b]	44.71 [b]	0.06 [d]	**0.08** [a]	0.08 [c]	**0.10** [a]
PLSR	1.18 [e]	**0.69** [a]	11.16 [b]	11.6 [c,d]	53.25 [c]	**40.44** [a]	0.06 [e]	0.13 [d]	0.09 [e]	0.12 [b]
LSLR	1.18 [e]	**0.68** [a]	11.15 [b]	12.05 [d]	53.14 [c]	**40.86** [a]	0.06 [e]	0.13 [d]	0.09 [e]	0.12 [b]
BooT	1.14 [d]	0.98 [c,d]	11.61 [d]	10.94 [b,c]	52.08 [c]	57.51 [c,d]	0.06 [f]	0.09 [b]	0.08 [d]	0.11 [a]
RegT	1.30 [f]	1.17 [e]	14.99 [e]	11.63 [c,d]	60.26 [d]	61.22 [d,e]	0.07 [g]	0.09 [b,c]	0.10 [f]	0.12 [b]
RFFE	1.40 [g]	1.44 [f]	16.03 [f]	13.18 [d]	64.36 [d]	68.32 [e]	0.08 [h]	0.10 [c]	0.11 [g]	0.12 [b]
GPR + AL	**0.94** [a]	1.31 [f]	**10.55** [a]	10.22 [b]	**43.81** [a]	54.56 [c]	**0.05** [a]	**0.08** [a]	**0.07** [a]	0.11 [a]
NN (SNAP)		0.74		10.94		44.61		0.108		0.105

Table 5. Mean RMSE values of 10 replicates for the estimation of LAI, LCC, CCC, FVC and FAPAR using different machine learning algorithms (MLRA) for cross-validation (CV) and ground validation (GV), for the Shunyi test site. Abbreviations for the MLRAs are reported in Table 2. Values sharing the same letters are not significantly different according to Friedman's aligned ranks post-hoc test. The best results for each variable are highlighted in bold.

MLRA	LAI		LCC		CCC		FVC	
	CV	GV	CV	GV	CV	GV	CV	GV
BagT	0.99 [b]	1.74 [c,d]	11.67 [d]	**17.28** [a]	46.12 [b]	**56.72** [a]	0.05 [c]	0.19 [c]
NN(ARTMO)	1.02 [b]	1.42 [b]	11.27 [b]	24.41 [c]	46.52 [b]	98.14 [e]	0.05 [b]	0.17 [a]
RFTB	1.00 [b]	1.67 [c]	11.8 [e]	**16.77** [a]	46.69 [b]	**56.51** [a]	0.05 [d]	0.19 [c]
PLSR	1.18 [d]	**1.01** [a]	11.43 [c]	24.77 [c]	52.94 [c]	68.11 [c]	0.06 [f]	0.18 [b]
LSLR	1.18 [d]	**1.00** [a]	11.43 [b,c]	25.98 [b,c]	52.86 [c]	74.33 [d]	0.06 [f]	0.18 [b]
BooT	1.12 [c]	1.46 [b]	11.91 [e]	25.91 [d]	51.69 [c]	94.77 [e]	0.05 [e]	0.18 [b]
RegT	1.32 [e]	1.76 [d]	15.26 [f]	20.75 [b]	60.63 [d]	62.71 [b]	0.07 [g]	0.21 [d]
RFFE	1.42 [e]	2.14 [e]	16.36 [g]	25.62 [b]	64.29 [d]	76.96 [d]	0.07 [h]	0.21 [d]
GPR + AL	**0.93** [a]	1.98 [e]	**10.79** [a]	22.75 [b]	**42.55** [a]	61.35 [b]	**0.05** [a]	0.18 [b]
NN (SNAP)		1.16		23.61		123.53		0.18

For Maccarese (Table 4), LSLR and PLSR were the best performing algorithms for LAI retrieval when validated with ground data, in terms of RMSE, with values of 0.68 ($R^2 = 0.78$, RRMSE = 19.48%) and 0.69 ($R^2 = 0.78$, RRMSE = 19.84%) respectively. These were also the fastest computing algorithms among all the MLRAs in terms of time required for training (data not shown). The cross-validation and ground validation results (Figure 3a) indicated that LSLR apparently did not show saturation, even up to LAI values of 5 or 6. The retrieval of LCC was best performed by RFTB (Figure 3b) and BagT, with RMSE values of 8.88 µg cm^{-2} ($R^2 = 0.26$ and RRMSE = 17.43%) and 8.90 µg cm^{-2} ($R^2 = 0.27$ and RRMSE = 17.46%). An overestimation of LCC was observed (Figure 3b). The lowest RMSE of 40.44 g cm^{-2} ($R^2 = 0.74$ RRMSE = 22.7%) (Figure 3e), was obtained for the retrieval of CCC with PLSR. The retrieval accuracy of LSLR was not significantly different from that of PLSR for CCC (RMSE = 40.86 g cm^{-2}, $R^2 = 0.74$ and RRMSE = 22.9%). Indeed, PLSR was not significantly different from LSLR in terms of RMSE for all the variables considered. The lowest RMSE for FVC was 0.08 ($R^2 = 0.9$ and RRMSE = 9.8%), obtained with the GPR algorithm (Figure 3c). For FVC, RFTB and BagT were also not significantly different from GPR in terms of RMSE. The FAPAR was best retrieved with a RMSE of 0.10 ($R^2 = 0.41$ and RRMSE = 12.06%) using the RFTB approach (Figure 3d), although GPR was not significantly different in terms of RMSE, which was only slightly higher. It was observed that, even in the best performing algorithms, an underestimation of low values of FVC and FAPAR was apparent (Figure 3c,d). It should also be noted that these latter variables show a smaller range of variation in the measured values as compared to the other biophysical variables. As can be seen, GPR was always the best performing algorithm in the cross-validation results, but this was not confirmed by the ground validation tests, with the exception of FVC and FAPAR estimation.

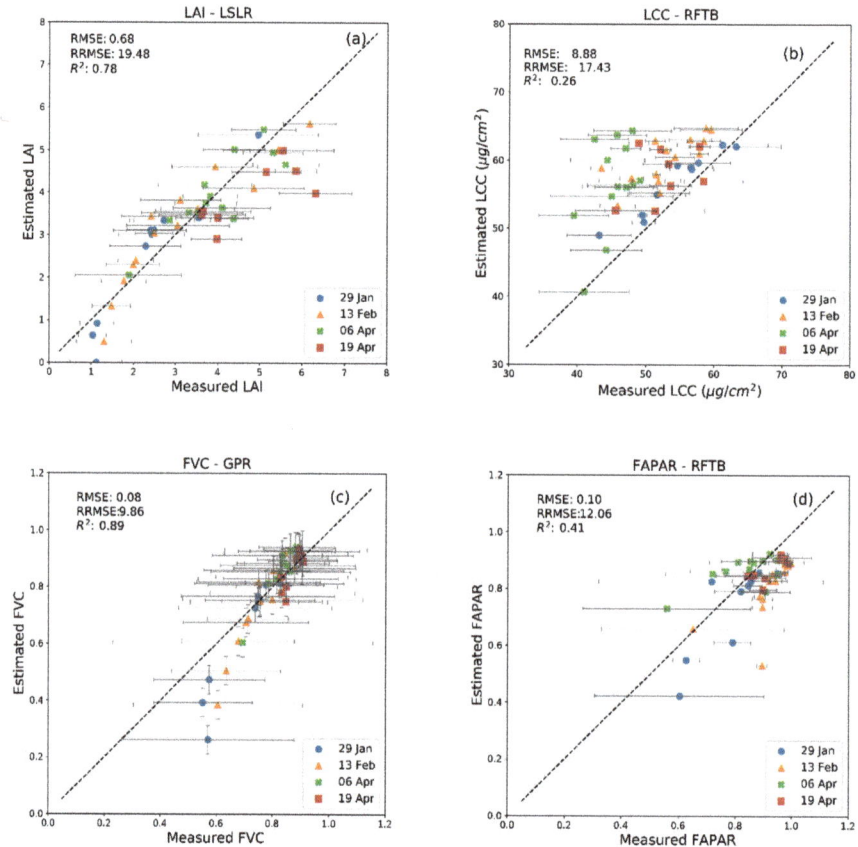

Figure 3. *Cont.*

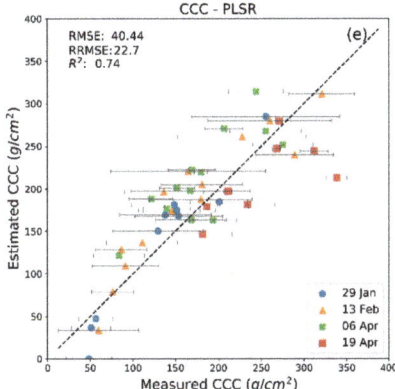

Figure 3. Best performing algorithms for the retrieval of (**a**) LAI, (**b**) LCC, (**c**) FVC, (**d**) FAPAR, and (**e**) CCC for the Maccarese test site. The horizontal error bars represent the standard deviation of measurements inside each ESU. Vertical error bars represent uncertainty estimates using GPR for FVC (**c**).

As can be observed from the horizontal error bars of Figure 3, considerable variability occurred among ground measurements inside ESUs, despite the visually apparent homogeneity of the crop canopy in the sampled area of each ESU.

For the Shunyi datasets (Table 5), the ground variability could not be reported for LAI, for which single-point measurements were carried out, but an estimate of the variability for chlorophyll could be made since multiple measurements were available for each point (Figure 4b).

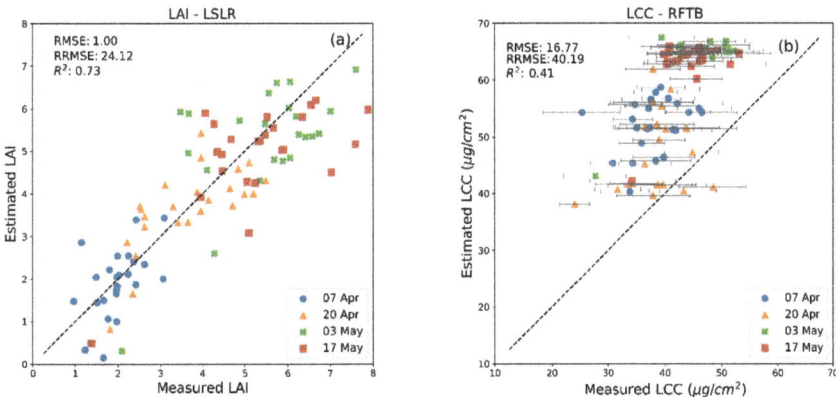

Figure 4. *Cont.*

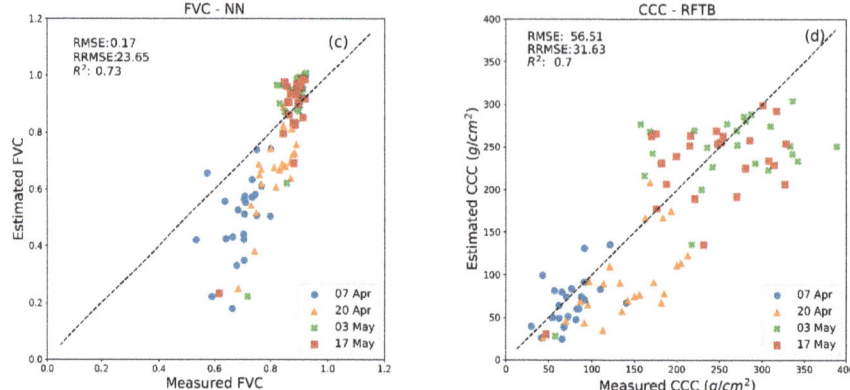

Figure 4. Best performing algorithms for (**a**) LAI, (**b**) LCC, (**c**) FVC, and (**d**) CCC retrieval for the dataset from Shunyi, China. The horizontal error bars for LCC represent the standard deviation of single chlorophyll measurements.

The best accuracy for LAI retrieval for the Shunyi site was a RMSE of 1.00 m^2 m^{-2} (R^2 = 0.73 and RRMSE = 24.12%) obtained with LSLR (Table 5), i.e., the same algorithm providing the best results also for Maccarese. For LAI, PLSR had a similar performance as that of LSLR, with a RMSE = 1.01 m^2 m^{-2} and (R^2 = 0.72 and RRMSE = 24.50%), i.e., also consistently with the results of LAI retrieval for Maccarese. Similarly, for Maccarese, LCC was best retrieved with an RMSE value of 16.77 µg cm^{-2}, (R^2 = 0.41 and RRMSE = 40.19%) using RFTB (Table 5). Although RFTB produced the lowest error for retrieving LCC, its retrieval using BagT approach was not significantly different, with RMSE = 17.28 µg cm^{-2}, R^2 = 0.41 and RRMSE = 41.41%. For CCC retrieval, the lowest RMSE, 56.51 g cm^{-2} (R^2 = 0.7, RRMSE = 31.63%), was achieved by the RFTB method. In this case, BagT was not significantly different from RFTB (R^2 = 0.71, RMSE = 56.72 and RRMSE = 31.75%). For this test site, FVC was retrieved with an RMSE of 0.17 (R^2 = 0.73 and RRMSE = 23.65%) with NNs (Figure 4c). Also, for the Shunyi test site, GPR was always the best performing algorithm in cross-validation but not in ground validation tests.

A summary of the results in terms of RRMSE for all the biophysical variables tested with different MLRAs for both sites is presented in Figure 5.

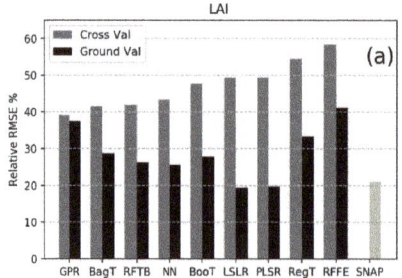
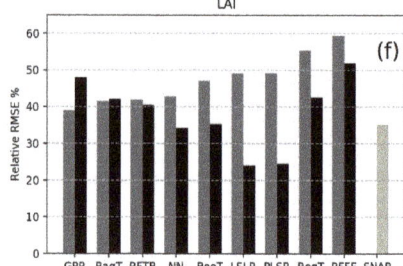

Figure 5. *Cont.*

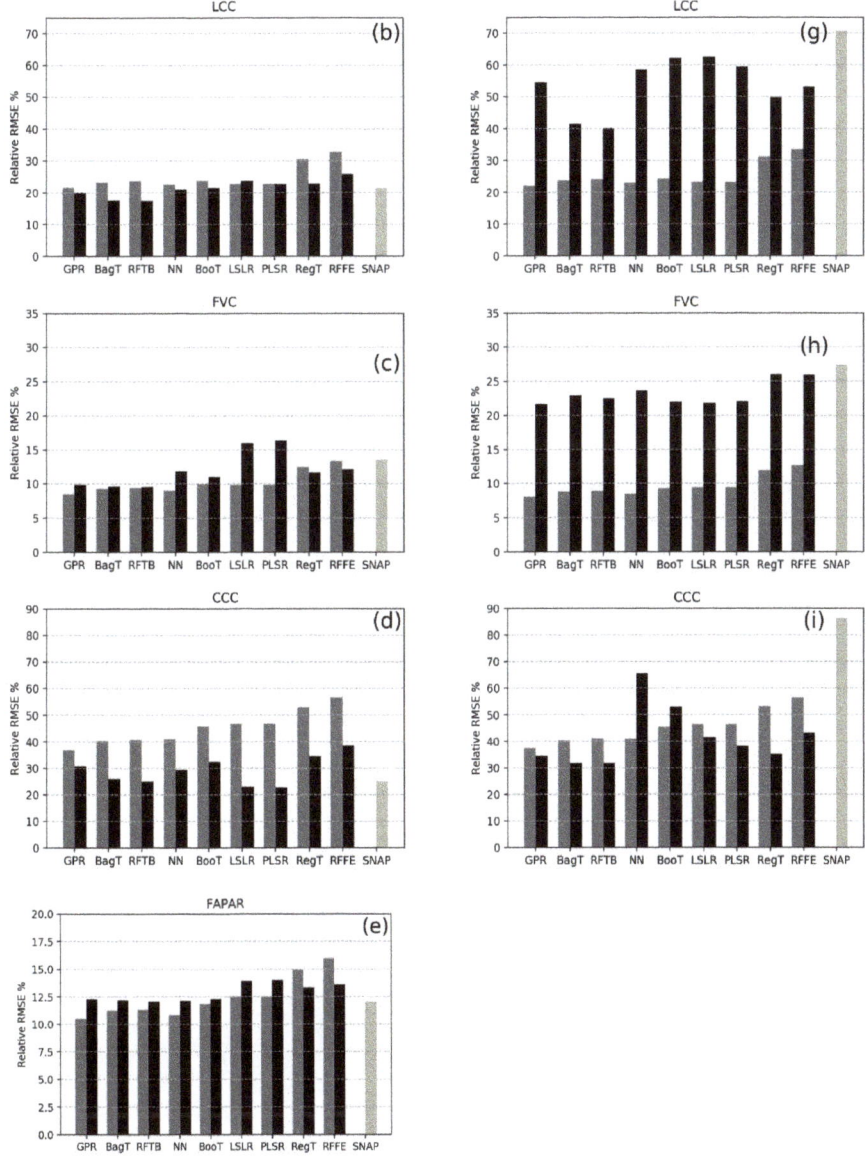

Figure 5. RRMSE (%) for LAI, LCC, FVC, CCC, and FAPAR for Maccarese, Italy (**left**) and LAI, LCC, FVC, and CCC for Shunyi, China (**right**).

It can be seen from Figure 5, that the k-fold cross-validation (k = 10) provided in some circumstances worse results (higher RRMSE) than the validation with ground data (e.g., Figure 5a). This is particularly evident for LAI and CCC for the Maccarese site. Usually, since it was carried out with subset of the same dataset as used for the training of the MLRAs, the cross-validation error was lower than the ground validation error, but this was not the case here, as can be seen from Figure 5a–e. The different statistical distributions of training and ground validation data might be a possible reason. It appears that the range of variation of the training sets was much larger than that of the ground validation sets, with more extreme values. This could have possibly led to more unreliable retrievals in the cross-validation when extreme values were sampled, which did not happen in the ground validation.

For the Shunyi test site, RRMSE values were generally higher than for the Maccarese (Figure 5), probably because of the larger differences in the statistical distributions of the ground validation dataset from the training set, as compared for the Maccarese and possibly for the larger error introduced by the sampling protocol.

In general, for LAI, algorithms such as LSLR and PLSR, seemed to outperform more complex MLRAs such as NNs, for example providing better results than those found by applying the biophysical processor implemented in the ESA SNAP toolbox.

For leaf chlorophyll (LCC), much worse results were obtained for Shunyi than for Maccarese, with a clear advantage of RFTB for both sites for ground validation tests. The performance of the NN implementation of SNAP was particularly poor for Shunyi, though it should be noted that this variable was back-calculated from the CCC variable generated by SNAP.

Fractional ground cover (FVC) showed the smallest error among all the biophysical variables tested, alongside FAPAR which was only available for Maccarese. Neural networks provided the best accuracies with the lowest values of relative RMSEs for the latter variable.

Although different algorithms from different families of MLRAs performed well for different biophysical variables, the GPR was considered particularly interesting, as it provided uncertainty estimates with the mean value of prediction (Figure 3c). However, despite its good cross-validation results, GPR ranked as the top algorithm only for FVC and FAPAR for the Maccarese test site (Table 4). The AL procedure only selected the most informative samples of the dataset, thus the training of GPR was performed with a smaller set of PROSAIL simulations compared to the other algorithms, showing that it was quite efficient and robust.

The time taken for training the models ranged from 0.003 seconds for LSLR to 47.5 seconds for NNs and a maximum 548.3 seconds for GPR. Although, GPR took longer to train, it should be noted that it performed AL using EBD and used fewer samples to train and achieved high accuracy in comparison to other models.

When considering single ground sampling dates separately, generally the retrieval performances were improved (Table 6) compared to the bulked data (Table 4). In some cases, e.g., LAI and CCC, the worst results were obtained at the latest date, due to the fact that it was generally more difficult to estimate higher LAI values.

Table 6. Ground validation estimation results at each ground sampling date, for the best performing algorithms for the Maccarese test site.

MLRA	29 January 2018		13 February 2018		6 April 2018		20 April 2018	
	R^2	RMSE	R^2	RMSE	R^2	RMSE	R^2	RMSE
LSLR (LAI)	0.89	0.59	0.88	0.55	0.73	0.51	0.55	1.09
RFTB (LCC)	0.91	3.01	0.55	7.75	0.25	12.56	0.11	7.03
GPR (FVC)	0.90	0.12	0.89	0.08	0.90	0.05	0.71	0.05
RFTB (FAPAR)	0.74	0.10	0.45	0.14	0.44	0.09	0.56	0.06
PLSR (CCC)	0.94	26.32	0.87	33.37	0.65	44.46	0.45	55.97

4. Discussion

There are different methods for the retrieval of biophysical variables from remote sensing data and all the methods have their pros and cons [16]. Currently, hybrid methods have the capability of combining physical and statistical methods and are considered state of the art. In this paper, different MLRAs (kernel-based and non-kernel based) were trained and applied for the retrieval of the crop biophysical variables LAI, LCC, FAPAR, FVC, and CCC with the same configuration and settings of RTM PROSAIL simulated spectra as those implemented in ESA SNAP biophysical variables retrieval toolbox, with the exception of the observation geometry. With the same of configuration of PROSAIL parameters (Section 2.1), a total of 41,472 simulated spectra was generated, which turned out to be rather redundant and inefficient for performing the training steps. Subsets of 2500 randomly extracted simulated spectra were used to perform the training of the algorithms and this procedure was repeated ten times with different subsets. This was done in order to allow a comparison of alternative MLRAs to a well-established operational algorithm [29].

The SNAP algorithm relies on the use of PROSAIL simulations for training NNs, which are the most widely-used tools for operational biophysical variables retrieval [25–28]. The downside of the NNs are that they require a relatively long time for training, tuning of parameters is a difficult task, they are black box in nature, and can be unpredictable if training and validation data deviate from each other even slightly [5]. In this paper, various alternative MLRAs outperformed NNs. For example, LAI retrieval was best performed with LSLR and PLSR, consistently for both tests, although with different accuracies, i.e., RMSE of 0.68 for Maccarese and 1 for Shunyi [5], compared with different retrieval strategies for LAI and reporting the best performing algorithms for each category. In the case of parametric regression, the best performing algorithm was Tian 3-band formulation (RMSE = 0.615 and R^2 = 0.823), whereas VH-GPR performed best among non-parametric regression algorithms (RMSE = 0.436 and R^2 = 0.902). However, it should be noted that the accuracies found by these authors, somehow better than those of the present study, could be explained by the fact that they used cross-validation in which the training was carried out using the ground dataset, not independent model simulations such as in the present work.

The LCC was best retrieved by the RFTB method for both sites, revealing a general overestimation, but the error of estimation was higher for the Shunyi test site compared to the Maccarese (Figures 3b and 4b), particularly for SNAP. Also, the error of CCC estimates was very high for the SNAP tool compared to the other algorithms tested in this work for the Shunyi ground validation (Figure 5i). It was previously shown by Reference [64] that Sentinel-2 bands at a 10 m spatial resolution are suitable for estimating LAI, LCC, and CCC. They retrieved LAI (R^2 = 0.809), LCC (R^2 = 0.696) and CCC (R^2 = 0.818) with vegetation indices approach. Multiple vegetation indices were compared for identification of potential vegetation index for the retrieval of LAI, LCC, and CCC in [65], the best correlation for LCC was with the Meris Terrestrial Chlorophyll Index (MTCI) (R^2 = 0.77) and Sentinel-2 red-edge position index (S2REP) (R^2 = 0.91), for LAI, inverted red-edge chlorophyll index (IRECI) (R^2 = 0.77) and NDI45 (R^2 = 0.62), Normalised Difference Vegetation Index (NDVI) for CCC (R^2 = 0.70). These results are better than those found in the present work, but again, empirical approaches employing the ground datasets were employed by these authors for the calibration of the models.

The FVC variable was more accurately predicted than LAI or LCC, though with slightly higher error for Shunyi. With hierarchical tree-based methods such as BagT, RFTB, and GPR, the error reached was even lower than 10%, which is in line with Global Monitoring for Environment and Security (GMES) goal accuracy [66] (Figure 5c). In the case of CCC retrieval, it can be noted that the lowest RRMSE was provided by RFTB and BagT, though also GPR performs similarly to RFTB. The FAPAR variable was only estimated for the Maccarese test site, due to the unavailability of the ground measurements for validation, it was not estimated for the Shunyi test site. The results showed that this variable was best retrieved with the RFTB method, with the lowest RMSE (Table 3).

The GPR was proved to be an efficient and powerful regressor for the biophysical variables retrieval in previous reports [18]. A study conducted by Reference [18], retrieved LAI, Chl, and FVC using different methods, such as NNs, kernel ridge regression (KRR), support vector regression, and GPR for different Sentinel-2 and Sentinel-3 configurations. They found that an overall good performance throughout all Sentinel configurations was provided by GPR. A study carried out by Reference [10] also found that GPR using AL techniques is an efficient and robust method for the retrieval of LAI and LCC from Sentinel-3 OLCI spectra. However, this was not the case in our study. The GPR method, coupled with AL procedure, provided the best results in terms of computational time and low error only for FVC and FAPAR ground validation tests (Figure 5c,e,h). On the other hand, GPR generally performed as the best algorithm for the retrieval of all the biophysical variables, only for the cross-validations tests (Tables 4 and 5). Hence, of importance for further analysis to investigate how accurately GPR performs when validated against independent ground data. As Reference [10] pointed out, GPR is based on non-parametric regression in a Bayesian framework, it provides insights in bands carrying relevant information and also in theoretical uncertainty estimates, thus partially overcoming the black box problem [5]. These uncertainties are a useful tool for the assessment of upscaling capabilities of biophysical variables from airborne or spaceborne platforms and their respective scales [32]. A study conducted by [9] introduced the AL approach with GPR and SVM regressions to deal with the problems of training sample collection for biophysical variables estimation. Their results obtained on simulated MEdium Resolution Imaging Spectrometer (MERIS) and real SeaWiFS Bio-optical Algorithm Mini-Workshop (SeaBAM) datasets were characterized by higher performances in terms of both accuracy and stability with respect to a completely random selection strategy. The present work seems to support these results highlighting the efficiency of the AL procedure, since, when using a smaller set of selected training data, comparable results were obtained than with a random selection of larger size (Figure 2).

5. Conclusions

Hybrid methods of biophysical variables retrieval rely on the generation of simulated spectra using physically-based RTM for the training of MLRAs, under the assumption of a more general applicability as compared to the training carried out using measured data, since RTMs allow to simulate a wider range of leaf and canopy properties. Operationally, so far, only NNs have been implemented as a hybrid method, e.g., in the ESA SNAP toolbox. Due to complexity of parameter tuning and the black box nature of the algorithm, in recent years, several studies have focused on alternative MLRAs such as kernel-based methods. They offer an interesting alternative to NNs in terms of performance and computational demand, sometimes partially overcoming the black box nature of NNs. In the present study, it was shown, using the same set of PROSAIL simulations, that some MLRAs could provide better results than NN-based algorithms implemented in SNAP. This was particularly evident for chlorophyll and fractional vegetation cover. It resulted that the best performing algorithms varied according to the biophysical variables to be estimated: e.g., LSLR and PLSR worked better for LAI, whereas RFTB and BagT for LCC. This study also showed that these variables' results were consistent for two different datasets respectively collected in Italy and China on wheat, though the retrieval accuracies depended somehow on errors introduced during the ground sampling protocol. The absolute values of the retrieval accuracies are in some cases higher than the best performances reported in previous studies, but most of these previous studies did not train the regression algorithms with completely independent datasets, as was done in the present study. Further studies should be conducted to investigate alternative strategies, e.g., of active learning procedures, using different kernel functions, and validate the retrievals using a wider range of agricultural crops, also assessing their performance for canopies that depart more from the assumptions of the turbid medium used in the PROSAIL RTM.

Author Contributions: D.U. implemented the methods, collected and analyzed the data for the Italian test site, and wrote the manuscript; R.C. designed the field experiments and collected the data in Italy, developed the idea and methodology, supervised its implementation and wrote the manuscript; S.P. (Simone Pascucci) and S.P. (Stefano Pignatti) collected and analyzed the data in Italy and edited the manuscript. W.H., X.Z., H.Y., and W.K. collected and analyzed the ground data for the China test site and edited the manuscript. All authors read and approved the final manuscript.

Funding: This research was funded by the European Space Agency (ESA) and the Ministry of Science and Technology of China (MOST), under the Dragon 4 Programme, Project ID32275-Topic1; it was also supported by the National Natural Science Foundation of China (41871339)and by the Italian National Research Council within the CNR-CAS Bilateral Agreement 2017–2019.

Acknowledgments: The authors gratefully acknowledge MIUR (Italian Ministry for Education, University and Research) for financial support (Law 232/216, Department of excellence) to the DAFNE Department. The authors would like to thank Marie Weiss (INRA Avignon, France) for providing the Matlab code of the BV-Net tool and the information on the SNAP biophysical processor. The authors would also like to thank Davide Tahani and Fabiola Fontana for helping in data collection and the Maccarese farm for their support and for granting access to the fields, and Jochem Verrelst for his support with the ARTMO toolbox.

Conflicts of Interest: The authors declare no conflict of interest.

References

1. Zhang, C.; Kovacs, J.M. The application of small unmanned aerial systems for precision agriculture: A review. *Precis. Agric.* **2012**, *13*, 693–712. [CrossRef]
2. Jay, S.; Maupas, F.; Bendoula, R.; Gorretta, N. Retrieving LAI, chlorophyll and nitrogen contents in sugar beet crops from multi-angular optical remote sensing: Comparison of vegetation indices and PROSAIL inversion for field phenotyping. *Field Crops Res.* **2017**, *210*, 33–46. [CrossRef]
3. Goffart, J.P.; Olivier, M.; Frankinet, M. Potato crop nitrogen status assessment to improve N fertilization management and efficiency: Past–present–future. *Potato Res.* **2008**, *51*, 355–383. [CrossRef]
4. Cilia, C.; Panigada, C.; Rossini, M.; Meroni, M.; Busetto, L.; Amaducci, S.; Boschetti, M.; Picchi, V.; Colombo, R. Nitrogen status assessment for variable rate fertilization in maize through hyperspectral imagery. *Remote Sens.* **2014**, *6*, 6549–6565. [CrossRef]
5. Verrelst, J.; Rivera, J.P.; Veroustraete, F.; Muñoz-Marí, J.; Clevers, J.G.; Camps-Valls, G.; Moreno, J. Experimental Sentinel-2 LAI estimation using parametric, non-parametric and physical retrieval methods–A comparison. *ISPRS J. Photogramm. Remote Sens.* **2015**, *108*, 260–272. [CrossRef]
6. Jordan, C.F. Derivation of leaf-area index from quality of light on the forest floor. *Ecology* **1969**, *50*, 663–666. [CrossRef]
7. Rouse, J.; Haas, R.; Schell, J.; Deering, D.; Harlan, J. Monitoring the Vernal Advancement of Retrogradation of Natural Vegetation, NASA/GSFC, Type III, Final Report. Available online: https://ntrs.nasa.gov/archive/nasa/casi.ntrs.nasa.gov/19730017588.pdf (accessed on 20 February 2019).
8. Rondeaux, G.; Steven, M.; Baret, F. Optimization of soil-adjusted vegetation indices. *Remote Sens. Environ.* **1996**, *55*, 95–107. [CrossRef]
9. Pasolli, E.; Melgani, F.; Alajlan, N.; Bazi, Y. Active learning methods for biophysical parameter estimation. *IEEE Trans. Geosci. Remote Sens.* **2012**, *50*, 4071–4084. [CrossRef]
10. Verrelst, J.; Dethier, S.; Rivera, J.P.; Muñoz-Marí, J.; Camps-Valls, G.; Moreno, J. Active learning methods for efficient hybrid biophysical variable retrieval. *IEEE Geosci. Remote Sens. Lett.* **2016**, *13*, 1012–1016. [CrossRef]
11. Verrelst, J.; Camps-Valls, G.; Muñoz-Marí, J.; Rivera, J.P.; Veroustraete, F.; Clevers, J.G.; Moreno, J. Optical remote sensing and the retrieval of terrestrial vegetation bio-geophysical properties–A review. *ISPRS J. Photogramm. Remote Sens.* **2015**, *108*, 273–290. [CrossRef]
12. Pignatti, S.; Acito, N.; Amato, U.; Casa, R.; Castaldi, F.; Coluzzi, R.; De Bonis, R.; Diani, M.; Imbrenda, V.; Laneve, G.; et al. Environmental products overview of the Italian hyperspectral prisma mission: The SAP4PRISMA project. In Proceedings of the 2015 IEEE International Geoscience and Remote Sensing Symposium (IGARSS), Milan, Italy, 26–31 July 2015.
13. Guanter, L.; Kaufmann, H.; Segl, K.; Foerster, S.; Rogass, C.; Chabrillat, S.; Kuester, T.; Hollstein, A.; Rossner, G.; Chlebek, C. The EnMAP spaceborne imaging spectroscopy mission for earth observation. *Remote Sens.* **2015**, *7*, 8830–8857. [CrossRef]
14. Durbha, S.S.; King, R.L.; Younan, N.H. Support vector machines regression for retrieval of leaf area index from multiangle imaging spectroradiometer. *Remote Sens. Environ.* **2007**, *107*, 348–361. [CrossRef]
15. Lázaro-Gredilla, M.; Titsias, M.K.; Verrelst, J.; Camps-Valls, G. Retrieval of biophysical parameters with heteroscedastic Gaussian processes. *IEEE Geosci. Remote Sens. Lett.* **2014**, *11*, 838–842. [CrossRef]
16. Verrelst, J.; Malenovský, Z.; Van der Tol, C.; Camps-Valls, G.; Gastellu-Etchegorry, J.-P.; Lewis, P.; North, P.; Moreno, J. Quantifying Vegetation Biophysical Variables from Imaging Spectroscopy Data: A Review on Retrieval Methods. *Surv. Geophys.* **2018**, 1–41. (in press). [CrossRef]
17. Knudby, A.; LeDrew, E.; Brenning, A. Predictive mapping of reef fish species richness, diversity and biomass in Zanzibar using IKONOS imagery and machine-learning techniques. *Remote Sens. Environ.* **2010**, *114*, 1230–1241. [CrossRef]
18. Verrelst, J.; Muñoz, J.; Alonso, L.; Delegido, J.; Rivera, J.P.; Camps-Valls, G.; Moreno, J. Machine learning regression algorithms for biophysical parameter retrieval: Opportunities for Sentinel-2 and -3. *Remote Sens. Environ.* **2012**, *118*, 127–139. [CrossRef]
19. Wang, L.; Zhou, X.; Zhu, X.; Dong, Z.; Guo, W. Estimation of biomass in wheat using random forest regression algorithm and remote sensing data. *Crop J.* **2016**, *4*, 212–219. [CrossRef]
20. Benítez, J.M.; Castro, J.L.; Requena, I. Are artificial neural networks black boxes? *IEEE Trans. Neural Netw.* **1997**, *8*, 1156–1164. [CrossRef] [PubMed]

21. Rivera, J.P.; Verrelst, J.; Leonenko, G.; Moreno, J. Multiple cost functions and regularization options for improved retrieval of leaf chlorophyll content and LAI through inversion of the PROSAIL model. *Remote Sens.* **2013**, *5*, 3280–3304. [CrossRef]
22. Combal, B.; Baret, F.; Weiss, M.; Trubuil, A.; Mace, D.; Pragnere, A.; Myneni, R.; Knyazikhin, Y.; Wang, L. Retrieval of canopy biophysical variables from bidirectional reflectance: Using prior information to solve the ill-posed inverse problem. *Remote Sens. Environ.* **2003**, *84*, 1–15. [CrossRef]
23. Atzberger, C. Object-based retrieval of biophysical canopy variables using artificial neural nets and radiative transfer models. *Remote Sens. Environ.* **2004**, *93*, 53–67. [CrossRef]
24. Houborg, R.; Boegh, E. Mapping leaf chlorophyll and leaf area index using inverse and forward canopy reflectance modeling and SPOT reflectance data. *Remote Sens. Environ.* **2008**, *112*, 186–202. [CrossRef]
25. Pozdnyakov, D.; Shuchman, R.; Korosov, A.; Hatt, C. Operational algorithm for the retrieval of water quality in the Great Lakes. *Remote Sens. Environ.* **2005**, *97*, 352–370. [CrossRef]
26. Schiller, H.; Doerffer, R. Improved determination of coastal water constituent concentrations from MERIS data. *IEEE Trans. Geosci. Remote Sens.* **2005**, *43*, 1585–1591. [CrossRef]
27. Verger, A.; Baret, F.; Weiss, M. Performances of neural networks for deriving LAI estimates from existing CYCLOPES and MODIS products. *Remote Sens. Environ.* **2008**, *112*, 2789–2803. [CrossRef]
28. Verger, A.; Baret, F.; Camacho, F. Optimal modalities for radiative transfer-neural network estimation of canopy biophysical characteristics: Evaluation over an agricultural area with CHRIS/PROBA observations. *Remote Sens. Environ.* **2011**, *115*, 415–426. [CrossRef]
29. Weiss, M.; Baret, F. S2ToolBox Level 2 Products: LAI, FAPAR, FCOVER. Available online: https://step.esa.int/docs/extra/ATBD_S2ToolBox_L2B_V1.1.pdf (accessed on 21 February 2019).
30. Chai, L.; Qu, Y.; Zhang, L.; Wang, J. Lai retrieval from cyclopes and modis products using artificial neural networks. In Proceedings of the 2008 Geoscience and Remote Sensing Symposium, Boston, MA, USA, 7–11 July 2008.
31. Camps-Valls, G.; Bruzzone, L. *Kernel Methods for Remote Sensing Data Analysis*; John Wiley & Sons: Hoboken, NJ, USA, 2009.
32. Verrelst, J.; Alonso, L.; Caicedo, J.P.R.; Moreno, J.; Camps-Valls, G. Gaussian process retrieval of chlorophyll content from imaging spectroscopy data. *IEEE J. Sel. Top. Appl. Earth Obs. Remote Sens.* **2013**, *6*, 867–874. [CrossRef]
33. Verrelst, J.; Rivera, J.P.; Moreno, J.; Camps-Valls, G. Gaussian processes uncertainty estimates in experimental Sentinel-2 LAI and leaf chlorophyll content retrieval. *ISPRS J. Photogramm. Remote Sens.* **2013**, *86*, 157–167. [CrossRef]
34. Verrelst, J.; Rivera, J.P.; Gitelson, A.; Delegido, J.; Moreno, J.; Camps-Valls, G. Spectral band selection for vegetation properties retrieval using Gaussian processes regression. *Int. J. Appl. Earth Obs. Geoinf.* **2016**, *52*, 554–567. [CrossRef]
35. Settles, B. Active Learning Literature Survey. Available online: http://burrsettles.com/pub/settles.activelearning.pdf (accessed on 21 January 2019).
36. Gascon, F.; Bouzinac, C.; Thépaut, O.; Jung, M.; Francesconi, B.; Louis, J.; Lonjou, V.; Lafrance, B.; Massera, S.; Gaudel-Vacaresse, A. Copernicus Sentinel-2A calibration and products validation status. *Remote Sens.* **2017**, *9*, 584. [CrossRef]
37. Jacquemoud, S.; Verhoef, W.; Baret, F.; Bacour, C.; Zarco-Tejada, P.J.; Asner, G.P.; François, C.; Ustin, S.L. PROSPECT+ SAIL models: A review of use for vegetation characterization. *Remote Sens. Environ.* **2009**, *113*, S56–S66. [CrossRef]
38. Feret, J.-B.; François, C.; Asner, G.P.; Gitelson, A.A.; Martin, R.E.; Bidel, L.P.; Ustin, S.L.; Le Maire, G.; Jacquemoud, S. PROSPECT-4 and 5: Advances in the leaf optical properties model separating photosynthetic pigments. *Remote Sens. Environ.* **2008**, *112*, 3030–3043. [CrossRef]
39. Verhoef, W. Light scattering by leaf layers with application to canopy reflectance modeling: The SAIL model. *Remote Sens. Environ.* **1984**, *16*, 125–141. [CrossRef]
40. Meroni, M.; Colombo, R.; Panigada, C. Inversion of a radiative transfer model with hyperspectral observations for LAI mapping in poplar plantations. *Remote Sens. Environ.* **2004**, *92*, 195–206. [CrossRef]
41. Richter, K.; Atzberger, C.; Vuolo, F.; Weihs, P.; d'Urso, G. Experimental assessment of the Sentinel-2 band setting for RTM-based LAI retrieval of sugar beet and maize. *Can. J. Remote Sens.* **2009**, *35*, 230–247. [CrossRef]

42. Richter, R.; Schlapfer, D.; Muller, A. Operational atmospheric correction for imaging spectrometers accounting for the smile effect. *IEEE Trans. Geosci. Remote Sens.* **2011**, *49*, 1772–1780. [CrossRef]
43. Palacharla, P.K.; Durbha, S.S.; King, R.L.; Gokaraju, B.; Lawrence, G.W. A hyperspectral reflectance data based model inversion methodology to detect reniform nematodes in cotton. In Proceedings of the 6th International Workshop on the Analysis of Multi-Temporal Remote Sensing Images (Multi-Temp), Trento, Italy, 12–14 July 2011.
44. Le Maire, G.; Marsden, C.; Verhoef, W.; Ponzoni, F.J.; Seen, D.L.; Bégué, A.; Stape, J.-L.; Nouvellon, Y. Leaf area index estimation with MODIS reflectance time series and model inversion during full rotations of Eucalyptus plantations. *Remote Sens. Environ.* **2011**, *115*, 586–599. [CrossRef]
45. Si, Y.; Schlerf, M.; Zurita-Milla, R.; Skidmore, A.; Wang, T. Mapping spatio-temporal variation of grassland quantity and quality using MERIS data and the PROSAIL model. *Remote Sens. Environ.* **2012**, *121*, 415–425.
46. Liang, L.; Qin, Z.; Zhao, S.; Di, L.; Zhang, C.; Deng, M.; Lin, H.; Zhang, L.; Wang, L.; Liu, Z. Estimating crop chlorophyll content with hyperspectral vegetation indices and the hybrid inversion method. *Int. J. Remote Sens.* **2016**, *37*, 2923–2949. [CrossRef]
47. Bacour, C.; Jacquemoud, S.; Tourbier, Y.; Dechambre, M.; Frangi, J.-P. Design and analysis of numerical experiments to compare four canopy reflectance models. *Remote Sens. Environ.* **2002**, *79*, 72–83. [CrossRef]
48. Baret, F.; Hagolle, O.; Geiger, B.; Bicheron, P.; Miras, B.; Huc, M.; Berthelot, B.; Niño, F.; Weiss, M.; Samain, O.; et al. LAI, fAPAR and fCover CYCLOPES global products derived from VEGETATION. *Remote Sens. Environ.* **2007**, *110*, 275–286. [CrossRef]
49. Bajwa, S.G.; Gowda, P.H.; Howell, T.A.; Leh, M. Comparing Artificial Neural Network and Least Square Regression Techniques for LAI Retrieval from Remote Sensing Data. Available online: http://www.asprs.org/a/publications/proceedings/pecora17/0006.pdf (accessed on 21 February 2019).
50. Geladi, P.; Kowalski, B.R. Partial least-squares regression: A tutorial. *Anal. Chim. Acta* **1986**, *185*, 1–17. [CrossRef]
51. Hagan, M.T.; Menhaj, M.B. Training feedforward networks with the Marquardt algorithm. *IEEE Trans. Neural Netw.* **1994**, *5*, 989–993. [CrossRef] [PubMed]
52. Breiman, L.; Friedman, J.H.; Olshen, R.A.; Stone, C.J. *Classification and Regression Trees*; Routledge: New York, NY, USA, 1984.
53. Friedman, J.; Hastie, T.; Tibshirani, R. Additive logistic regression: A statistical view of boosting (with discussion and a rejoinder by the authors). *Ann. Stat.* **2000**, *28*, 337–407. [CrossRef]
54. Breiman, L. Bagging predictors. *Mach. Learn.* **1996**, *24*, 123–140. [CrossRef]
55. Breiman, L. Random forests. *Mach. Learn.* **2001**, *45*, 5–32. [CrossRef]
56. Williams, C.K.; Rasmussen, C.E. Gaussian Processes for Machine Learning. Available online: http://www.gaussianprocess.org/gpml/chapters/RW.pdf (accessed on 20 February 2019).
57. Williams, C.K.; Rasmussen, C.E. Gaussian Processes for Regression. Available online: https://pdfs.semanticscholar.org/e251/664b68fe1b2bdbf39d938b96adfabe54c4fd.pdf (accessed on 20 February 2019).
58. Verrelst, J.; Rivera, J.; Alonso, L.; Moreno, J. ARTMO: An Automated Radiative Transfer Models Operator Toolbox for Automated Retrieval of Biophysical Parameters Through Model Inversion. Available online: http://ipl.uv.es/artmo/files/JV_ARTMO_paper_Earsel.pdf (accessed on 20 February 2019).
59. Camacho, F.; Baret, F.; Weiss, M.; Fernandes, R.; Berthelot, B.; Sánchez, J.; Latorre, C.; García-Haro, J.; Duca, R. Validación de algoritmos para la obtención de variables biofísicas con datos Sentinel2 en la ESA: Proyecto VALSE-2. In Proceedings of the XV Congreso de la Asociación Española de Teledetección INTA, Torrejón de Ardoz (Madrid), 22–24 October 2013. [CrossRef]
60. Li-Cor, I. *LAI-2000 Plant Canopy Analyzer Instruction Manual*; LI-COR Inc.: Lincoln, NE, USA, 1992.
61. Casa, R.; Castaldi, F.; Pascucci, S.; Pignatti, S. Chlorophyll estimation in field crops: An assessment of handheld leaf meters and spectral reflectance measurements. *J. Agric. Sci.* **2015**, *153*, 876–890. [CrossRef]
62. Nielsen, D.C.; Miceli-Garcia, J.J.; Lyon, D.J. Canopy cover and leaf area index relationships for wheat, triticale, and corn. *Agron. J.* **2012**, *104*, 1569–1573. [CrossRef]
63. Das, B.; Nair, B.; Reddy, V.K.; Venkatesh, P. Evaluation of multiple linear, neural network and penalised regression models for prediction of rice yield based on weather parameters for west coast of India. *Int. J. Biometeorol.* **2018**, *62*, 1809–1822. [CrossRef] [PubMed]
64. Clevers, J.G.; Kooistra, L.; van den Brande, M.M. Using Sentinel-2 data for retrieving LAI and leaf and canopy chlorophyll content of a potato crop. *Remote Sens.* **2017**, *9*, 405. [CrossRef]

65. Frampton, W.J.; Dash, J.; Watmough, G.; Milton, E.J. Evaluating the capabilities of Sentinel-2 for quantitative estimation of biophysical variables in vegetation. *ISPRS J. Photogramm. Remote Sens.* **2013**, *82*, 83–92. [CrossRef]
66. Drusch, M.; Gascon, F.; Berger, M. GMES Sentinel-2 Mission Requirements Document. Available online: https://earth.esa.int/pub/ESA_DOC/GMES_Sentinel2_MRD_issue_2.0_update.pdf (accessed on 20 February 2019).

© 2019 by the authors. Licensee MDPI, Basel, Switzerland. This article is an open access article distributed under the terms and conditions of the Creative Commons Attribution (CC BY) license (http://creativecommons.org/licenses/by/4.0/).

Article

Physically-Based Retrieval of Canopy Equivalent Water Thickness Using Hyperspectral Data

Matthias Wocher *, Katja Berger, Martin Danner, Wolfram Mauser and Tobias Hank

Department of Geography, Ludwig-Maximilians-University Munich, Luisenstraße 37, 80333 Munich, Germany; katja.berger@lmu.de (K.B.); martin.danner@iggf.geo.uni-muenchen.de (M.D.); w.mauser@lmu.de (W.M.); tobias.hank@lmu.de (T.H.)
* Correspondence: m.wocher@lmu.de; Tel.: +49-892-180-6695

Received: 19 October 2018; Accepted: 27 November 2018; Published: 30 November 2018

Abstract: Quantitative equivalent water thickness on canopy level (EWT_{canopy}) is an important land surface variable and retrieving EWT_{canopy} from remote sensing has been targeted by many studies. However, the effect of radiative penetration into the canopy has not been fully understood. Therefore, in this study the Beer-Lambert law is applied to inversely determine water content information in the 930 to 1060 nm range of canopy reflectance from measured winter wheat and corn spectra collected in 2015, 2017, and 2018. The spectral model was calibrated using a look-up-table (LUT) of 50,000 PROSPECT spectra. Internal model validation was performed using two leaf optical properties datasets (LOPEX93 and ANGERS). Destructive in-situ measurements of water content were collected separately for leaves, stalks, and fruits. Correlation between measured and modelled water content was most promising for leaves and ears in case of wheat, reaching coefficients of determination (R^2) up to 0.72 and relative RMSE (rRMSE) of 26% and in case of corn for the leaf fraction only ($R^2 = 0.86$, rRMSE = 23%). These findings indicate that, depending on the crop type and its structure, different parts of the canopy are observed by optical sensors. The results from the Munich-North-Isar test sites indicated that plant compartment specific EWT_{canopy} allows us to deduce more information about the physical meaning of model results than from equivalent water thickness on leaf level (EWT) which is upscaled to canopy water content (CWC) by multiplication of the leaf area index (LAI). Therefore, it is suggested to collect EWT_{canopy} data and corresponding reflectance for different crop types over the entire growing cycle. Nevertheless, the calibrated model proved to be transferable in time and space and thus can be applied for fast and effective retrieval of EWT_{canopy} in the scope of future hyperspectral satellite missions.

Keywords: hyperspectral; spectroscopy; equivalent water thickness; canopy water content; agriculture; EnMAP

---

1. Introduction

The quantification of water stored in agricultural plants plays an essential role in understanding the impact of cultivated areas on the earth's water cycle. Due to its close association to biochemical factors, such as vegetation transpiration [1] and net primary production [2], the knowledge of quantities of water contained within agricultural crops is crucial, particularly for the development of environmental process models [3,4]. Moreover, quantifying canopy water content is important in regards to the water use efficiency of plants [5], evaluation of plant physiological status and health [6,7], and crop ripening monitoring [8].

Within the optical spectral domain (400 nm–2500 nm), absorption by vegetation liquid water occurs in the near-infrared (NIR) at 970 nm and 1200 nm and in the shortwave infrared (SWIR) at 1450 nm and 1950 nm [9,10]. Due to a higher absorption coefficient in the SWIR [11] most of the early studies combined those wavelengths with water insensitive wavelengths in the NIR to define empirical narrow-band indices for water content retrieval [12–15]. However, the strong absorption by water may saturate those bands at high water contents in optically thick canopies [16]. Moreover, absorption by atmospheric water vapor at 1450 nm and 1900 nm results in noisy measurements which renders these spectral regions unsuitable for further analysis [9], both for top-of-atmosphere (TOA) and top-of-canopy (TOC) spectroscopy. Some vegetation biophysical variables may disturb the signal of water: for instance Jacquemoud, et al. [17] noted that the leaf area index (LAI) masks the water signal between 1000 nm and 1400 nm and advised caution when using such indices for water retrieval. The comparatively low 970 nm water absorption depth is embedded in an area of generally high vegetation reflectance in the NIR. Due to low absorption it is expected that radiation at 970 nm penetrates deeper into the canopy reflecting a larger portion of its total water content without a tendency to saturation [18–22]. Therefore, Peñuelas, et al. [6,7] developed the 970 nm water index (WI) to retrieve relative plant water concentration (PWC). In the following, other studies also focused on the 970 nm absorption to estimate canopy water content [3,5,23,24].

Methodologically, the definition of a narrow-band spectral index to retrieve vegetation water content information constitutes the parametric regression type of methods. Their simplicity and thus computational feasibility make them highly desirable for large-scale remote sensing applications. However, a fundamental problem of parametric regression methods is their lack of generality and transferability [25]. Since indices are not solely influenced by liquid water, but also affected by leaf internal structure and leaf dry matter [26] or canopy structure, LAI and soil background [15,27,28], the established regression-based relationships and estimated quantities of water stored in a canopy are limited to local conditions [29]. Accordingly, the obtained results are site-, time- and crop-specific [30]. Moreover, as more hyperspectral image data with a continuous spectral coverage become accessible, the limited use of a small number of bands does not correspond to the up to date possibilities in view of the available data information density.

For the implications given, physically based model inversion methods have been introduced as a promising alternative to retrieve biochemical and biophysical vegetation variables. Radiative transfer models (RTM) describe interactions between solar radiation and vegetation constituents using physical laws. Their ability to generate an infinite number of simulated spectra with known input parameters conversely allows their inversion in order to estimate the underlying parameters. For the inversion of RTMs, a variety of strategies have been applied. These include numerical optimization algorithms, look-up table (LUT) approaches, artificial neural networks (ANN) and other machine learning algorithms (for an overview please refer to Verrelst, et al. [25,31]). Although RTM-based inversion methods are considered to be physically sound, the techniques require profound knowledge, are often computationally demanding and are mathematically highly non-linear [31,32]. Another limitation of RTM-inversion is the ill-posed nature or equifinality of model inversion. Many different parameter sets may be equally valid in terms of their ability to reproduce a measured reflectance spectrum (for a discussion of this topic see Atzberger and Richter [33]).

In view of future hyperspectral satellite missions like Italian PRISMA [34], US HyspIRI [35], Israeli-Italian SHALOM [36], European CHIME [37], and German EnMAP [38] fast and efficient retrieval methods for large datasets are required. Mathematically simpler physically-based approaches have been applied before to circumvent the equifinality problem and to reduce the computational effort of RTM-based model inversion. Green, et al. [39,40] originally incorporated the Beer-Lambert law to separate liquid water from atmospheric water vapor and determine both to allow the retrieval of surface reflectance from measured AVIRIS radiance. Thereby, the Beer-Lambert law was applied to directly infer the path length through optically active liquid water, i.e., the equivalent water thickness (EWT), from a measured reflectance spectrum using water absorption coefficients for pure liquid water [41,42]. Since multiple NIR scattering, and the attendant increase in optical path length at both the leaf and the canopy scale are not accounted for in the simple Beer-Lambert law [43,44], absolute quantification of EWT can only be achieved by calibration. Subsequently, validation has to be performed on accurate in-situ measurements. Studies that aimed at separating all three phases of water were not designed to quantify canopy water content in absolute terms and therefore accurate measurements were not carried out [45–48]. On the other hand, studies which derived water content explicitly by applying the Beer-Lambert law often relied on the assumption that upscaling leaf EWT to canopy water content (CWC) could be done by a simple multiplication with the leaf area index (LAI) (see references [5,23,25,49,50]). In other publications, biomass sampling strategies have not been designed to deduce the single water components of a canopy that an optical sensor can actually detect (e.g., references [21,51–53]). Consequently, validation of these approaches could not approve translation into transferable and generally applicable retrieval tools [42].

Therefore, the objective of the present study was to test the performance of the Beer-Lambert law to retrieve crop water content from spectra with a contiguous spectral coverage around 970 nm and perform validation separately for leaves, stalks, and fruits by means of the two very different crop types: corn and winter wheat.

2. Materials

2.1. Munich-North-Isar Test Site

2.1.1. Biomass Sampling and Water Content Determination

Biomass collection was performed in 2015, 2017, and 2018 at three winter wheat fields (*triticum aestivum*) and two corn fields (*zea mays*) of communal farmland 30 km north of Munich (southern Germany) east of the river Isar (Table 1, Figure 1).

Figure 1. Munich-North-Isar test sites overview (**left**) and exemplary 2017 simulated 30 × 30 m corn EnMap-pixel with 9 ESUs (**right**).

Table 1. Munich-North-Isar winter wheat and corn test sites, locations, periods of sample collection, number of biomass samples, and number of spectral measurements at cloud-free days.

Crop Type	Coordinates	Sampling Period	No. of Samplings	No. of Spectral Measurements
Winter wheat	48°14′51.46″N 11°42′24.10″E	10 April–29 July 2015	17	7
Winter wheat	48°14′56.70″N 11°43′03.60″E	29 March–17 July 2017	16	12
Winter wheat	48°14′52.27″N 11°42′57.06″E	04 April–13 July 2018	12	7
Corn	48°17′06.56″N 11°42′49.98″E	8 June–15 September 2017	11	8
Corn	48°14′56.70″N 11°43′03.60″E	25 May–29 August 2018	13	6

Within the fields, three different sampling points were selected based on long-term biomass distribution pattern observations (TalkingFields Base Map: www.talkingfields.de) representing low, medium, and high persistent relative fertility.

Plant leaf water content is commonly expressed as equivalent water thickness (EWT, Equation (1)) corresponding to a hypothetical thickness of a single layer of water averaged over the whole leaf area [10]:

$$\text{EWT} = \frac{\text{FW} - \text{DW}}{\text{A}} \left[\text{g cm}^{-2}\right] \text{ or [cm]}, \left[\text{kg m}^{-2}\right] \text{ or [mm]} \quad (1)$$

where FW is the fresh sample weight, DW is the oven dry weight and A is the leaf area. While EWT refers to the water content on leaf level, canopy water content (CWC, Equation (2)) is commonly derived through extrapolation by means of the LAI:

$$\text{CWC} = \text{EWT} * \text{LAI} \quad (2)$$

Due to the linkage of LAI to the whole canopy, CWC may be biased towards the leaf fraction. Furthermore, CWC does not allow inferring the actual water detectability of plant components in a canopy from total detected water. Consequently, in this study total $\text{EWT}_{\text{canopy}}$ (EWT_{leaf} + $\text{EWT}_{\text{stalk}}$ + $\text{EWT}_{\text{fruit}}$) will be defined as the above-ground total equivalent water layer averaged over one square meter of ground surface (Equation (3)).

$$\text{total EWT}_{\text{canopy}} = \sum (\text{FW}_{\text{leaves+stalks+fruits}} - \text{DW}_{\text{leaves+stalks+fruits}}) * A_g^{-1} \quad (3)$$

where A_g denotes the ground area. To monitor the development of total amounts of water stored in the canopy throughout the growing season, biomass samples were collected on a weekly basis. In case of wheat, a minimum transect of 50 cm along a sowing track or an area of 0.25 m² was cut at soil level. For corn, 2–3 plants were cut. In-field plant density was obtained by counting plants and rows per meter. The samples were separated into leaf, stalk and fruit compartments, weighed in fresh state and oven-dried until constant weight for 24 h at 105 °C before dry weight was determined. EWT_{leaf}, $\text{EWT}_{\text{stalk}}$, $\text{EWT}_{\text{fruit}}$ (EWT_{ear} and EWT_{cob}, respectively) and total $\text{EWT}_{\text{canopy}}$ per cm² (Table 2) were calculated from laboratory results (specific water contents per ground area) and from farm management metadata (plants per meter and row spacing). The phenology was determined according to secondary growth stages of the BBCH-scale [54].

Table 2. Statistics (range, mean, standard deviation) for in-situ measured EWT_{leaf}, $\text{EWT}_{\text{stalk}}$, $\text{EWT}_{\text{fruit}}$, total $\text{EWT}_{\text{canopy}}$ and BBCH-range. Values correspond to measurements with available spectral reflectance data.

Year	2015	2017		2018	
Crop Type	Winter Wheat	Winter Wheat	Corn	Winter Wheat	Corn
BBCH range [-]	22–87	25–87	30–85	28–87	32–83
EWT_{leaf}: range [cm]	0.007–0.179	0.005–0.182	0.009–0.104	0.045–0.121	0.095–0.161
mean (std) [cm]	0.066 (0.058)	0.082 (0.050)	0.059 (0.035)	0.082 (0.027)	0.132 (0.023)
$\text{EWT}_{\text{stalk}}$: range [cm]	0.012–0.256	0.003–0.275	0.008–0.295	0.019–0.268	0.252–0.619
mean (std) [cm]	0.123 (0.084)	0.144 (0.089)	0.161 (0.115)	0.126 (0.099)	0.472 (0.126)
$\text{EWT}_{\text{fruit}}$: range [cm]	0.000–0.100	0.000–0.112	0.000–0.248	0.000–0.148	0.000–0.306
mean (std) [cm]	0.044 (0.045)	0.045 (0.045)	0.068 (0.100)	0.048 (0.068)	0.171 (0.123)
Total $\text{EWT}_{\text{canopy}}$: range [cm]	0.041–0.417	0.019–0.490	0.017–0.606	0.064–0.503	0.347–1.019
mean (std) [cm]	0.233 (0.141)	0.271 (0.145)	0.289 (0.221)	0.256 (0.170)	0.775 (0.227)

2.1.2. Spectroscopic Measurements

At each study site, a 30 × 30 m grid of nine 10 × 10 m squares was marked out delineating the elementary sampling units (ESU). This grid layout was designed to trace the geometric properties of one EnMAP pixel; hence, regarding the viewing geometry and grid location, the sensors descending orbit and inclination angle of 97.96° was accounted for (Table 1). At each sampling date with clear sky conditions (Table 1) all nine ESUs were revisited and spectral measurements were taken using an Analytical Spectral Devices Inc. (ASD, Boulder, CO, USA) FieldSpec3 Jr. spectroradiometer with an effective spectral resolution of 3 nm in the VIS (≤700 nm) and 10 nm in the NIR and SWIR (≤2500 nm). Five nadir measurements were conducted per ESU at a height of 25 cm above the canopy, the same height at which the white reference panel (OptoPolymer, Munich, Germany) could be fully observed with the instruments field of view of 25°. Throughout the measurements, the sensor was slightly moved back and forth manually while maintaining the observation angle to obtain a representative spectral sample of the canopy. The five recorded spectra were averaged per ESU and a spatial mean of the nine ESUs was calculated to provide reflectance of the complete 30 × 30 m grid. Further, post-processing included splice-correction, white reference baseline calibration, and slight smoothing using a Savitzky-Golay-Filter with a frame size of 13 nm.

Note that it was not possible to conduct destructive sampling at exactly the same locations where the continuous spectral measurements were taken. However, due to averaging of spectral sampling points over the 30 × 30 m sampling area, it was possible to capture the in-field variability and therefore to represent average field water conditions.

Altogether, the collected dataset comprised destructively measured, plant compartment specific water content samples with corresponding spectral measurements at 26 dates for wheat and 14 dates for corn over three and two years, respectively (see Tables 1 and 2).

2.2. Leaf Optical Data

Preliminary tests of the EWT_{canopy} retrieval model presented in this study were performed on two different leaf optical datasets. The LOPEX93 database was established in 1993 by the Joint Research Centre (JRC, Ispra, Italy). It associates transmittance and reflectance in the range of 400–2500 nm with biophysical and biochemical measurements of 66 leaf samples from 45 species [55]. In total, the database comprises 330 spectra with corresponding measurements of EWT. Secondly, tests were performed on the ANGERS database containing 276 reflectance spectra and EWT measurements of 43 species [56]. While woody species make up the majority of the ANGERS database, both datasets represent a large variety of leaf internal structure and spectra.

2.3. Radiative Transfer Models and Look-Up Tables

To check consistency between leaf optical data and modelled spectra, large look-up tables (LUT) using the RTMs PROSPECT and PROSAIL were created. PROSAIL [17] is coupling the *Leaf Optical Properties Spectra* model PROSPECT [57] and the turbid medium canopy reflectance model 4SAIL (*Scattering by Arbitrary Inclined Leaves*) [58,59]. The latest recalibrated version PROSPECT-D [60,61] simulates bidirectional-hemispherical reflectance and transmittance in the optical domain as a function of leaf pigments (chlorophyll a+b content C_{ab}, carotenoids C_{ar}, and anthocyanins C_{anth}), dry matter C_m, and brown pigments C_{brown} as well as a leaf mesophyll structure parameter N, and EWT(C_w). The canopy model SAIL calculates a bidirectional reflectance factor of 1-D turbid medium plant canopies. With regard to leaf optical properties and reflectance of the underlying soil (p_{soil}), it implements canopy structure (LAI), average leaf inclination angle (ALA) or optionally, ellipsoidal leaf inclination distribution (LIDF), and hot spot size parameter (h_{spot}) for a given illumination and viewing geometry (observation zenith angle (OZA), relative azimuth angle (rAA) between sun and sensor, and the solar zenith angle (SZA)).

Considering the impact of plant foliar water on the 970 nm absorption band, during LUT generation, all parameters with sensitivity in the NIR region were uniformly distributed over a wide value range (Table 3). Leaf pigments, having no effect on reflectance in the NIR [26,44], remained constant.

Table 3. Parameter ranges for PROSPECT-D and PROSPECT-D + 4SAIL (PROSAIL) LUT. Specified ranges are uniformly distributed, single values are fixed.

PROSPECT-D-Parameters	Range	Notation [Unit]	4SAIL-Parameters	Range	Notation [Unit]
N	1.0–3.0	[-]	LAI	0.5–8.0	[m² m^{-2}]
C_{ab}	55	[µg cm^{-2}]	ALA	0–90	[deg]
C_w	0.0002–0.07	[g cm^{-2}]	h_{spot}	0.01–0.5	[-]
C_m	0.001–0.02	[g cm^{-2}]	OZA	0	[deg]
C_{brown}	0.0–1.0	[-]	SZA	35–50	[deg]
C_{ar}	15	[µg cm^{-2}]	rAA	0	[deg]
C_{anth}	5	[µg cm^{-2}]	P_{soil}	0.0–1.0	[-]

3. Methods

3.1. The Beer-Lambert Law and Retrieval Method Development

The Beer-Lambert law is mathematically formulated as Equation (4):

$$\Phi = \Phi_0 e^{-\alpha(\lambda)d}. \tag{4}$$

Passing through a medium of thickness d the incident radiation intensity Φ_0 is exponentially attenuated with increasing penetration depth. The absorption characteristics of a medium are defined by its wavelength-dependent absorption coefficients $\alpha(\lambda)$. In this study, due to the accurate spectral resolution in the 970 nm domain [61], water absorption coefficients for pure liquid water as determined by Kou, et al. [11] are used. Furthermore, it is assumed that within the absorption band at 970 nm, water is the only quantity-depending, varying active absorber and that variance within absorption of further components is neglectable. Thus, concluding from Equation (4), the absorption depth of measured fresh leaves or canopies at 970 nm is uniquely dependent on the thickness of the optically active water layer (see also discussion in Section 3.2). For dry leaves or senescent canopies, absorption by liquid water is neglectable, resulting in a strictly linear reflectance signature at 970 nm. For the retrieval of EWT_{canopy}, Equation (4) is rearranged in accordance with Bach ([51]; Equation (5)), where R_0 is the measured reflectance, d is the thickness of the optically active water layer, and R' is the d-dependent reflectance:

$$R' = \frac{R_0}{e^{-\alpha(\lambda)d}} \tag{5}$$

Using Equation (5), d is iteratively optimized so that an objective function—the sum of absolute residuals between the modelled reflectance and the linear connection between the descending and ascending vertices of the 970 nm absorption—is minimal (Figure 2). The wavelength range considered by the plant water retrieval (PWR) model has been limited to 930–1060 nm based on preliminary minimization of the standard deviation of yielded EWT results from the PROSPECT LUT. Describing the thickness of the optically active water layer, the results can directly be compared to measured EWT on leaf level (Equation (1)), CWC (Equation (2)) and EWT_{canopy} (Equation (3)). The algorithm was implemented in Python, where retrieval of EWT for 50,000 spectra was completed in 69 s on an Intel Core i5-3570K @ 3.40 GHz.

3.2. Global Sensitivity Analysis

The PWR model expects the thickness of optically active water to reflect the vegetation water content detected by a hyperspectral sensor. Both the PROSPECT and PROSAIL LUT were subjected to a global sensitivity analysis (GSA) to identify and evaluate the impact of contributing parameters in the 970 nm domain and to validate the performance of the model. The Fourier amplitude sensitivity test (FAST) identifies the main effects (first-order sensitivity effects), i.e., the contribution (S_{Ti}) to the variance of the model output by each input variable and interactions with other variables [62]. The contribution of parameters to the 970 nm absorption depth and shape is assessed by its distribution width using the variance-to-mean ratio (VMR).

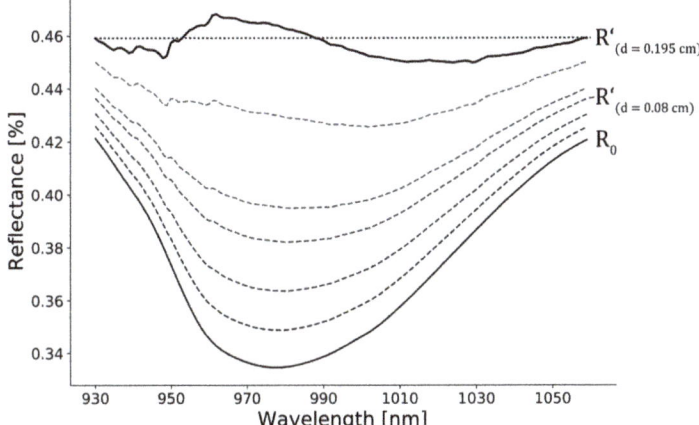

Figure 2. Determination of optically active water thickness d from a measured spectrum R_0 through minimization of residuals to assumed dry reflectance line (dotted line).

Within PROSPECT (Figure 3a) N contributes to 98% of leaf reflectance at the vertices left and right of the 970 nm water absorption band. At 970 nm C_w is the highest contributing parameter in terms of VMR (10^{-2}). Minor influence on the shape of the absorption band is caused by C_{brown} at the descending vertex (VMR = 10^{-3}). C_m and parameter interactions also affect overall reflectance at 970 nm but interference with its shape is smaller by more than two orders of magnitude (10^{-4}). Regarding PROSAIL (Figure 3b), C_w likewise is the strongest shape-determining factor at 970 nm in terms of VMR (10^{-2}). However, the 970 nm absorption shape is affected by canopy structural parameters (ALA, LAI, h_{spot}, p_{soil}) and parameter interactions. The sum of influential parameter VMR may exceed C_w VMR, which may result in masking of the water signal when unfavorable parameter combinations occur.

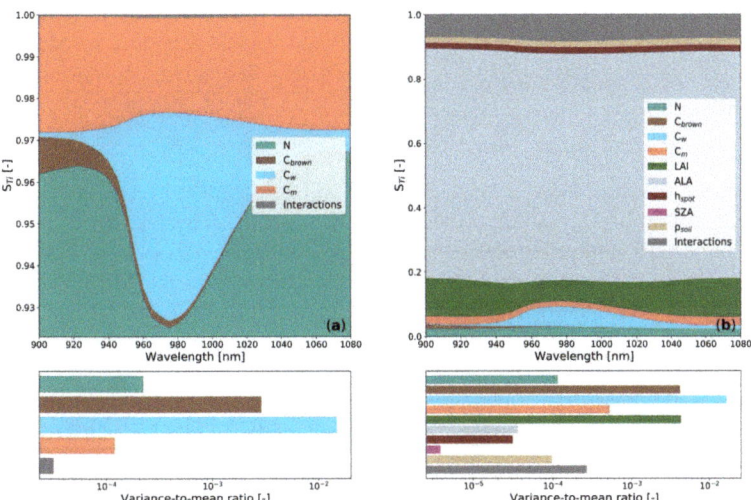

Figure 3. FAST first-order sensitivity coefficients and interactions (S_{Ti}) to reflectance (900–1080 nm) for PROSPECT (**a**) and PROSAIL (**b**) parameters. Due to high contribution of the leaf structure coefficient N within PROSPECT, only the upper contribution range ≥ 0.935 is shown. Below, influences of parameters that affect the shape of the water absorption band considered within the PWR model (930–1060 nm) are quantified by the variance-to-mean ratio (VMR).

The retrieval method was first tested on all the spectra within both the PROSPECT and the PROSAIL LUT (Figure 4). With a coefficient of determination (R^2) of 0.96 the approach indicates a strong correlation between PROSPECT modelled water content C_w and retrieved optically active water content EWT (Figure 4a). However, the high relative root mean square error ($rRMSE = RMSE * mean_{observations}^{-1}$) of 286% with an intercept close to zero revealed a strong systematical offset. The growing spread of results towards higher values of C_w suggests a simultaneously increasing influence of other parameters due to the exponential radiative transfer from specific absorption coefficients to transmission and reflectance [57].

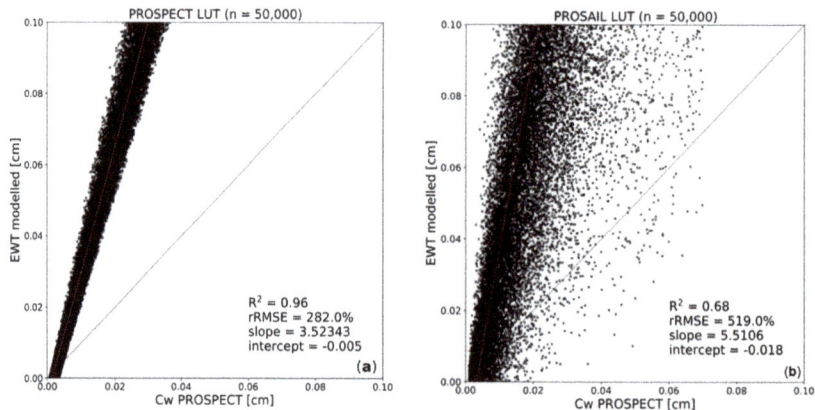

Figure 4. Modelled EWT results from synthetic LUT containing PROSPECT (**a**) and PROSAIL (**b**) spectra.

Applied to PROSAIL spectra (Figure 4b) the R^2-results (0.68) are significantly lower and although model results correspond to LUT C_w-values, both regression residuals and intercept do not show a systematic bias. However, within the created LUT, several parameter combinations can be considered unrealistic [63], masking or flattening the water signal due to model parameter related interference with the shape of the 970 nm absorption band. The resulting outliers and overall spread of modelled C_w-values render the PROSAIL LUT unsuitable for further calibration of the model.

3.3. Using PROSPECT for Calibration of the PWR Model

The model was further tested on the LOPEX93 [55] and ANGERS [56] datasets, which showed a similar systematical bias as model results from PROSPECT spectra (Figure 5a,c,e). Since the overestimation seemed to be solely defined by the slope of the regression line, the water absorption coefficients in Equation (5) were adjusted by multiplying the slope of the PROSPECT LUT linear regression model as a constant (Equation (6)):

$$R' = \frac{R_0}{e^{-\alpha(\lambda)d*3.52343}} \quad (6)$$

The calibration procedure accounts for unknown effects of the leaf surface and of leaf internal structure on reflectance in the 970 nm domain [61] and for potential multiple leaf internal scattering [44,64]. Using the calibrated water absorption coefficients (Equation (6)), minimization of the objective function is achieved more quickly, resulting in lower modelled values of EWT that are consistent with the measured order of magnitude. Subsequently, the altered absorption coefficients in the 930 to 1060 nm range were used for an improved water content retrieval. Applied to the PROSPECT LUT (Figure 5b), EWT was estimated with a much smaller error (rRMSE = 12%). Applying the algorithm with updated coefficients to LOPEX93 data, measured EWT was estimated with R^2 = 0.93 and rRMSE = 22% and for ANGERS data with 0.93 and 39% respectively (Figure 5d,f).

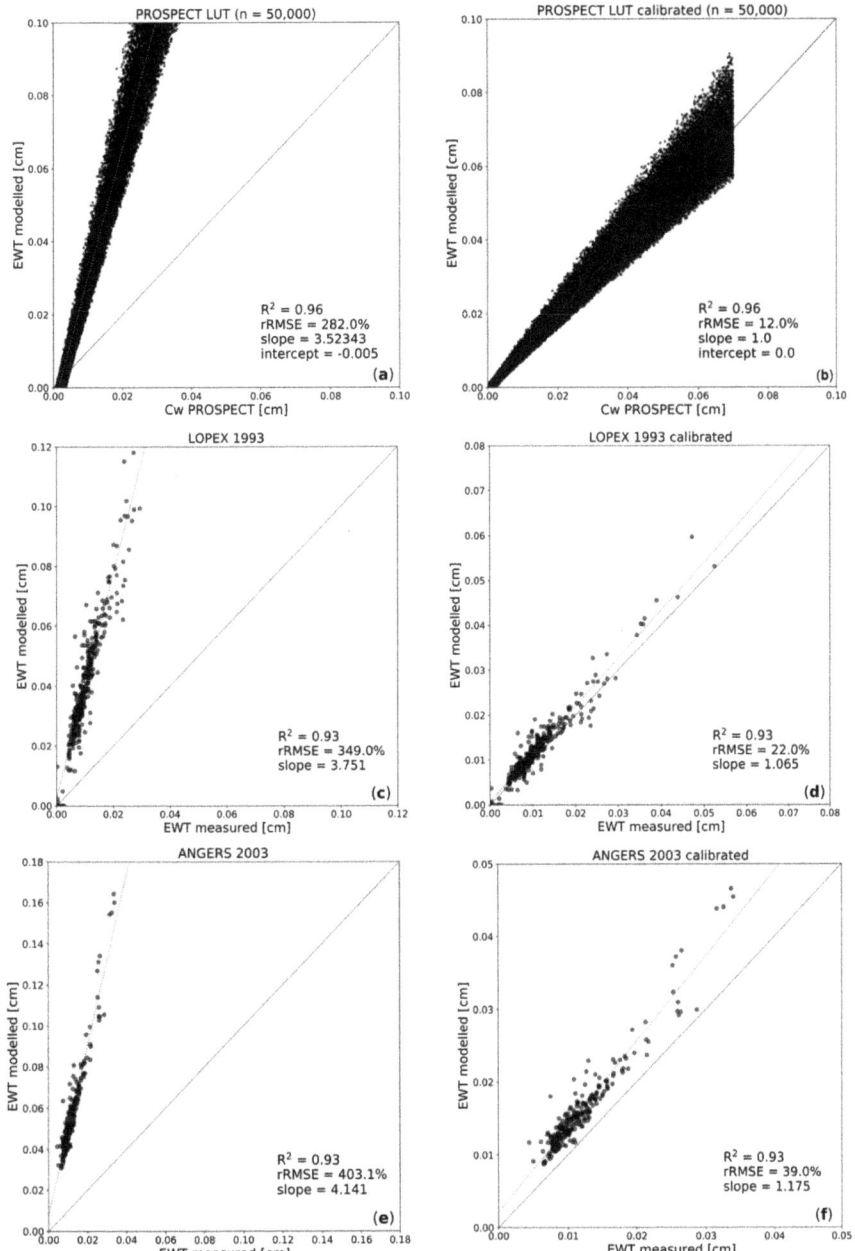

Figure 5. Uncalibrated PWR-results for the PROSPECT LUT (**a**); LOPEX93 data (**c**); and ANGERS data (**e**); compared to results after calibration: PROSPECT LUT (**b**) LOPEX93 (**d**); and ANGERS (**f**).

4. Results

The minimization process for retrieving optically active water using the PWR model with recalibrated absorption coefficients (factor 3.52342, Equation (6)) was applied to both in-situ winter wheat and corn spectral data. The results were compared to combinations of destructively measured leaf, stalk, and ear or cob water contents. For further analysis, the BBCH-scale was included to relate to growth stage dependencies of the model results.

4.1. Winter Wheat Data

Considering only the measured water content of wheat leaves, the results showed low correlation (Figure 6a: $R^2 = 0.27$; rRMSE = 81%); however, annotated BBCH-values showed good results for tillering (20+) and stalk elongation stages (30+) and progressing senescence (87). On the other hand, heading (47+) and flowering stages (60+) as well as ear development and ripening stages (70+) were invariably overestimated by the model. The sum of leaf and stalk water content yielded better results (Figure 6b: $R^2 = 0.68$; rRMSE = 52%) in particular for early growth stages. However, as growth proceeds, strong underestimation of $EWT_{leaf} + EWT_{stalk}$ occurs due to saturation.

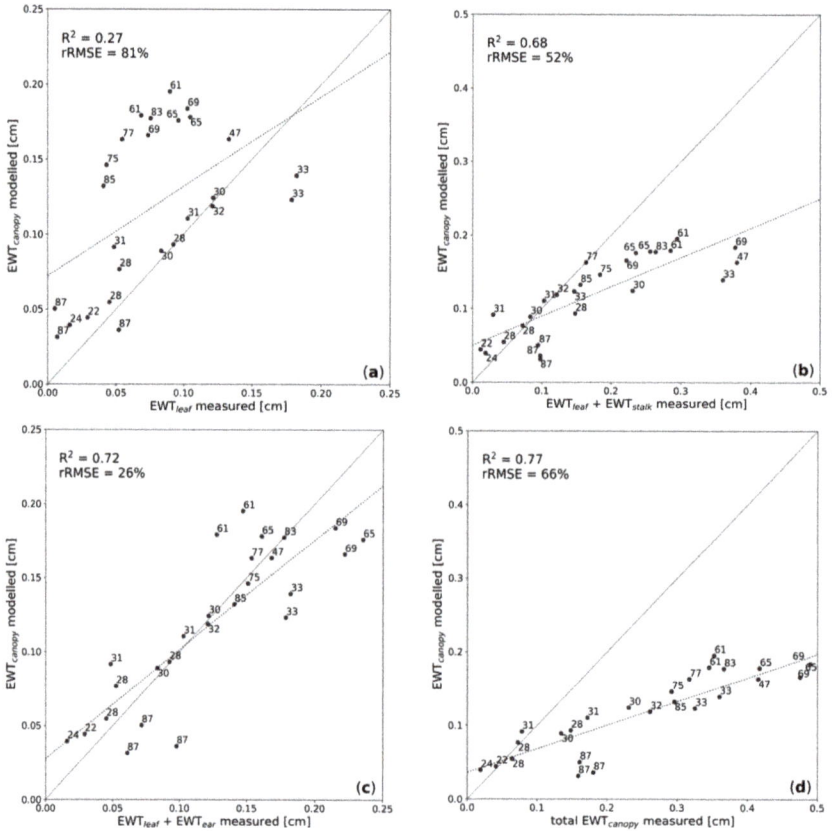

Figure 6. Modelled optically active water in relation to measured water contents of wheat compartments. Annotated numbers refer to secondary growth stages according to BBCH-scale. Results compared to EWT_{leaf} (a); $EWT_{leaf} + EWT_{stalk}$ (b); $EWT_{leaf} + EWT_{ear}$ (c); and total EWT_{canopy} (d).

The best results were obtained when combining the measured water contents of leaves and ears (Figure 6c: $R^2 = 0.72$; rRMSE = 26%). Thereby, model results adequately reflected measured $EWT_{leaf} + EWT_{ear}$ across all phenological stages over three years. Aggregating measured EWT_{leaf}, EWT_{stalk} and EWT_{ear} (= total EWT_{canopy}) yet again largely resulted in an underestimation (Figure 6d: $R^2 = 0.77$; rRMSE = 66%); only tillering stages were modelled with reasonable accuracy.

4.2. Corn Data

Regarding the two-year corn dataset, best results were achieved for leaf water contents (Figure 7a: $R^2 = 0.86$; rRMSE = 23%) with a minor tendency to underestimation towards higher growth stages. Despite good correlation, the combination of leaf and stalk measured water content was largely underestimated by the model (Figure 7b: $R^2 = 0.91$; rRMSE = 95%). Aggregated EWT_{leaf} and EWT_{cob} resulted in both lower correlation and error (Figure 7c: $R^2 = 0.61$; rRMSE = 84%) due to underestimation when cobs were registered.

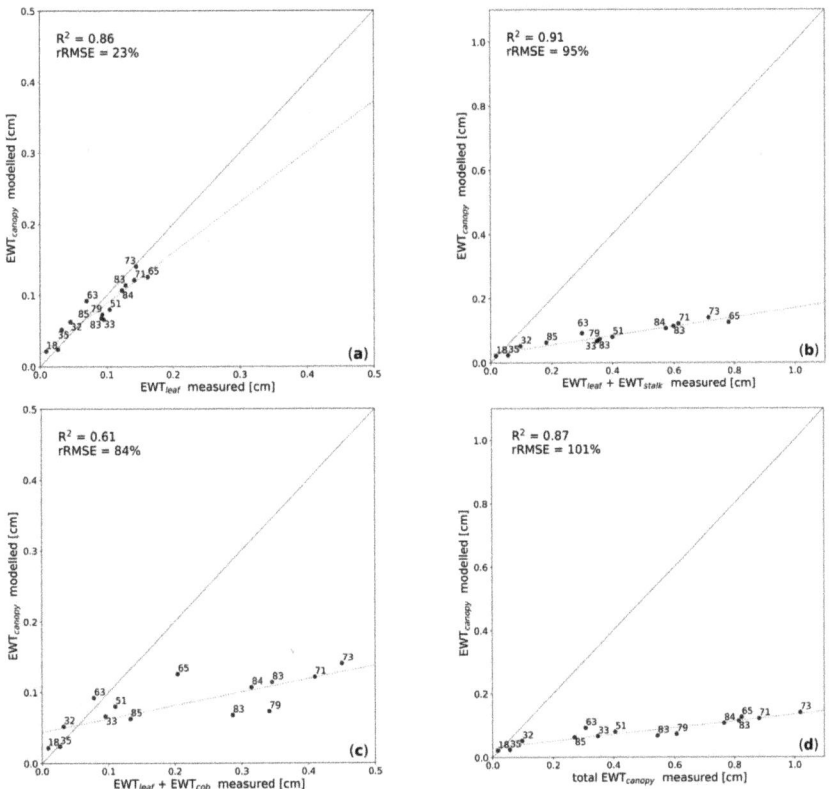

Figure 7. Modelled optically active water in relation to measured water contents of corn compartments. Annotated numbers refer to secondary growth stages according to BBCH-scale. Results compared to EWT_{leaf} (a); $EWT_{leaf} + EWT_{stalk}$ (b); $EWT_{leaf} + EWT_{cob}$ (c); and total EWT_{canopy} (d).

In relation to total measured EWT_{canopy}, both correlation and underestimation are large (Figure 7d: $R^2 = 0.87$; rRMSE = 101%). In view of phenological dependencies, low water contents were consistently modelled with high accuracy during early leaf development stage (BBCH = 18) and beginning stalk elongation (30–32). Furthermore, unlike for wheat, no clear growth stage related dependencies were recognizable.

5. Discussion

5.1. Inversion of the Beer-Lambert Law for Water Content Retrieval

A simple physically based model was developed which applies the Beer-Lambert law to inversely retrieve optically active water content on leaf and canopy scale. In view of the fact that only one parameter—the thickness of the optically active water layer *d*—needs to be inverted, the algorithm allows a fast processing of large hyperspectral datasets. As shown by a GSA of the PROSPECT

LUT in the 970 nm domain, interference of other parameters is marginal (Figure 3a) rendering C_w to be the dominant driver of leaf reflectance in this spectral region ($R^2 = 0.96$; Figure 4a). However, this does not apply to the PROSAIL LUT where, according to GSA, the cumulative influence of leaf and canopy structural parameters may mask the water signal (Figure 3b). Hereby, two issues interact: first, the 4SAIL model assumes a horizontally homogenous canopy, which may not be valid for complex canopy architectures and clumped vegetation through, e.g., formation in rows [65–67]. Second, unrealistic parameter combinations may occur in LUTs [63]. Both issues may unfavorably affect modelled reflectance in the 970 nm domain, reducing the predictive power of C_w for water content information ($R^2 = 0.57$; Figure 4b) and rendering the PROSAIL LUT unsuitable for calibration of the presented PWR model. When applied to PROSPECT spectra the linear offset of the regression model indicates that the absorption coefficients of pure liquid water differ from those of leaves, because reflectance in interaction with the leaf surface and multiple leaf internal reflections are not accounted for [44,64]. Using the slope of the regression from the PROSPECT LUT results as a factor to calibrate the absorption coefficients, the absolute quantification of PROSPECT C_w, LOPEX93 and ANGERS EWT significantly improved (Figure 5). The high correlation of $R^2 = 0.96$ between PROSPECT C_w and modelled EWT approved application of the model to in-situ measured TOC data. Nevertheless, using PROSPECT for calibration implies that potentially occurring canopy architectural effects on radiation [68] are being neglected. Hence, the PWR model considers the 970 nm absorption to be caused solely by liquid water. In addition, since reflected radiance in the 930–1060 nm range is also affected by atmospheric water vapor [16,46], the process of accurate atmospheric correction is a critical prerequisite when the PWR model is applied to future available hyperspectral TOC reflectance acquired from space.

5.2. Dependency of Canopy Water Detection on Canopy Structure

Absolute measures of EWT_{canopy} were inversely extracted from a three-year TOC winter wheat and two-year corn spectral dataset by means of the proposed PWR model. The results indicated a strong correlation between water absorption centered around 970 nm and measured EWT_{canopy}. However, the comparison of retrieved EWT_{canopy} from in-situ spectra with measured aggregations of plant compartment specific water contents raises the question, how deep radiation at 970 nm penetrates into the canopy and thus, which amounts of water actually can be observed by optical sensors [21]. Although absorption by water and vegetation in the NIR is low and penetration depth of radiation is higher in this wavelength range [6,69], the presented results showed that not all of the contained canopy water is detected by the sensor. Our results suggest that in the case of winter wheat modelled EWT_{canopy} largely reflects the absolute water contained in the leaves and present ears (Figure 6c). Taking only EWT_{leaf} as a reference, EWT_{canopy} is overestimated due to the presence of EWT_{ear}, which manifests in the spectral response but is not reflected by the in-situ data (Figure 6b,d). This also implies potential water content overestimation for wheat when referencing is done based on CWC records, which in the case of barley can be seen in the results of Vohland [3]. On the other hand, including measured EWT_{stalk}, the underestimation resulting with advanced growth stage indicates that radiation at 970 nm cannot penetrate increasingly hardened stalk tissue and thus cannot transport information about the water contained within. This has also been noted by Sims and Gamon [21] and Champagne, et al. [53]. This finding is further supported by the fact that residual water in ripe wheat (BBCH = 87) is consistently underestimated, rendering the PWR model unable to detect residual water in senescent wheat.

Despite good results of modelled EWT_{canopy} for EWT_{leaf} of corn, the results indicate a tendency to underestimation towards higher water contents (Figure 7a) due to maximum radiation transmission through stacked leaves [18]. Once stalks and cobs have developed, the underestimation of total corn EWT_{canopy} reveals the limited ability of NIR radiation to penetrate the thick stalk/cob tissues or the canopy depth or both (Figure 7b–d).

In summary, the retrieval results of winter wheat and corn vary because—depending on canopy structure—different plant components manifest in the 970 nm water absorption band. Several other studies also raised the fact of canopy structural influence on crop variables retrieval [68,70–73]. Cereal crops with prominent ears will not be satisfactorily modelled if the ear water content is not included in the in-situ measurements as it can be seen in the study of Champagne, et al. [53]. On the contrary, when modelling corn water content it may be sufficient to only collect leaf samples since only the water fraction of the leaves can be estimated directly from optical sensors. The specific structure of corn mostly covers the stalks and cobs, masking the water stored in these plant compartments.

In recent studies, mostly parametric regression models based on vegetation indices [3,49], derivative- [5,23], or integration-based [50] indices have been applied to retrieve crop canopy water content information from hyperspectral data. Verrelst, et al. [74] obtained very good CWC correlation on SPARC03 data (R^2 = 0.95) by applying Gaussian process regression with integrated sequential backward band removal. Cernicharo, et al. [24] used both an ANN and a LUT approach to estimate CWC from CHRIS/PROBA data (R^2 = 0.82 and R^2 = 0.64, respectively). Earlier studies which presented retrieval methods based on the Beer-Lambert law include Champagne, et al. [53] with good results for corn but an overestimation of wheat canopy water content, because EWT_{ear} has not been sampled separately (index of agreement D = 0.80 and D = 0.38, respectively). The findings of this study are also confirmed by the Beer-Lambert law based study of Sims and Gamon [21], in which best results for water content of thin tissues were obtained (R^2 = 0.66), whereas total canopy water content was underestimated (R^2 = 0.35).

The presented PWR model is considered superior to empirical regression models by its physical basis, allowing insights into the physical meaning of results, while outperforming other Beer-Lambert law based approaches by the possibility to infer absolute measures of canopy water content from measured TOC reflectance spectra. This absolute quantifiability of canopy water content represents a novelty among available retrieval approaches. Besides, the accurate underlying data basis proved transferability of the model to different sites and crop types and, given that a sensor detects the maximal depth of the 970 nm water absorption, promises applicability also to hyperspectral data on an operational basis.

6. Conclusions

The proposed PWR model based on the inversion of the Beer-Lambert law effectively succeeds in the determination of wheat EWT_{leaf} and EWT_{ear} with consistent results over a three-year dataset (R^2 = 0.72; rRMSE = 26%). For corn EWT_{leaf} was estimated from two-year data with even better results (R^2 = 0.86; rRMSE = 23%). Since the detectability of canopy water content fractions seems to be largely dependent on the crop type, its canopy structure, depth, and growth stage, it is recommended to collect EWT_{leaf}, EWT_{stalk} and EWT_{fruit} data and corresponding reflectance for different crop types over all phenological stages along the growing cycle. However, an evaluation is needed to assess limits of canopy water content retrieval in terms of optical radiation penetration depth through thick canopies and tissues, also in view of a possibly improved retrieval from off-nadir spectroscopy [75]. Our study could proof the transferability of the developed PWR model to other sites and crop types and represents a novelty in crop water content absolute quantifiability. The PWR model will be provided as a slim and applicable tool within the open source software EnMAP-Box [76] to accurately and efficiently retrieve water content information from ground-based, airborne and spaceborne hyperspectral data, as it will become available through future missions.

Author Contributions: Conceptualization, M.W., K.B. and T.H.; Data curation, M.W.; Formal analysis, M.W. and K.B.; Funding acquisition, W.M. and T.H.; Investigation, M.W.; Methodology, M.W.; Project administration, W.M. and T.H.; Resources, M.W.; Software, M.W. and M.D.; Supervision, W.M. and T.H.; Validation, M.W. and K.B.; Visualization, M.W.; Writing—original draft, M.W.; Writing—review & editing, M.W., K.B., M.D. and T.H.

Funding: This research received no external funding.

Acknowledgments: The study was supported by the Space Agency of the German Aerospace Center (DLR) in the frame of the project "EnMAP Scientific Advisory Group Phase III—Developing the EnMAP Managed Vegetation Scientific Processor" through funding by the German Federal Ministry of Economic Affairs and Energy (BMWi) based on enactment of the German Bundestag under the grant code number 50EE1623. The responsibility for the content of this publication lies with the authors.

Conflicts of Interest: The authors declare no conflict of interest. The funders had no role in the design of the study; in the collection, analyses, or interpretation of data; in the writing of the manuscript, or in the decision to publish the results.

References

1. Running, S.W.; Gower, S. Forest-BGC, a general model of forest ecosystem processes for regional applications. Ii. Dynamic carbon allocation and nitrogen budgets. *Tree Physiol.* **1991**, *9*, 147–160. [CrossRef] [PubMed]
2. Running, S.W.; Nemani, R.R. Regional hydrologic and carbon balance responses of forests resulting from potential climate change. *Clim. Chang.* **1991**, *19*, 349–368. [CrossRef]
3. Vohland, M. Using imaging and non-imaging spectroradiometer data for the remote detection of vegetation water content. *J. Appl. Remote Sens.* **2008**, *2*, 023520. [CrossRef]
4. Hank, T.; Bach, H.; Mauser, W. Using a remote sensing-supported hydro-agroecological model for field-scale simulation of heterogeneous crop growth and yield: Application for wheat in central Europe. *Remote Sens.* **2015**, *7*, 3934–3965. [CrossRef]
5. Clevers, J.G.P.W.; Kooistra, L.; Schaepman, M.E. Estimating canopy water content using hyperspectral remote sensing data. *Int. J. Appl. Earth Obs. Geoinf.* **2010**, *12*, 119–125. [CrossRef]
6. Peñuelas, J.; Filella, I.; Biel, C.; Serrano, L.; Savé, R. The reflectance at the 950–970 nm region as an indicator of plant water status. *Int. J. Remote Sens.* **1993**, *14*, 1887–1905. [CrossRef]
7. Peñuelas, J.; Pinol, J.; Ogaya, R.; Filella, I. Estimation of plant water concentration by the reflectance water index wi (r900/r970). *Int. J. Remote Sens.* **1997**, *18*, 2869–2875. [CrossRef]
8. Hank, T.B.; Berger, K.; Bach, H.; Clevers, J.G.P.W.; Gitelson, A.; Zarco-Tejada, P.; Mauser, W. Spaceborne imaging spectroscopy for sustainable agriculture: Contributions and challenges. *Surv. Geophys.* **2018**. [CrossRef]
9. Tucker, C.J. Remote sensing of leaf water content in the near infrared. *Remote Sens. Environ.* **1980**, *10*, 23–32. [CrossRef]
10. Danson, F.M.; Steven, M.D.; Malthus, T.J.; Clark, J.A. High-spectral resolution data for determining leaf water content. *Int. J. Remote Sens.* **1992**, *13*, 461–470. [CrossRef]
11. Kou, L.; Labrie, D.; Chylek, P. Refractive indices of water and ice in the 0.65- to 2.5-µm spectral range. *Appl. Opt.* **1993**, *32*, 3531–3540. [CrossRef] [PubMed]
12. Hardisky, M.; Klemas, V.; Smart, R.M. The influence of soil salinity, growth form, and leaf moisture on the spectral radiance of spartina alterniflora canopies. *Photogramm. Eng. Remote Sens.* **1983**, *49*, 77–83.
13. Hunt, E.R.; Rock, B.N.; Nobel, P.S. Measurement of leaf relative water content by infrared reflectance. *Remote Sens. Environ.* **1987**, *22*, 429–435. [CrossRef]
14. Hunt, E.R.; Rock, B.N. Detection of changes in leaf water content using near- and middle-infrared reflectances. *Remote Sens. Environ.* **1989**, *30*, 43–54.
15. Gao, B.-C. NDWI—A normalized difference water index for remote sensing of vegetation liquid water from space. *Remote Sens. Environ.* **1996**, *58*, 257–266. [CrossRef]
16. Datt, B. Remote sensing of water content in eucalyptus leaves. *Aust. J. Bot.* **1999**, *47*, 909–923. [CrossRef]
17. Jacquemoud, S.; Verhoef, W.; Baret, F.; Bacour, C.; Zarco-Tejada, P.J.; Asner, G.P.; François, C.; Ustin, S.L. Prospect+SAIL models: A review of use for vegetation characterization. *Remote Sens. Environ.* **2009**, *113*, S56–S66. [CrossRef]
18. Lillesaeter, O. Spectral reflectance of partly transmitting leaves: Laboratory measurements and mathematical modeling. *Remote Sens. Environ.* **1982**, *12*, 247–254. [CrossRef]
19. Newton, J.E.; Blackman, G.E. The penetration of solar radiation through leaf canopies of different structure. *Ann. Bot.* **1970**, *34*, 329–348. [CrossRef]
20. Bull, C.R. Wavelength selection for near-infrared reflectance moisture meters. *J. Agric. Eng. Res.* **1991**, *49*, 113–125. [CrossRef]
21. Sims, D.A.; Gamon, J.A. Estimation of vegetation water content and photosynthetic tissue area from spectral reflectance: A comparison of indices based on liquid water and chlorophyll absorption features. *Remote Sens. Environ.* **2003**, *84*, 526–537. [CrossRef]
22. Ghulam, A.; Li, Z.-L.; Qin, Q.; Yimit, H.; Wang, J. Estimating crop water stress with ETM+ NIR and SWIR data. *Agric. For. Meteorol.* **2008**, *148*, 1679–1695. [CrossRef]
23. Clevers, J.G.P.W.; Kooistra, L.; Schaepman, M.E. Using spectral information from the NIR water absorption features for the retrieval of canopy water content. *Int. J. Appl. Earth Obs. Geoinf.* **2008**, *10*, 388–397. [CrossRef]
24. Cernicharo, J.; Verger, A.; Camacho, F. Empirical and physical estimation of canopy water content from CHRIS/PROBA data. *Remote Sens.* **2013**, *5*, 5265. [CrossRef]

25. Verrelst, J.; Camps-Valls, G.; Muñoz-Marí, J.; Rivera, J.P.; Veroustraete, F.; Clevers, J.G.P.W.; Moreno, J. Optical remote sensing and the retrieval of terrestrial vegetation bio-geophysical properties—A review. *ISPRS J. Photogramm. Remote Sens.* **2015**, *108*, 273–290. [CrossRef]
26. Ceccato, P.; Flasse, S.; Tarantola, S.; Jacquemoud, S.; Grégoire, J.-M. Detecting vegetation leaf water content using reflectance in the optical domain. *Remote Sens. Environ.* **2001**, *77*, 22–33. [CrossRef]
27. Zarco-Tejada, P.J.; Rueda, C.A.; Ustin, S.L. Water content estimation in vegetation with MODIS reflectance data and model inversion methods. *Remote Sens. Environ.* **2003**, *85*, 109–124. [CrossRef]
28. Yilmaz, M.T.; Hunt, E.R.; Jackson, T.J. Remote sensing of vegetation water content from equivalent water thickness using satellite imagery. *Remote Sens. Environ.* **2008**, *112*, 2514–2522. [CrossRef]
29. Baret, F.; Guyot, G. Potentials and limits of vegetation indices for LAI and APAR assessment. *Remote Sens. Environ.* **1991**, *35*, 161–173. [CrossRef]
30. Houborg, R.; Soegaard, H.; Boegh, E. Combining vegetation index and model inversion methods for the extraction of key vegetation biophysical parameters using terra and aqua MODIS reflectance data. *Remote Sens. Environ.* **2007**, *106*, 39–58. [CrossRef]
31. Verrelst, J.; Malenovský, Z.; Van der Tol, C.; Camps-Valls, G.; Gastellu-Etchegorry, J.-P.; Lewis, P.; North, P.; Moreno, J. Quantifying vegetation biophysical variables from imaging spectroscopy data: A review on retrieval methods. *Surv. Geophys.* **2018**. [CrossRef]
32. Jacquemoud, S.; Bacour, C.; Poilvé, H.; Frangi, J.P. Comparison of four radiative transfer models to simulate plant canopies reflectance: Direct and inverse mode. *Remote Sens. Environ.* **2000**, *74*, 471–481. [CrossRef]
33. Atzberger, C.; Richter, K. Spatially constrained inversion of radiative transfer models for improved LAI mapping from future sentinel-2 imagery. *Remote Sens. Environ.* **2012**, *120*, 208–218. [CrossRef]
34. Labate, D.; Ceccherini, M.; Cisbani, A.; De Cosmo, V.; Galeazzi, C.; Giunti, L.; Melozzi, M.; Pieraccini, S.; Stagi, M. The PRISMA payload optomechanical design, a high performance instrument for a new hyperspectral mission. *Acta Astronaut.* **2009**, *65*, 1429–1436. [CrossRef]
35. Lee, C.M.; Cable, M.L.; Hook, S.J.; Green, R.O.; Ustin, S.L.; Mandl, D.J.; Middleton, E.M. An introduction to the NASA hyperspectral InfraRed imager (HyspIRI) mission and preparatory activities. *Remote Sens. Environ.* **2015**, *167*, 6–19. [CrossRef]
36. Feingersh, T.; Eyal, B.D. SHALOM—A commercial hyperspectral space mission. In *Optical Payloads for Space Missions*; Qian, S.-E., Ed.; John Wiley & Sons: Hoboken, NJ, USA, 2015.
37. Nieke, J.; Rast, M. Towards the copernicus hyperspectral imaging mission for the environment (CHIME). In Proceedings of the IGGARS 2018, Valencia, Spain, 22–27 July 2018; pp. 157–159.
38. Guanter, L.; Kaufmann, H.; Segl, K.; Foerster, S.; Rogass, C.; Chabrillat, S.; Kuester, T.; Hollstein, A.; Rossner, G.; Chlebek, C.; et al. The enmap spaceborne imaging spectroscopy mission for earth observation. *Remote Sens.* **2015**, *7*, 8830–8857. [CrossRef]
39. Green, R.O.; Conel, J.E.; Margolis, J.; Bruegge, J.C.; Hoover, L.G. An inversion algorithm for retrieval of atmospheric and leaf water absorption from AVIRIS radiance with compensation for atmospheric scattering. In *Third Airborne Visible/Infrared Imaging Spectrometer (AVIRIS) Workshop*; Green, O.R., Ed.; NASA: Pasadena, CA, USA, 1991; pp. 51–61.

40. Green, R.O.; Conel, J.E.; Roberts, D.A. Estimation of aerosol optical depth, pressure elevation, water vapor, and calculation of apparent surface reflectance from radiance measured by the airborne visible/infrared imaging spectrometer (AVIRIS). In Proceedings of the Summaries of the 4th Annual JPL Airborne Geoscience Workshop, AVJRIS Workshop, Washington, DC, USA, 25–29 October 1993; Volume 1937, pp. 73–76.
41. Ustin, S.L.; Riaño, D.; Hunt, E.R. Estimating canopy water content from spectroscopy. *Isr. J. Plant Sci.* **2012**, *60*, 9–23. [CrossRef]
42. Hunt, E.R.; Ustin, S.L.; Riaño, D. Remote sensing of leaf, canopy, and vegetation water contents for satellite environmental data records. In *Satellite-Based Applications on Climate Change*; Qu, J., Powell, A., Sivakumar, M.V.K., Eds.; Springer: Dordrecht, The Netherlands, 2013; pp. 335–357.
43. Knipling, E.B. Physical and physiological basis for the reflectance of visible and near-infrared radiation from vegetation. *Remote Sens. Environ.* **1970**, *1*, 155–159. [CrossRef]
44. Carter, G.A. Primary and secondary effects of water content on the spectral reflectance of leaves. *Am. J. Bot.* **1991**, *78*, 916–924. [CrossRef]
45. Gao, B.-C.; Goetz, A.F.H. Column atmospheric water vapor and vegetation liquid water retrievals from airborne imaging spectrometer data. *J. Geophys. Res. Atmos.* **1990**, *95*, 3549–3564. [CrossRef]
46. Gao, B.-C.; Goetz, A.F.H. Retrieval of equivalent water thickness and information related to biochemical components of vegetation canopies from AVIRIS data. *Remote Sens. Environ.* **1995**, *52*, 155–162. [CrossRef]
47. Green, R.O.; Painter, T.H.; Roberts, D.A.; Dozier, J. Measuring the expressed abundance of the three phases of water with an imaging spectrometer over melting snow. *Water Resour. Res.* **2006**, *42*. [CrossRef]
48. Thompson, D.R.; Gao, B.-C.; Green, R.O.; Roberts, D.A.; Dennison, P.E.; Lundeen, S.R. Atmospheric correction for global mapping spectroscopy: ATREM advances for the HyspIRI preparatory campaign. *Remote Sens. Environ.* **2015**, *167*, 64–77. [CrossRef]
49. Yi, Q.; Wang, F.; Bao, A.; Jiapaer, G. Leaf and canopy water content estimation in cotton using hyperspectral indices and radiative transfer models. *Int. J. Appl. Earth Obs. Geoinf.* **2014**, *33*, 67–75. [CrossRef]
50. Pasqualotto, N.; Delegido, J.; Van Wittenberghe, S.; Verrelst, J.; Rivera, J.P.; Moreno, J. Retrieval of canopy water content of different crop types with two new hyperspectral indices: Water absorption area index and depth water index. *Int. J. Appl. Earth Obs. Geoinf.* **2018**, *67*, 69–78. [CrossRef]
51. Bach, H. *Die Bestimmung Hydrologischer und Landwirtschaftlicher Oberflächenparameter aus Hyperspektralen Fernerkundungsdaten*; Geobuch-Verlag: München, Germany, 1995.
52. Ustin, S.L.; Roberts, D.A.; Pinzón, J.; Jacquemoud, S.; Gardner, M.; Scheer, G.; Castañeda, C.M.; Palacios-Orueta, A. Estimating canopy water content of chaparral shrubs using optical methods. *Remote Sens. Environ.* **1998**, *65*, 280–291. [CrossRef]
53. Champagne, C.M.; Staenz, K.; Bannari, A.; McNairn, H.; Deguise, J.-C. Validation of a hyperspectral curve-fitting model for the estimation of plant water content of agricultural canopies. *Remote Sens. Environ.* **2003**, *87*, 148–160. [CrossRef]
54. Meier, U. *Growth Stages of Mono- and Dicotyledonous Plants: BBCH Monograph*; Open Agrar Repositorium: Quedlinburg, Germany, 2018.
55. Hosgood, B.; Jacquemoud, S.; Andreoli, J.; Verdebout, A.; Pedrini, A.; Schmuck, G. *Leaf Optical Properties Experiment 93 (LOPEX93)*; European Commission: Brussels, Belgium, 1995.
56. Jacquemoud, S.; Bidel, C.; Pavan, F.G. Angers Leaf Optical Properties Database. 2003. Available online: http://ecosis.org (accessed on 14 November 2017).
57. Jacquemoud, S.; Baret, F. Prospect: A model of leaf optical properties spectra. *Remote Sens. Environ.* **1990**, *34*, 75–91. [CrossRef]
58. Verhoef, W. Light scattering by leaf layers with application to canopy reflectance modeling: The sail model. *Remote Sens. Environ.* **1984**, *16*, 125–141. [CrossRef]
59. Verhoef, W.; Bach, H. Coupled soil–leaf-canopy and atmosphere radiative transfer modeling to simulate hyperspectral multi-angular surface reflectance and TOA radiance data. *Remote Sens. Environ.* **2007**, *109*, 166–182. [CrossRef]
60. Féret, J.B.; Gitelson, A.A.; Noble, S.D.; Jacquemoud, S. Prospect-D: Towards modeling leaf optical properties through a complete lifecycle. *Remote Sens. Environ.* **2017**, *193*, 204–215. [CrossRef]
61. Feret, J.-B.; François, C.; Asner, G.P.; Gitelson, A.A.; Martin, R.E.; Bidel, L.P.R.; Ustin, S.L.; le Maire, G.; Jacquemoud, S. Prospect-4 and 5: Advances in the leaf optical properties model separating photosynthetic pigments. *Remote Sens. Environ.* **2008**, *112*, 3030–3043. [CrossRef]

62. Cannavó, F. Sensitivity analysis for volcanic source modeling quality assessment and model selection. *Comput. Geosci.* **2012**, *44*, 52–59. [CrossRef]
63. Wang, Z.; Skidmore, A.K.; Darvishzadeh, R.; Wang, T. Mapping forest canopy nitrogen content by inversion of coupled leaf-canopy radiative transfer models from airborne hyperspectral imagery. *Agric. For. Meteorol.* **2018**, *253–254*, 247–260. [CrossRef]
64. Zhang, Q.; Li, Q.; Zhang, G. Scattering impact analysis and correction for leaf biochemical parameter estimation using vis-NIR spectroscopy. *Spectroscopy* **2011**, *26*, 28–39.
65. Dorigo, W.A. Improving the robustness of cotton status characterisation by radiative transfer model inversion of multi-angular CHRIS/PROBA data. *IEEE J. Sel. Top. Appl. Earth Obs. Remote Sens.* **2012**, *5*, 18–29. [CrossRef]
66. Zou, X.; Hernandez Clemente, R.; Tammeorg, P.; Lizarazo, C.; Stoddard, F.; Mäkelä, P.; Pellikka, P.; Mõttus, M. Retrieval of leaf chlorophyll content in field crops using narrow-band indices: Effects of leaf area index and leaf mean tilt angle. *Int. J. Remote Sens.* **2015**, *36*, 6031–6055. [CrossRef]
67. Combal, B.; Baret, F.; Weiss, M.; Trubuil, A.; Macé, D.; Pragnère, A.; Myneni, R.; Knyazikhin, Y.; Wang, L. Retrieval of canopy biophysical variables from bidirectional reflectance: Using prior information to solve the ill-posed inverse problem. *Remote Sens. Environ.* **2003**, *84*, 1–15. [CrossRef]
68. Kuester, T.; Spengler, D. Structural and spectral analysis of cereal canopy reflectance and reflectance anisotropy. *Remote Sens.* **2018**, *10*, 1767. [CrossRef]
69. Serrano, L.; Ustin, S.L.; Roberts, D.A.; Gamon, J.A.; Peñuelas, J. Deriving water content of chaparral vegetation from AVIRIS data. *Remote Sens. Environ.* **2000**, *74*, 570–581. [CrossRef]
70. Trombetti, M.; Riaño, D.; Rubio, M.A.; Cheng, Y.B.; Ustin, S.L. Multi-temporal vegetation canopy water content retrieval and interpretation using artificial neural networks for the continental USA. *Remote Sens. Environ.* **2008**, *112*, 203–215. [CrossRef]
71. Berger, K.; Hank, T.; Vuolo, F.; Mauser, W.; D'Urso, G. Optimal exploitation of the sentinel-2 spectral capabilities for crop leaf area index mapping. *Remote Sens.* **2012**, *4*, 561–582.
72. Richter, K.; Hank, T.B.; Mauser, W.; Atzberger, C. Derivation of biophysical variables from earth observation data: Validation and statistical measures. *J. Appl. Remote Sens.* **2012**, *6*, 063557. [CrossRef]
73. Transon, J.; d'Andrimont, R.; Maugnard, A.; Defourny, P. Survey of hyperspectral earth observation applications from space in the sentinel-2 context. *Remote Sens.* **2018**, *10*, 157. [CrossRef]
74. Verrelst, J.; Rivera, J.P.; Gitelson, A.; Delegido, J.; Moreno, J.; Camps-Valls, G. Spectral band selection for vegetation properties retrieval using Gaussian processes regression. *Int. J. Appl. Earth Obs. Geoinf.* **2016**, *52*, 554–567. [CrossRef]
75. Danner, M.; Berger, K.; Wocher, M.; Mauser, W.; Hank, T. Retrieval of biophysical crop variables from multi-angular canopy spectroscopy. *Remote Sens.* **2017**, *9*, 726. [CrossRef]
76. van der Linden, S.; Rabe, A.; Held, M.; Jakimow, B.; Leitão, P.; Okujeni, A.; Schwieder, M.; Suess, S.; Hostert, P. The EnMAP-Box—A toolbox and application programming interface for EnMAP data processing. *Remote Sens.* **2015**, *7*, 11249. [CrossRef]

© 2018 by the authors. Licensee MDPI, Basel, Switzerland. This article is an open access article distributed under the terms and conditions of the Creative Commons Attribution (CC BY) license (http://creativecommons.org/licenses/by/4.0/).

Review

Estimation of LAI with the LiDAR Technology: A Review

Yao Wang [1,2,*] **and Hongliang Fang** [1,2]

[1] LREIS, Institute of Geographic Sciences and Natural Resources Research, Chinese Academy of Sciences, Beijing 100101, China; fanghl@lreis.ac.cn
[2] College of Resources and Environment, University of Chinese Academy of Sciences, Beijing 100049, China
* Correspondence: wangy.18b@igsnrr.ac.cn

Received: 3 September 2020; Accepted: 19 October 2020; Published: 21 October 2020

Abstract: Leaf area index (LAI) is an important vegetation parameter. Active light detection and ranging (LiDAR) technology has been widely used to estimate vegetation LAI. In this study, LiDAR technology, LAI retrieval and validation methods, and impact factors are reviewed. First, the paper introduces types of LiDAR systems and LiDAR data preprocessing methods. After introducing the application of different LiDAR systems, LAI retrieval methods are described. Subsequently, the review discusses various LiDAR LAI validation schemes and limitations in LiDAR LAI validation. Finally, factors affecting LAI estimation are analyzed. The review presents that LAI is mainly estimated from LiDAR data by means of the correlation with the gap fraction and contact frequency, and also from the regression of forest biophysical parameters derived from LiDAR. Terrestrial laser scanning (TLS) can be used to effectively estimate the LAI and vertical foliage profile (VFP) within plots, but this method is affected by clumping, occlusion, voxel size, and woody material. Airborne laser scanning (ALS) covers relatively large areas in a spatially contiguous manner. However, the capability of describing the within-canopy structure is limited, and the accuracy of LAI estimation with ALS is affected by the height threshold and sampling size, and types of return. Spaceborne laser scanning (SLS) provides the global LAI and VFP, and the accuracy of estimation is affected by the footprint size and topography. The use of LiDAR instruments for the retrieval of the LAI and VFP has increased; however, current LiDAR LAI validation studies are mostly performed at local scales. Future research should explore new methods to invert LAI and VFP from LiDAR and enhance the quantitative analysis and large-scale validation of the parameters.

Keywords: leaf area index (LAI); vertical foliage profile (VFP); terrestrial laser scanning (TLS); airborne laser scanning (ALS); spaceborne laser scanning (SLS)

1. Introduction

Leaf area index (LAI) is defined as one half the total green leaf area per unit ground surface area [1]. It is listed as an essential climate variable by the global climate change research community (GCOS) and is a critical variable in processes such as photosynthesis, respiration, and interception [2,3]. The field LAI can be measured using direct sampling or indirect optical methods [4–7]. With a direct sampling method, the LAI can be directly obtained by harvesting vegetation leaves through the collection of leaf litter or destructive sampling [8]. With an indirect optical method, the LAI is estimated from the canopy gap fraction or transmittance using the Beer–Lambert law. The LAI values obtained from ground measurement are often used as references for remote sensing validation. However, these methods are labor-intensive and time-consuming, and the deployment over large areas is difficult.

The LAI estimations from remote sensing data show the most promise for accurate estimations in large scales. Existing techniques can be divided into two main categories, that is, passive optical remote sensing and active light detection and ranging (LiDAR) remote sensing. Passive optical remote sensing has been widely used to estimate the LAI [9–12]. Based on both theoretical models and observations, the LAI and vegetation indices (VI) strongly correlate [13]. One major issue in estimating the LAI from the vegetation index calculated from passive optical sensors is the LAI saturation [7,14,15].

LiDAR is an active remote sensing technology for indirect LAI measurements, which alleviates the saturation problem because of the direct detection of the vertical structure [16]. LiDAR has been applied in many studies for the retrieval of the forest LAI [17–21]. The LAI is estimated from LiDAR data based on the correlation with the gap fraction, which is derived from various laser penetration metrics (LPMs) [19,22]. The LAI can also be estimated through allometric relationships using forest biophysical parameters derived from LiDAR data such as the canopy height and foliage density [23–25]. A few review papers have pointed out that LAI can be effectively estimated from the LiDAR technology [6,7], but further understanding is still required regarding the LAI retrieval methods from different platforms and the basic rationales of the retrieval methods.

This work provides a review of LiDAR technology and the LAI estimation with LiDAR, LAI validation studies, and factors affecting the LAI estimation. Different LiDAR systems, data, and measurement principles are described in Section 2. The methods and applications of LiDAR to LAI estimation are specified in Section 3. The literature on LAI validation studies is discussed in Section 4. In Section 5, factors affecting the LAI estimation are reviewed based on LiDAR data. The prospects of the application of LiDAR to LAI estimation are discussed in Section 6, and the conclusions are summarized in Section 7.

2. LiDAR Technology

2.1. Types of LiDAR Systems

The LiDAR systems can be divided into three categories, that is, discrete return, full waveform, and photon-counting, based on the detection methods [26]. A conceptual diagram of different LiDAR systems is shown in Figure 1. When the laser signal is reflected back to the sensor, discrete return systems record clouds of points representing intercepted features. For example, when light hits different parts of the tree in a forest, the first, second, and third returns are recorded (Figure 1b). Discrete return systems systematically sample and record the X, Y, and Z (elevation) values, producing a high-density three-dimensional (3D) point cloud. In addition to the time, intensity information is recorded [27].

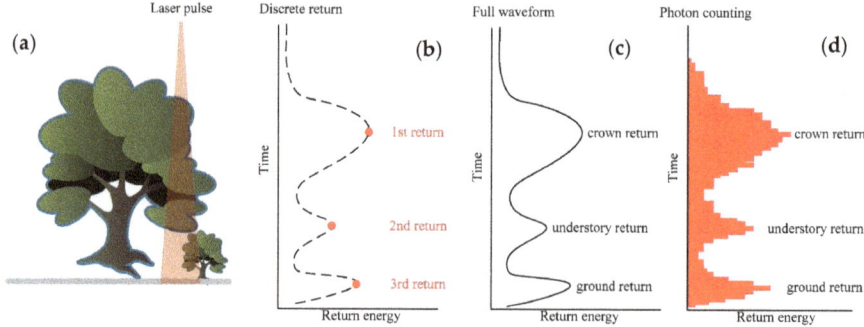

Figure 1. Conceptual diagram of different light detection and ranging (LiDAR) systems (signals returned from trees and the ground). (**a**) Intersection of the laser illumination area with the tree crown and signals received with the (**b**) discrete return, (**c**) full waveform, and (**d**) photon-counting LiDAR systems.

In contrast, full-waveform LiDAR systems digitize the entire reflected energy, resulting in complete submeter vertical vegetation profiles. Within a forest environment, the full waveform indicates the forest structure (i.e., from the top through the crown and understory to the ground) (Figure 1c) [16,26]. In contrast to discrete return and full-waveform systems, photon-counting LiDAR (PCL) systems are unique. They operate based on the concept that a low-power laser pulse is transmitted, and the utilized detectors are sensitive at the single-photon level. Based on this type of detector, any returned photon, whether from the reflected signal or solar background, can trigger an event within the detector [26,28]. An individual photon could be reflected from anywhere within the vertical canopy, the probability distribution function (PDF) of that reflected photon would be the full waveform (Figure 1d).

For different platforms, LiDAR can be categorized into three groups: terrestrial laser scanning (TLS), airborne laser scanning (ALS), and spaceborne laser scanning (SLS). The main characteristics of LiDAR systems used for the estimation of the vegetation LAI are presented in Table 1.

Table 1. Main characteristics of different LiDAR systems for LAI estimation.

Platform	Detection Method	Footprint Size	Capability	Limitation	References
Terrestrial	Discrete return Full waveform	1–5 cm	Derive leaf area density and vertical LAI distribution.	Different sampling frequency for upper and lower canopy, complicated data processing.	[29–32]
Airborne	Discrete return Full waveform	0.1–3 m 10–30 m	Estimate understory and overstory LAI.	Poor penetration, expensive data acquisition.	[17–19]
Spaceborne	Full waveform Photon counting	12–70 m	Derive vertical LAI distribution over a large scale.	Terrain impact, data gaps.	[33–35]

TLS is a ground-based LiDAR technology that acquires 3D details. Typically, these TLS systems emit laser pulses having footprints ranging from 1 cm to 5 cm. TLS systems can be classified into discrete return and full-waveform models by how they record the energy returning to the sensor (Table 1). Discrete return TLS measures the surrounding 3D space using millions to billions of 3D points, which are commonly presented in a point cloud [36]. Most discrete return TLSs record a single range for each laser shot. The TLS point cloud data provide the 3D distribution of canopy at individual tree or stand level and have been widely used for the estimation of the vegetation LAI [20,37]. The full waveform TLS records the light reflected from objects along the laser path, which can be calibrated to power units. Full waveform data can be used to improve the vegetation foliage extraction in forest stands with the fully digitized return pulse [38]. Such data have been used to measure the LAI, foliage profile, and stand height [30,39,40]. Based on the scanning geometry of TLS, near-range objects are more frequently sampled than far-range objects, which may limit its broad application. Figure 2 shows two examples of the TLS data: the point cloud data of a forest sample plot based on single-scan data colored by height (Figure 2a), and the TLS intensity image of the Harvard Forest, observed from Echidna Validation Instrument (EVI) full-waveform data (Figure 2b).

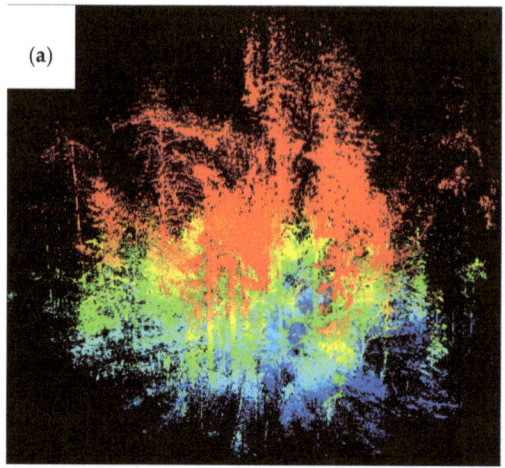

Figure 2. Terrestrial laser scanning (TLS) data examples. (**a**) A forest plot based on single-scan TLS data colored by height; (**b**) Example of Echidna Validation Instrument (EVI) full-waveform-intensity image of the Harvard Forest Plot 01 (42°31'51" N, 72°10'55" W), 2007 [41].

ALS is deployed on fixed or rotary-wing aircraft, which is a data source for multiscale LAI estimation [42]. ALS covers relatively large areas in a spatially contiguous manner. The difference between TLS and ALS is that the ground laser pulse returns from TLS are limited due to the scanning geometry [43]. ALS systems can be classified as discrete and full-waveform. The ALS point cloud data are acquired at altitudes between 500 and 3000 m using a small laser pulse footprint ranging from 0.1 m to 3 m. The ALS point cloud data are the most common type of data used in forest inventory applications and several system developers and an increasing number of commercial vendors support the acquisition and analysis [44]. The ALS full-waveform data are typically acquired at higher altitudes (up to 20,000 m) using a large footprint ranging from 10 m to 30 m. A common ALS full-waveform data system is the airborne Land, Vegetation, and Ice Sensor (LVIS) [45]. Figure 3 shows the point cloud data of a forest plot (colored by height), with an average point density of 4.3 points/m^2 obtained by an ALS (upper), and the LVIS waveform data with a footprint size of 20 m (lower).

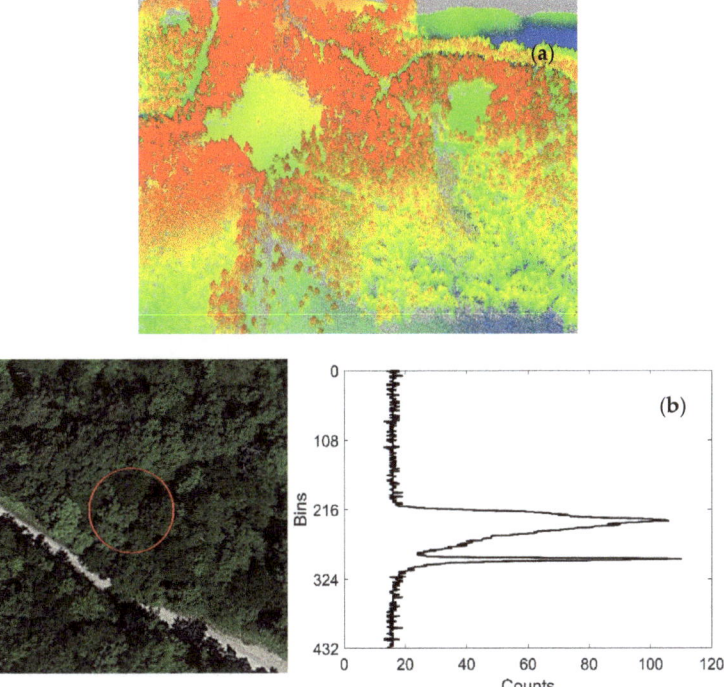

Figure 3. Airborne laser scanning (ALS) data examples. (**a**) Airborne point cloud data colored by height (4.3 points/m^2); (**b**) airborne large-footprint full-waveform data—Land, Vegetation, and Ice Sensor (LVIS)—~20 m in diameter (44°2′59″ N, 71°17′18″ W), 2009 [46]. The image in the left panel is from Google Earth®, and the right panel is a waveform return where the upper and lower peaks come from the forest canopy and the ground surface, respectively.

SLS is deployed onboard satellites and can be classified into full-waveform and photo counting. Compared with the ALS and TLS, which have insufficient spatial coverage, the SLS is more suitable for forest surveys at global scales [47]. The Geoscience Laser Altimeter System (GLAS) is a spaceborne full-waveform LiDAR system onboard the Ice, Cloud, and land Elevation Satellite-1 (ICESat-1) [48,49]. The GLAS full-waveform data provide the first spaceborne laser altimetry data for Earth observations and have been applied for LAI estimations [33]. Figure 4 shows an example of the GLAS data with a footprint size of 70 m. The follow-on to the ICESat mission, ICESat-2, launched in September 2018, carries the Advanced Topographic Laser Altimeter System (ATLAS), a LiDAR system that utilizes the photon-counting technology [35]. The ICESat-2/ATLAS data have potential for the estimation of the forest canopy cover and LAI [50]. In addition, the Global Ecosystem Dynamics Investigation (GEDI) installed on the International Space Station (ISS) was launched in December 2018 [34]. The GEDI measures the forest canopy height and canopy vertical structure. GEDI provides the raw waveforms (Level 1B) and the LAI and vertical profile data (L2B) from the Land Processes Distributed Active Archive Center (LPDAAC) (https://lpdaac.usgs.gov/data/get-started-data/collection-overview/missions/gedi-overview/).

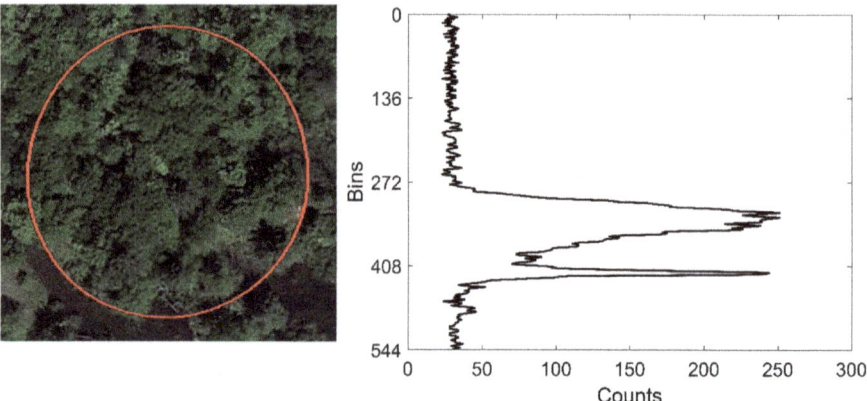

Figure 4. Example Geoscience Laser Altimeter System (GLAS) data, ~70 m in diameter (44°5'10" N, 71°17'43" W), in 2004. Image in the left panel is from Google Earth®, and the right panel is a waveform return where the upper and lower peaks come from the forest canopy and the ground surface, respectively.

2.2. Data Preprocessing

Preprocessing of the point cloud data includes registration, noise removal, and ground filtering (Figure 5) [51,52]. During the multiple scans, the data of all scan locations must be registered into a single-point cloud dataset with a common coordinate to clip overlapping points and resolve the point redundancy [53,54]. Two registration methods can be used: the reflector registration and the reflector-free registration method [55]. Commercial TLS scanners provide reflectors to be placed in each scan. However, this method is limited in forestry applications because few natural tie points can be found. Therefore, artificial reflectors are used as tie points for registration [31]. The placement of artificial reflectors is labor-intensive because of obstructions. Therefore, several reflector-free registration methods have been applied to overcome difficulties associated with the placement of artificial reflectors [56–59]. After the data registration, noise removal and ground filtering were handled together. The noises in point cloud data, that is, the unreasonably high or low elevation values, are often randomly distributed. The simplest way to identify such outliers is to examine the frequency distribution of elevation values [60,61]. Subsequently, non-ground points and ground are separated by ground filtering methods [61,62]. Filtering methods can be mainly categorized into three groups: slope-based, mathematical morphology-based, and surface-based methods [63]. Slope-based methods are based on the assumption that the change of the slope of the terrain is generally gradual in a neighborhood, while the change of the slope between buildings or trees and the ground is very abrupt and large [64,65]. Slope-based methods are affected by complex terrain. Mathematical morphology-based methods use mathematical morphology to remove non-ground points [62]. The window size is critical for mathematical morphology-based methods. In contrast to slope-based and mathematical morphology-based methods, surface-based methods gradually approximate the ground surface by iteratively selecting ground points from the original dataset. The core of surface-based methods is to create a surface that is close to the bare ground [55,66]. After filtering the data, ground points are used to produce a digital elevation model (DEM). The height of LiDAR returns is normalized with respect to the corresponding DEM. The relative height above the DEM can be used to remove understory vegetation [37,67] and the height threshold can be used to separate ground returns and canopy returns [23].

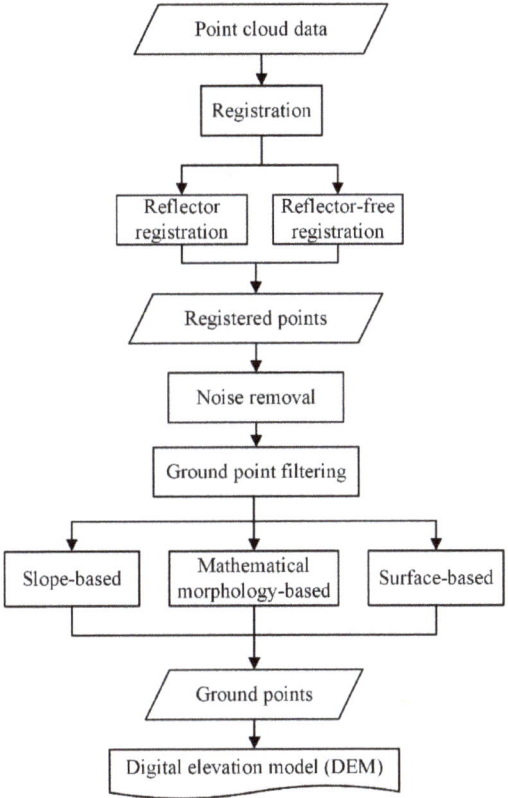

Figure 5. Preprocessing of the point cloud data.

Preprocessing of the full waveform data includes smoothing, the identification of signal start and end points, and Gaussian decomposition (Figure 6). Because of the rough shapes of waveforms, the estimation of initial values results in a large number of modes with narrow widths and low amplitudes. Therefore, it is necessary to smooth the waveforms to obtain a smaller number of modes [68]. To identify the signal start and end points, each received waveform is first smoothed using a Gaussian filter with a width similar to that of the transmitted laser pulse [69,70]. A threshold above the background noise level is then used to obtain the signal start and end points. The signal start and end points are the first and last bin locations at which the waveform intensity is larger than the threshold above the mean background noise in the waveform [71]. Different thresholds have been used in the literature including 3 times the standard deviation [70,72,73], 4 times the standard deviation [74], or 4.5 times the standard deviations [75–77]. There is no consistent optimal threshold that can be applied to all sites [69]. Finally, a Gaussian decomposition algorithm (Equation (1)) is used to decompose the filtered waveform into a series of Gaussian peaks, where the last peak of decomposed components corresponds to the ground surface [78]:

$$min\left\{\left\|f(x) - \sum_{i=1}^{n} |a_i| exp\left[-\frac{(x-\mu_i)^2}{2\sigma_i^2}\right]\right\|\right\} \quad (1)$$

where $f(x)$ is the LiDAR waveform; a least-squared method is used to compute the model parameters; a_i, μ_i, and σ_i are the amplitude, location, and width of a decomposed Gaussian peak; and n is the number of decomposed Gaussian peaks. The absolute value of a_i is used in Equation (1) to avoid decomposed Gaussian peaks with negative amplitudes.

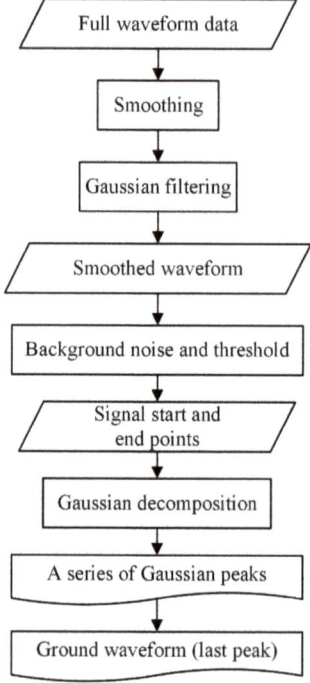

Figure 6. Preprocessing of the full waveform data.

3. LAI Retrieval Methods

3.1. Gap-Based Methods

The LAI is mainly estimated from LiDAR data by means of the correlation with the gap fraction [6,21,22,30,31,79,80]. The theoretical basis of gap-based methods is the Beer–Lambert law [81]:

$$P(\theta) = e^{-G(\theta) \cdot LAI / \cos \theta} \quad (2)$$

where $P(\theta)$ is the canopy gap fraction at zenith angle θ.

Based on the above equation, the light attenuation in vegetation canopies can be represented by the light extinction as a function of the LAI [22,82]:

$$LAI = -\frac{1}{k} \ln(I/I_0) \quad (3)$$

where I is the below-canopy light intensity, I_0 is the above-canopy light intensity, k is the extinction coefficient, and I/I_0 is the fraction of light transmitted through the canopy.

However, the gap fraction cannot directly be measured by LiDAR but is derived from LiDAR metrics. Various LPMs are used as proxies for I/I_0 to estimate LAI. For point cloud data, the LPM can be calculated as the ratio of the number of ground returns to the number of total returns or the number of sky pixels to the number of total pixels (number-based ratio) and the ratio of the sum of intensity values of ground returns to the sum of intensity values of total returns (intensity-based ratio) [19,83]. For the waveform data, the LPM can be calculated by the ratio of the ground return energy to the total return energy (ground-to-total energy ratio) [18,84,85]. Based on different LiDAR data characteristics, a variety of LPMs can be used for the estimation of the LAI (Table 2).

Table 2. Summary of different laser penetration metrics (LPMs) used for LAI estimation.

Data	LPM	Description	Symbol
Point cloud	Number-based ratio	Ratio of ground (or sky) return number to the total return number	$\frac{N_{ground}}{N_{total}}$ or $\frac{N_{sky}}{N_{total}}$
	Intensity-based ratio	Ratio of ground return point intensity to the total intensity value	$\frac{I_{ground}}{I_{total}}$
Full waveform	Ground-to-total energy ratio	Ratio of ground return energy to the total return energy	$\frac{E_{ground}}{E_{total}}$

For point cloud data from TLS, the gap fraction can be calculated by the ratio of the number of sky pixels to the number of total pixels [79]. Two methods are commonly used for the LAI estimation from point cloud data based on the gap fraction theory. One is a two-dimensional (2D) approach and the other is a 3D approach [20]. The main steps of the 2D method are as follows (Figure 7): (1) The 3D point cloud data are first converted to the spherical coordination system using spherical projection; (2) data from the hemisphere are then converted to a plane; that is, the 3D point cloud data are converted to 2D raster images using geometrical projection; (3) 2D raster images are divided into sky and foliage elements, and the gap fraction is calculated from the ratio of the number of sky pixels to the number of total pixels; and (4) the gap fraction is used to retrieve the LAI. Danson et al. [79] estimated the forest canopy gap fraction from TLS point cloud data based on the 2D method. The results showed that the gap fraction obtained from TLS data is similar to the gap fraction measured in the field. Zheng et al. [20] converted the 3D point cloud data to 2D raster images using two geometrical projection techniques and estimated the effective LAI (LAI$_{eff}$). They reported that the stereographic projection-based TLS LAI$_{eff}$ model is more robust than the Lambert azimuthal equal-area projection TLS LAI$_{eff}$ model.

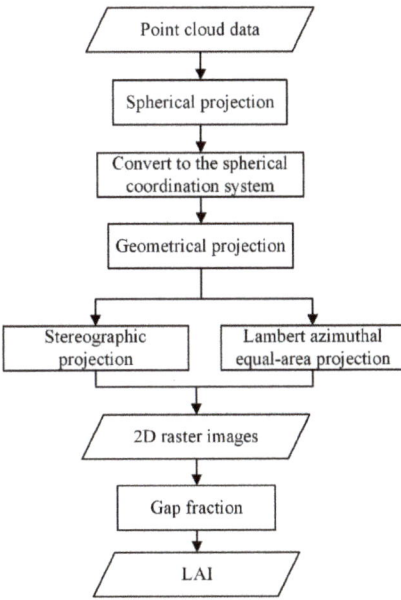

Figure 7. Estimation of leaf area index (LAI) from point cloud data with the 2D method.

LAI derived from TLS point cloud data using 2D method is in good agreement with the field LAI data. However, the 3D structural information of the scanned canopy is lost and the gap fraction must be obtained from the original 3D data. In 3D-based methods, the main steps to estimate LAI include (Figure 8): (1) the point cloud data are voxelized and a voxel-based tree model is produced from the registered point cloud dataset; (2) gap fraction of the layer is calculated by the number of empty voxels and total number of incident laser beams that reach each horizontal layer; (3) the LAI of each horizontal layer is calculated by using the gap fraction in the layer and extinction coefficient; and (4) the LAI of tree is cumulated from the first horizontal layer to the last layer. Takeda et al. [86] divided the TLS point cloud data into horizontal and vertical elements, calculated the gap fraction for each voxel, and estimated the plant area index (PAI). The PAI estimate from TLS is in good agreement with the field data (R^2 = 0.69). Zheng and Moskal [31] proposed a voxel-based algorithm to quantitatively identify the canopy structure and directly predict the LAI_{eff} from TLS. The results showed that the voxel-based method explains 88.7% of the LAI_{eff} of the field measurement. These results show that TLS provides direct, nondestructive estimates of the LAI.

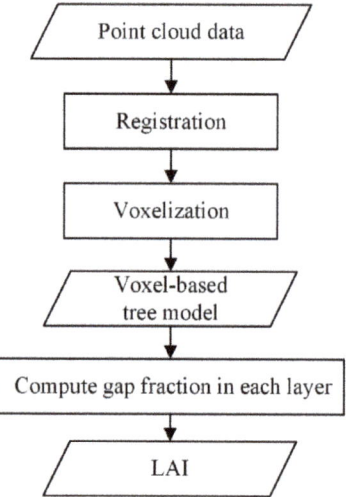

Figure 8. Estimation of LAI from point cloud data with the 3D method.

The LPM can also be calculated as the ratio of the number of ground returns to the number of total returns [19,83,87]. Figure 9 shows the flowchart of LAI estimation based on return number or intensity: (1) the ground and canopy returns are separated according to the height threshold; (2) the gap fraction is calculated as the ratio of the number of ground returns to the number of total returns (or the ratio of the sum of intensity values of the ground returns to the sum of intensity values of the total returns); and (3) the LAI is determined using the gap fraction based on the Beer–Lambert law. The LAI obtained from ALS point cloud data using the number-based ratio is in good agreement with the field LAI [22,23,88]. Aside from the number of returns, the ALS point cloud data intensity is increasingly used in remote sensing applications. Zhao and Popescu [19] used the ratio of the sum of intensity values of the ground returns to the sum of intensity values of the total returns to estimate the LAI. In addition, the forest LAI can be reliably estimated using the LPM based on intensity alone, with R^2 = 0.610 [83].

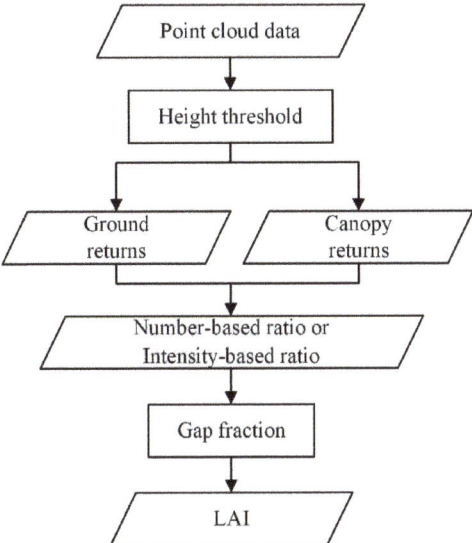

Figure 9. Estimation of LAI based on return number or intensity ratios.

The full-waveform LiDAR records a continuous distribution of returned energy and can be used to characterize the LAI and vertical LAI profile. For the full waveform, the LPM can be calculated as the ratio of the ground return energy to the total return energy (ground-to-total energy ratio) [18,84,85]. The main steps to estimate LAI based on ground-to-total energy ratio are as follows: (1) the ground and canopy returns are separated using height threshold/Gaussian decomposition; (2) the ground-to-total energy ratio is used to calculate the LPM; and (3) the LPM is used to estimate the LAI (Figure 10).

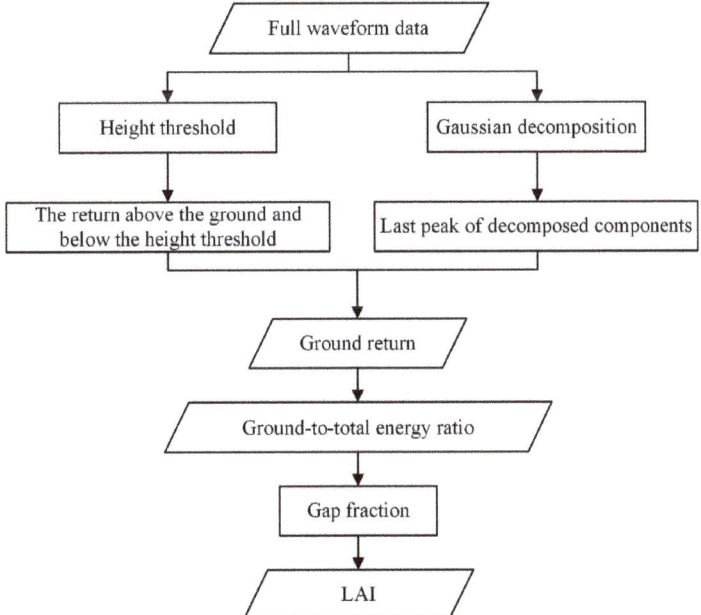

Figure 10. Estimation of LAI based on the ground-to-total energy ratio.

The separation of the ground and canopy returns is a key step to calculate the ground-to-total energy ratio. Different separation methods have been applied. A simple method is based on the height threshold; that is, the ground return energy is calculated using the height above the ground and below the height threshold. A height threshold of 2.0 m has been used to separate the ground and canopy when calculating the ground-to-total energy ratio. Based on the method, the GLAS-predicted and field-measured LAI are in good agreement [84,89]. However, the height threshold varies depending on the location and species. Therefore, the Gaussian decomposition method is incorporated to derive the LAI and vertical foliage profile (VFP) using LiDAR. Tang et al. [18] used the Gaussian decomposition method [90] to decompose full-waveform data into multiple Gaussian functions. The last peak of all decomposed Gaussian components is assumed to be the ground peak and the last Gaussian fit represents the ground return [78,91]. The gap fraction was calculated using laser energy returns from the canopy and ground and used to estimate the LAI and VFP [15,18,33]. The ability to derive the LAI and VFP facilitates large-scale LAI mapping using LiDAR, and it frees the requirement for associated field data.

3.2. Contact-Based Methods

The basis of contact-based methods is the contact frequency, calculated as the probability of a beam penetrating the canopy coming into contact with a vegetative element [32,92]. The contact frequency $N(\theta)$ between a light beam and vegetation element in the direction (θ) is given by

$$N(\theta) = G(\theta)\frac{LAI}{\cos\theta} \quad (4)$$

where $G(\theta)$ is the projection function that corresponds to the fraction of foliage projected on the plane normal to the solar direction.

For TLS point cloud data, the main steps to estimate LAI based on contact frequency include the following (Figure 11): (1) the point cloud data are voxelized, and a voxel-based tree model is produced from the registered point cloud dataset; (2) the contact frequency of the layer is calculated from the number of intercepted laser beams and total number of incident laser beams that reach each horizontal layer (Equation (5)); (3) the LAI of each horizontal layer is calculated by using the contact frequency of the laser beams in the layer and projection function (Equation (6)); and (4) the LAI of the entire tree is cumulated from the first horizontal layer to the last layer.

$$N(k) = \frac{n_I(k)}{n_I(k) + n_P(k)} \quad (5)$$

where $N(k)$ is the contact frequency of laser beams in the kth layer, $n_I(k)$ is the number of laser beams intercepted by the kth layer, $n_P(k)$ is the number of laser beams passing through the kth layer, and $n_I(k) + n_P(k)$ is the total number reach the kth layer.

$$LAI(k) = N(k) \cdot \frac{\cos\theta}{G(\theta)} \quad (6)$$

where $LAI(k)$ is the LAI of the kth horizontal layer within the canopy.

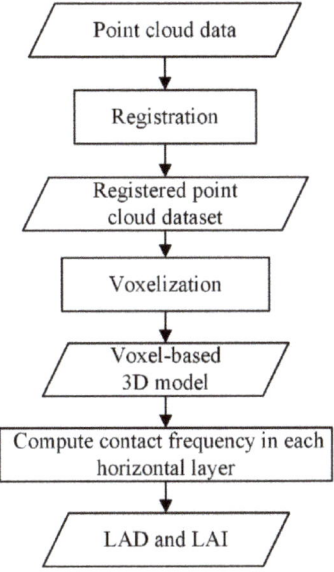

Figure 11. Estimation of LAI based on the contact frequency.

Hosoi and Omasa [32] developed a voxel-based canopy profiling (VCP) method based on the contact frequency for accurate LAD and LAI estimation using TLS point cloud data. First, they produced a voxel-based tree model, where the voxel is defined as a volume element in a 3D array. In each horizontal layer, the voxel was then assigned a different attribute value. For example, a voxel with attribute 1 indicates that the laser beams are intercepted, and a voxel with attribute 2 reflects that the laser beams pass through. The contact frequency of the layer is calculated by the number of intercepted laser beams and total number of incident laser beams. Finally, the contact frequency of each horizontal layer is calculated from the bottom canopy layer (lowest horizontal layer) to the top canopy layer (highest horizontal layer), and the LAIs of each horizontal layer and entire tree are obtained. The authors demonstrated that the LAD and LAI can be computed by directly counting the contact frequency based on the precise voxel model. The error of the best LAD and LAI estimations was 0.7%. The contact frequency calculated from TLS point cloud data utilizing the voxel-based approach can also be used to estimate different LAI layers in savanna trees [93]. The VCP method is one of the methods used to estimate the LAI and VFP from TLS point cloud data. The method converts point cloud data into a voxel-based 3D model that can reproduce each tree. The voxel model computes LAD and LAI by directly counting the contact frequency for each layer and the whole canopy. However, voxel-based approaches are usually associated with time-consuming data acquisition and registration. In addition, the accuracy of these approaches depends on the voxel size.

3.3. Biophysical Regression Methods

The LAI can be estimated from the regression of forest biophysical parameters derived from LiDAR, such as LiDAR height and foliage density metrics [19,94,95]. The main steps are as follows: (1) LiDAR metrics are extracted from LiDAR data; and (2) LiDAR-derived metrics are used to estimate the vegetation LAI using allometric relationships. A multivariate linear regression model [94] and partial least squares regression model [96] utilize the height and foliage density metrics to estimate the LAI of the forest. In addition, LiDAR intensity data are increasingly used in estimation of LAI, and intensity metrics are related to the LAI [25]. The combination of the height and intensity metrics has a higher predictive power [19,83].

3.4. Model Inversion Methods

Physical radiative transfer (RT) models have been implemented to simulate LiDAR data under specific forest stand representations and LiDAR specifications [97–101]. The LAI can be retrieved from LiDAR data using the model inversion method. Sun and Ranson [100] developed a 3D model that successfully simulates full-waveform data by building a 3D forest stand scene, which is divided into cells with specific characteristics. Koetz et al. [102] adapted the model of Sun and Ranson [100] and inverted a 3D LiDAR waveform model to estimate the fractional cover and overstory LAI of a coniferous forest. The LAI estimate agreed well with the field measurements (RMSE = 0.41). Ma et al. [103] used the canopy height derived from LiDAR and canopy structure information derived from the geometric-optical mutual shadowing (GOMS) model to estimate the large-scale forest LAI. Based on the comparison of their results with the field LAI, the highest R^2 values were 0.73. However, their study failed to retrieve the vertical distribution of canopy. Based on the assumption that the gap probability is the reverse of the vertical canopy profile, the vertical distribution of the gap probability can be derived [104], and the LAI and vertical foliage area volume density (FAVD) profile can be directly retrieved from ALS full-waveform data using an RT model [105].

3.5. Method Comparison

Table 3 summarizes the performance of different methods to estimate LAI from the LiDAR data. Among the gap-based category, the 3D method shows the best performance with R^2 = 0.89 and RMSE = 0.007 because the method could improve the accuracy of gap fraction for each layer and provide information about the light penetration condition within the canopy. In contrast, the accuracy based on return intensity is relatively low, because the LiDAR intensity is affected by many factors, such as laser power, incidence angle, object reflectivity, and range of the LiDAR sensor to the object [106]. The other three categories give moderate estimation accuracy with R^2 > 0.69. The regression methods are relatively simple to apply; however, these methods are not universally applicable and need ground LAI measurements. The voxel-based model not only computes LAD and LAI by directly counting the contact frequency for each layer, but also provides the contact frequency information for the whole canopy. However, the voxel-based approaches are usually associated with time-consuming data acquisition and registration processes. In addition, the accuracy of the approaches depends on the voxel size. The LiDAR RT model could simulate LiDAR data under specific forest stand representations and sensor specifications. However, the model is usually complicated and requires many data inputs.

Table 3. Comparison of different methods to estimate LAI from the LiDAR data. DHP: digital hemispherical photography.

Categories	Methods	Advantages and Disadvantages	R^2	RMSE	References
Gap-based	2D method	Based on commonly accepted theories adopted in DHP, easily applicable in practice. Lacks 3D structural information of the scanned canopy.	0.71	1.03	[20]
	3D method	Improves the accuracy of gap fraction for each layer and provides the light penetration information within the canopy. The voxel resolution directly determines the level of details for the canopy structure.	0.89	0.007	[31]
	Number-based ratio	The penetration metrics are related to LAI, whereas the selection of height threshold and plot size greatly affects the effectiveness of the metrics.	0.70	N/A	[23]
	Intensity-based ratio	The intensity metrics are related to LAI. The combination of intensity and other metrics could provide a higher predictive power. However, the LiDAR intensity value needs to be corrected.	0.61	0.66	[83]
	Ground to total energy ratio	An effective method to derive LAI and VFP from large footprint waveforms. Estimation of the canopy vertical structural information is affected by topography.	0.69	0.33	[91]
Contact-based	Voxel-based method	Compute LAD and LAI by directly counting the contact frequency for each layer, and provide the contact frequency for the whole canopy. The methods are usually associated with time-consuming data acquisition and registration processes, and the accuracy depends on the voxel size.	N/A	0.14	[32]
Biophysical regression	Regression of LiDAR metrics	Approximate LAI from LiDAR metrics, relatively simple to apply. Not universally applicable and need ground LAI measurements.	0.69	0.13	[94]
Model inversion	LiDAR RT model	Simulate LiDAR data under specific forest stand representations and sensor specifications. Complicated and require many data inputs.	0.73	0.67	[105]

4. LAI Validation

Different schemes that have been used to validate the LiDAR LAI include direct comparison methods, scaling-up strategies, and the intercomparison of multiple products.

Field measurements, typically limited to a point or very small area, are vital because they are the basis for all validation studies. Based on the direct comparison method, field measurements and the LAI from different LiDAR systems are directly compared. The field LAI obtained from destructive sampling was used to validate the TLS LAI and LVIS LAI; the LAI derived from LiDAR and destructive sampling were in excellent agreement [18,32]. In addition, the LAI from digital hemispherical photography (DHP) and LAI-2200 are commonly used to validate the LAI from different LiDAR platforms [19,23,30,107,108]. Because of the spatial scale mismatch between field measurements and remote sensing estimation, it is usually difficult to use this method for global validation.

Based on the scaling-up strategy, the field LAI is scaled up via different platforms for the validation with SLS LAI products, thus bridging the scale differences between the field LAI and the LAI derived from SLS. The TLS provides an additional indirect ground-based technique to estimate the LAI. The LAI derived from TLS can be used as field measurement [109–111]. The LAI can be validated using scaling-up strategy at multiple spatial scales through LiDAR remote sensing [91]. First, the ground-based (DHP, LAI-2200, TLS) LAI is related to aircraft observations of the LAI. Then, the ALS observations of the LAI are used to validate the LAI derived from SLS tracks that intersect the aircraft coverage. The upscaling validation method has been widely used in the remote sensing community [112]. However, this method may be affected by several factors. First, ground LAI derived from photos, TLS, and LAI sensors may be inconsistent among themselves. Second, errors are introduced by the scale mismatch between ground field data and ALS. Third, different data sources are based on varying spatial footprints and viewing geometries, which may complicate LAI validation.

Multiple products can be compared to determine the relative quality of land products. The intercomparison method has been used as a proxy to assess the temporal and spatial consistency. The LiDAR-derived LAI values are aggregated to the resolution of the passive satellite LAI products to evaluate all LAI data. The GLOBCARBON [19] and MODIS [113] LAI products have been used to compare with the LiDAR-derived LAI map. The registration between LiDAR and the satellite LAI maps is important because misregistration could severely bias the pixel-by-pixel comparison.

Current validation studies are mostly performed at local scales. The results indicate a significant correlation between airborne LiDAR and the field-derived LAI at the plot scale in a tropical forest, with $R^2 = 0.58$ and RMSE = 1.36 [96], and a moderate agreement ($R^2 = 0.63$, RMSE = 1.36) between LVIS and the field-derived LAI at tower footprint scales in tropical rainforest [18]. Based on a large-scale validation method, R^2 and RMSE values of 0.69 and 0.33 were obtained between the LVIS LAI and GLAS LAI at GLAS tracks in the Sierra National Forest [91]. The LiDAR-derived LAI was evaluated using the MODIS LAI product, yielding R^2 and RMSE values of 0.86 and 0.76 in mixed coniferous forest [113]. However, the LAIs derived from ALS or SLS still lack sufficient ground validation and intercomparison validation using existing global LAI products generated from passive optical sensors. The LiDAR also has the capability to provide the LAI vertical profile, from site [18] to regional and continental [15,114] scales. Due to the lack of ground-measured data on the LAI vertical profile of the forest, the LAI vertical distribution map has not been completely validated. Existing validation work is mainly based on limited site or observation tower data [114].

5. Impact Factors

Several factors should be considered in the estimation of LAI with LiDAR technology (Table 4). The gap fraction theory is based on the assumption of a random distribution of foliage elements [6]. However, in reality, leaves are generally clumped rather than randomly distributed, and thus the LAI estimation must be corrected by considering the error caused by foliage clumping, which leads to underestimations of the LAI [37]. Therefore, several traditional clumping algorithms based on the gap size analysis theory [115] and gap size distribution method [37] were implemented in TLS. The vertical distribution profile of the forest CI can also be calculated from TLS and used to correct the clumping effect of the LAI at different heights [116]. These studies proved that the CI can be successfully estimated using TLS [117]. In contrast, the estimation of the CI using ALS and SLS is rare. Yang et al. [78] presented a physical method with the gap fraction model to estimate the LAI by correcting the between-crown clumping in discontinuous forest using GLAS data. Based on the correction, R^2 and RMSE values of 0.83 and 0.39, respectively, were obtained. However, it is difficult to use SLS to quantify the clumping because the CI is highly variable and changes even in the same forest. The footprint size of ALS and SLS is too large for the detection of small gaps. Therefore, the quantification of the clumping effect remains an ongoing task in ALS and SLS.

The height threshold and sampling size of LiDAR data are key factors affecting the accuracy of the LAI estimation [83,118]. However, there is no optimal sampling size and height threshold when estimating the vegetation LAI. In some cases, the height threshold of LiDAR was set to a constant value, such as 2 m, to separate the ground and canopy returns in ALS point cloud data [22] and separate the ground and canopy when calculating a ground-to-total energy ratio in GLAS [84,89]. A threshold of 3.6 m was used to calculate the LPM value in a temperate forest [19]. In other cases, the height threshold of LiDAR was related to the setup height of the ground measurement. The setup height of the fisheye camera (i.e., 1.25 m) was used to separate ground and vegetation returns [119]. For the LiDAR plot radius, a common choice is to use the same LiDAR sampling size as that used to collect field plot data [94,120]. On the other hand, the sampling size of LiDAR is related to the canopy height. A radius size of the LiDAR sampling scale equivalent to the entire forest canopy height [24] and 0.75 times the tree height [17] was used to calculate the LPM value. However, the choice of a variable radius is difficult: optimum values range from 75% to 100% of the canopy height [19,119]. In summary, the accuracy of the LAI estimation using LiDAR strongly depends on the height threshold and sampling size [121]. However, the height threshold and sampling size are site-specific, and there are no clear guidelines for the determination of an appropriate value.

Table 4. Factors affecting the LAI estimation from LiDAR data.

Factor	Description	Mitigation Method	Advantages and Disadvantages	References
Clumping effect	Leaves are not randomly distributed but clumped, which may cause the underestimation of the LAI.	Estimate CI and make clumping correction.	The CI can be successfully estimated using TLS, whereas it is rare to estimate CI from ALS and SLS.	[37,78,115,116]
Footprint size	The relative contribution of the vegetation and terrain signal of waveforms is different under different footprint sizes.	Study the influence of footprint size by LiDAR model and obtain optimal footprint size.	A large footprint size contains sufficient gaps for the detection of the underlying ground. However, the ALS and SLS footprint sizes are too large for the small gap detection.	[34,122]
Height threshold and sampling size	The height threshold is critical for the separation of ground and vegetation returns. The sampling size affects the LiDAR and field LAI comparison.	Set the height threshold similar to the ground measurement, and the sampling size the size of field plot.	The accuracy of the LAI estimation is highest with the optimal height threshold and sampling size. However, the optimal value is site-specific.	[17,19,24,119,121]
Occlusion	Vegetation elements intercept laser beams and stop them from being in contact with further material along the path.	Acquire data from multiple TLS scans, or combine TLS and ALS data.	Easy to eliminate blind regions and overcome the occlusion effects. However, it will increase the measurement work and the data size.	[32,92]
Topography	The slope can blur the boundary between vegetation and ground return and affects the accuracy of the canopy vertical structure estimation.	Filter out larger slopes, or compensate the terrain effect using slope-adaptive waveform metrics.	Can compensate the terrain stretching on the forest waveform. However, the performance decreases for complex terrain.	[71,77,91,110,123]
Types of return	LiDAR returns are from different canopy layers; using all returns is more effective than using only the first and last returns in deriving LPM.	Calculate LPMs using all returns.	Increases the effective pulse density and the sensitivity to smaller gap sizes. However, the method is not applicable for LiDAR intensity data.	[22,87,88,95,124,125]
Voxel size	Different voxel sizes significantly affect the gap fraction and LAI estimation.	Determine voxel size based on the minimum element size of the object, or based on the TLS characteristics.	Can obtain higher LAI accuracy with the optimal voxel size. However, it is difficult to determine the optimal voxel size, which depends on many factors.	[92,126,127]
Woody material	Woody materials (i.e., stems and branches) may lead to the LAI overestimation.	Joint use of leaf-on and leaf-off LiDAR data; or make use of the geometric and radiometric features in TLS.	Foliar and woody materials can be effectively separated using TLS. However, the classification method used in TLS is not applicable to ALS and SLS.	[37,128–130]

The relative contribution of the vegetation and terrain signal of waveforms is different under different footprint sizes [71]. Pang et al. [122] studied the effects of footprint size on the precision of canopy height estimates by means of simulation. They found that footprints with a diameter between 25 m and 30 m would be ideal to level the effects of vegetation height and terrain slope on waveform length. Milenković et al. [131] studied the influence of footprint size on the precision of forest biomass estimates from spaceborne waveform LiDAR. They recommend an optimal footprint size that is similar to the size of the field plots, i.e., 20 m diameter in their case. Dubayah et al. [34] demonstrated that ~25 m is the optimal footprint size for GEDI. The suggestion is that the size is large enough to capture the entire canopy of larger diameter trees and small enough to limit the vertical mixing of vegetation and ground signals caused by surface slope.

Occlusion plays an important role in the spatial distribution of the density of the point cloud [92,132]. Occlusion effects are caused by material intercepting laser beams and stopping them from being in contact with material along their path, which results in a relatively lower inner-canopy point density and underestimation of the LAI [43]. Occlusion effects typically occur when single-scan LiDAR is applied to complex forest canopies. The addition of scans is an effective strategy to alleviate the occlusion [133]. Multiple scans can be used to obtain detailed information on the 3D distribution of the canopy, leading to an increase in the amount of data for forests and contributing to the solution of problems associated with a short zenith range [126,127,134]. However, multiple scans significantly increase the field work and size of the datasets, and biophysical parameter estimations cannot be directly compared with other instruments. The combination of TLS and ALS data is another strategy that may solve occlusion effects for very dense crowns because the profiles obtained by ALS and TLS complement each other, eliminating blind regions and yielding more accurate LAD profiles than by using each type of LiDAR alone [135].

Topographic effects can lead to significant errors in the vertical distribution of the plant area, with an RMSE up to 66.2% [134]. The slope affects the accuracy of the vertical LAI distribution because it can blur the boundary between vegetation and ground signals in a LiDAR waveform. The ground peak gradually decreases and finally disappears as the terrain slope increases, making their separation difficult and potentially leading to errors in the LAI and VFP estimates. To identify the accurate ground return and minimize the slope-induced error, Yang et al. [123] extended the geometric optical and radiative transfer (GORT) vegetation LiDAR model to consider the effect of the surface topography. The slope effect on waveforms was then assessed using model simulations. Filtering out a larger slope is an effective method to reduce the effect of the terrain [136,137]; for example, GLAS footprints were filtered using a cutoff slope threshold of 20 [91]. Although significant progress has been made regarding the use of waveforms on slopes below 20°, the effect of the terrain slope on the estimation of forest parameters has not been thoroughly addressed, especially for steep slopes (i.e., >20°). Wang et al. [71] proposed slope-adaptive waveform metrics for the estimation of the forest aboveground biomass (AGB). They used a model to calculate waveforms of bare ground with known terrain slopes to compensate the terrain effect. However, the model performance decreases for complex terrains including bumps or pits. The terrain conditions are always simpler for small footprints than for larger footprints. The waveform LiDAR with a small footprint may limit the vertical mixing of vegetation and ground signals caused by the slope [122].

LiDAR returns can be reflected back from different layers of canopies, e.g., crown surfaces, inside or below top crowns, and the ground [19]. For various LAI estimation methods, it is important to note that the LPM is sensitive to the gap fraction regardless of the gap size. The penetration rate based on the first return might be insensitive to small gaps because the ALS footprints are too large to penetrate such gaps [87]. Heiskanen et al. [124] reported that the LPM utilizing the first and last returns to derive the ALS penetration rate is more sensitive to the gap fraction variations because the returns from different part of canopy are considered. Therefore, the LPMs calculated from both the first and last returns may be the best predictors of LAI [95]. Richardson et al. [22] further proved that the ALS penetration rate calculated from the first and last returns strongly correlates with the field LAI_{eff} estimates ($R^2 > 0.9$). Hence, the LPM utilizing the first and last returns is recommended for LAI estimation.

The voxelization procedure involves the specification of the voxel size, which has a marked impact on the LAI estimation, because it significantly affects the gap fraction estimation [126]. Different voxel sizes have been used to produce voxel-based tree models. Van der Zande et al. [92] defined voxel size based on the minimum size of the object element, which ensures a number of point cloud data in each voxel. However, several small gaps between leaves can easily be overlooked. More recently, voxel size has been defined based on TLS characteristics [127]. For example, the voxel size is defined depending on the range and scan resolution of the TLS [32]. Grau et al. [138] assessed the effect of the voxel size on LAI estimation using a simulation framework based on the discrete anisotropic radiative transfer (DART) model. They found that voxel size is site-specific, and to obtain good accuracy of estimation, voxel size should be varied in different forest scenes.

Because woody components are sources of error in indirect LAI estimations, the separation of foliar and woody materials is crucial for the accurate estimation of the LAI [139]. The quantitative evaluation of the contribution of non-photosynthetic canopy components to the LAI will be helpful to improve the LAI estimation accuracy. A possible approach is the joint use of leaf-on and leaf-off LiDAR data. The wood area index (WAI) is generally estimated during leaf-off periods and subtracted from the total plant area index. First, the effective woody area index (WAI_{eff}) is estimated under leaf-off conditions, and the effective PAI (PAI_{eff}) is estimated under leaf-on conditions. Subsequently, the LAI_{eff} is obtained by subtracting WAI_{eff} from PAI_{eff} [110]. However, this method is not suitable for evergreen vegetation such as coniferous forests. Another approach is based on the use of leaf-on LiDAR data and makes use of the geometric and radiometric features. Geometric features (linear features, random features, and surface features) have been utilized to quantitatively evaluate the contribution of woody material to the LAI using TLS point cloud data [128,140], and the shape, normal vector distribution, and structure tensor of TLS data features have been used to separate various tree organs [129]. Zhu et al. [130] proposed an adaptive radius near-neighbor search method to accurately separate foliar and woody materials in a mixed forest using TLS point cloud data. They reported that the use of a combination of radiometric and geometric features outperforms either one of them alone, yielding an average overall accuracy of 84.4% for mixed forests. However, due to the lower point density and bigger footprint of ALS and SLS relative to TLS, the classification method used in TLS is not applicable to ALS and SLS. The quantitative evaluation of the contribution of non-photosynthetic canopy components to the LAI of a forest stand must be focused on in the future.

6. Future Prospects

The major advantage of LiDAR technology is its capability to characterize the vertical vegetation structure at different heights [15]. LiDAR-derived LAIs have been used in the validation of the passive satellite LAI products [19,112]. We expect the use of LiDAR LAI will increase with the growing availability of high-quality LAI data derived from LiDAR. Different LiDAR data provided by the different lidar systems have been used to estimate LAI. However, there is no universal LiDAR metric for LAI estimation; therefore, the selection of proper LiDAR metrics is crucial for LAI estimation. More field measurements and novel LiDAR metrics are necessary for improved LAI estimation in the future.

The ALS observations act as a validation link between field and satellite data [91]. However, the relatively high cost of ALS flight mission has significantly limited its applications. As an alternative platform for ALS, the unmanned aerial vehicle (UAV) costs less but can provide denser points. Therefore, UAV provides an effective platform for LAI estimation [141] and acts as a validation link between field and satellite data [21]. The TLS provides an additional indirect ground-based technique to estimate the LAI [109–111]. However, TLS data acquisition is highly time-consuming and labor-intensive. A new backpack LiDAR system was developed for efficient and accurate forest inventory applications, and the derived LAD fits well with the TLS estimates ($R^2 > 0.92$, RMSE = 0.01 m^2/m^3) [142]. A backpack LiDAR system may provide an alternative platform for TLS data acquisition.

The increasing availability of LiDAR data will greatly enhance the LAI estimation. Fusion of multiple LiDAR data from different systems, platforms, and temporal observations is also a continued research direction [14].

7. Conclusions

In this study, the LiDAR technology, LAI retrieval and validation methods, and factors affecting LAI estimation were reviewed. The use of LiDAR has become an operational data collection option for the retrieval of the LAI. All TLS, ALS, and SLS systems can provide the LAI and VFP.

LAI is mainly estimated from LiDAR data by means of the correlation with the gap fraction and contact frequency, and LAI is also estimated from the regression of forest biophysical parameters derived from LiDAR, such as LiDAR height and foliage density metrics. The TLS provides detailed information on the within-canopy structure at individual tree or stand levels and accurate LAI and VFP distribution within plots. However, TLS data processing is complex and the spatial coverage of TLS is limited. The ALS covers relatively large areas in a spatially contiguous manner and provides multiscale LAI estimation. However, the description of the within-canopy structure is limited because the penetration of the ALS point cloud data is poor in dense vegetation. The SLS provides information regarding the canopy structure with near-global coverage and thus has the potential to produce the global LAI and VFP. However, the accuracy of LAI estimation based on a large-footprint waveform is affected by the terrain. In addition to the limitation of the LiDAR instrument itself, several factors should also be considered in the estimation of LAI with LiDAR technology. Clumping and wood are common factors for all LiDAR systems. The LAI estimation is also affected by occlusion, and voxel size for TLS; sampling size, height threshold and sampling size, and types of return for ALS; and footprint size and topography for SLS. Quantification of these factors remains an ongoing task for LiDAR LAI estimation.

Direct comparison methods, scaling-up strategies, and the intercomparison of multiple products are used to validate the LiDAR LAI. The results show that the LAI derived from LiDAR and reference data were in good agreement. However, current LiDAR LAI validation studies are mostly performed at local scales, such as limited site or observation tower.

Remote sensing techniques have provided powerful and effective tools for estimating the spatial distribution of LAI for large areas. The LAI and VFP have been produced at a large scale using ICESat-1/GLAS, whereas the operation of ICESat-1/GLAS was stopped in 2009. The ICESat-2/ATLAS and GEDI are expected to provide vegetation structure information and large-scale LAI estimates. The usage of LiDAR is expected to increase based on the capability to provide the LAI vertical profile. Future research should explore LiDAR remote sensing inversion with respect to the LAI and VFP, quantitative analysis, and the large-scale validation of the LAI and VFP.

Author Contributions: Conceptualization, Y.W. and H.F.; writing—original draft preparation, Y.W.; writing—review and editing, Y.W. and H.F. All authors have read and agreed to the published version of the manuscript.

Funding: This research was funded by the National Key Research and Development Program of China, grant number 2016YFA0600201 (H.F.).

Acknowledgments: We thank the anonymous reviewers for helpful insights in improving this paper.

Conflicts of Interest: The authors declare no conflict of interest.

References

1. Chen, J.M.; Black, T.A. Defining leaf-area index for non-flat leaves. *Plant Cell Environ.* **1992**, *15*, 421–429. [CrossRef]
2. Alton, P.B. The sensitivity of models of gross primary productivity to meteorological and leaf area forcing: A comparison between a Penman-Monteith ecophysiological approach and the MODIS Light-Use Efficiency algorithm. *Agric. For. Meteorol.* **2016**, *218*, 11–24. [CrossRef]
3. Asner, G.P.; Braswell, B.H.; Schimel, D.S.; Wessman, C.A. Ecological research needs from multiangle remote sensing data. *Remote Sens. Environ.* **1998**, *63*, 155–165. [CrossRef]
4. Jonckheere, I.; Fleck, S.; Nackaerts, K.; Muys, B.; Coppin, P.; Weiss, M.; Baret, F. Review of methods for in situ leaf area index determination-Part I. Theories, sensors and hemispherical photography. *Agric. For. Meteorol.* **2004**, *121*, 19–35. [CrossRef]
5. Weiss, M.; Baret, F.; Smith, G.J.; Jonckheere, I.; Coppin, P. Review of methods for in situ leaf area index (LAI) determination Part II. Estimation of LAI, errors and sampling. *Agric. For. Meteorol.* **2004**, *121*, 37–53. [CrossRef]
6. Yan, G.; Hu, R.; Luo, J.; Weiss, M.; Jiang, H.; Mu, X.; Xie, D.; Zhang, W. Review of indirect optical measurements of leaf area index: Recent advances, challenges, and perspectives. *Agric. For. Meteorol.* **2019**, *265*, 390–411. [CrossRef]
7. Zheng, G.; Moskal, L.M. Retrieving Leaf Area Index (LAI) Using Remote Sensing: Theories, Methods and Sensors. *Sensors* **2009**, *9*, 2719–2745. [CrossRef]
8. Fang, H.L.; Li, W.J.; Wei, S.S.; Jiang, C.Y. Seasonal variation of leaf area index (LAI) over paddy rice fields in NE China: Intercomparison of destructive sampling, LAI-2200, digital hemispherical photography (DHP), and AccuPAR methods. *Agric. For. Meteorol.* **2014**, *198*, 126–141. [CrossRef]
9. Vina, A.; Gitelson, A.A.; Nguy-Robertson, A.L.; Peng, Y. Comparison of different vegetation indices for the remote assessment of green leaf area index of crops. *Remote Sens. Environ.* **2011**, *115*, 3468–3478. [CrossRef]
10. Kimura, R.; Okada, S.; Miura, H.; Kamichika, M. Relationships among the leaf area index, moisture availability, and spectral reflectance in an upland rice field. *Agric. Water Manag.* **2004**, *69*, 83–100. [CrossRef]
11. Gitelson, A.A. Wide dynamic range vegetation index for remote quantification of biophysical characteristics of vegetation. *J. Plant Physiol.* **2004**, *161*, 165–173. [CrossRef] [PubMed]
12. Broge, N.H.; Leblanc, E. Comparing prediction power and stability of broadband and hyperspectral vegetation indices for estimation of green leaf area index and canopy chlorophyll density. *Remote Sens. Environ.* **2001**, *76*, 156–172. [CrossRef]
13. Chen, J.M.; Cihlar, J. Retrieving leaf area index of boreal conifer forests using landsat TM images. *Remote Sens. Environ.* **1996**, *55*, 153–162. [CrossRef]
14. Shao, G.; Stark, S.C.; de Almeida, D.R.A.; Smith, M.N. Towards high throughput assessment of canopy dynamics: The estimation of leaf area structure in Amazonian forests with multitemporal multi-sensor airborne lidar. *Remote Sens. Environ.* **2019**, *221*, 1–13. [CrossRef]
15. Tang, H.; Ganguly, S.; Zhang, G.; Hofton, M.A.; Nelson, R.F.; Dubayah, R. Characterizing leaf area index (LAI) and vertical foliage profile (VFP) over the United States. *Biogeosciences* **2016**, *13*, 239–252. [CrossRef]
16. Lim, K.; Treitz, P.; Wulder, M.; St-Onge, B.; Flood, M. LiDAR remote sensing of forest structure. *Progress Phys. Geogr.* **2003**, *27*, 88–106. [CrossRef]
17. Solberg, S.; Brunner, A.; Hanssen, K.H.; Lange, H.; Naesset, E.; Rautiainen, M.; Stenberg, P. Mapping LAI in a Norway spruce forest using airborne laser scanning. *Remote Sens. Environ.* **2009**, *113*, 2317–2327. [CrossRef]
18. Tang, H.; Dubayah, R.; Swatantran, A.; Hofton, M.; Sheldon, S.; Clark, D.B.; Blair, B. Retrieval of vertical LAI profiles over tropical rain forests using waveform lidar at La Selva, Costa Rica. *Remote Sens. Environ.* **2012**, *124*, 242–250. [CrossRef]
19. Zhao, K.G.; Popescu, S. Lidar-based mapping of leaf area index and its use for validating GLOBCARBON satellite LAI product in a temperate forest of the southern USA. *Remote Sens. Environ.* **2009**, *113*, 1628–1645. [CrossRef]
20. Zheng, G.; Moskal, L.M.; Kim, S.H. Retrieval of Effective Leaf Area Index in Heterogeneous Forests With Terrestrial Laser Scanning. *IEEE Trans. Geosci. Remote Sens.* **2013**, *51*, 777–786. [CrossRef]
21. Fang, H.L.; Baret, F.; Plummer, S.; Schaepman-Strub, G. An Overview of Global Leaf Area Index (LAI): Methods, Products, Validation, and Applications. *Rev. Geophys.* **2019**, *57*, 739–799. [CrossRef]
22. Richardson, J.J.; Moskal, L.M.; Kim, S.H. Modeling approaches to estimate effective leaf area index from aerial discrete-return LIDAR. *Agric. For. Meteorol.* **2009**, *149*, 1152–1160. [CrossRef]

23. Luo, S.Z.; Wang, C.; Pan, F.F.; Xi, X.H.; Li, G.C.; Nie, S.; Xia, S.B. Estimation of wetland vegetation height and leaf area index using airborne laser scanning data. *Ecol. Indic.* **2015**, *48*, 550–559. [CrossRef]
24. Riano, D.; Valladares, F.; Condes, S.; Chuvieco, E. Estimation of leaf area index and covered ground from airborne laser scanner (Lidar) in two contrasting forests. *Agric. For. Meteorol.* **2004**, *124*, 269–275. [CrossRef]
25. Pope, G.; Treitz, P. Leaf Area Index (LAI) Estimation in Boreal Mixedwood Forest of Ontario, Canada Using Light Detection and Ranging (LiDAR) and WorldView-2 Imagery. *Remote Sens.* **2013**, *5*, 5040–5063. [CrossRef]
26. Wulder, M.A.; White, J.C.; Nelson, R.F.; Naesset, E.; Orka, H.O.; Coops, N.C.; Hilker, T.; Bater, C.W.; Gobakken, T. Lidar sampling for large-area forest characterization: A review. *Remote Sens. Environ.* **2012**, *121*, 196–209. [CrossRef]
27. Lim, K.; Treitz, P.; Baldwin, K.; Morrison, I.; Green, J. Lidar remote sensing of biophysical properties of tolerant northern hardwood forests. *Can. J. Remote Sens.* **2003**, *29*, 658–678. [CrossRef]
28. Zhang, J.S.; Kerekes, J.P. First-Principle Simulation of Spaceborne Micropulse Photon-Counting Lidar Performance on Complex Surfaces. *IEEE Trans. Geosci. Remote Sens.* **2014**, *52*, 6488–6496. [CrossRef]
29. Beland, M.; Parker, G.; Sparrow, B.; Harding, D.; Chasmer, L.; Phinn, S.; Antonarakis, A.; Strahler, A. On promoting the use of lidar systems in forest ecosystem research. *For. Ecol. Manag.* **2019**, *450*, 117484. [CrossRef]
30. Zhao, F.; Yang, X.Y.; Schull, M.A.; Roman-Colon, M.O.; Yao, T.; Wang, Z.S.; Zhang, Q.L.; Jupp, D.L.B.; Lovell, J.L.; Culvenor, D.S.; et al. Measuring effective leaf area index, foliage profile, and stand height in New England forest stands using a full-waveform ground-based lidar. *Remote Sens. Environ.* **2011**, *115*, 2954–2964. [CrossRef]
31. Zheng, G.; Moskal, L.M. Computational-Geometry-Based Retrieval of Effective Leaf Area Index Using Terrestrial Laser Scanning. *IEEE Trans. Geosci. Remote Sens.* **2012**, *50*, 3958–3969. [CrossRef]
32. Hosoi, F.; Omasa, K. Voxel-based 3-D modeling of individual trees for estimating leaf area density using high-resolution portable scanning lidar. *IEEE Trans. Geosci. Remote Sens.* **2006**, *44*, 3610–3618. [CrossRef]
33. Tang, H.; Dubayah, R.; Brolly, M.; Ganguly, S.; Zhang, G. Large-scale retrieval of leaf area index and vertical foliage profile from the spaceborne waveform lidar (GLAS/ICESat). *Remote Sens. Environ.* **2014**, *154*, 8–18. [CrossRef]
34. Dubayah, R.; Blair, J.B.; Goetz, S.; Fatoyinbo, L.; Hansen, M.; Healey, S.; Hofton, M.; Hurtt, G.; Kellner, J.; Luthcke, S.; et al. The Global Ecosystem Dynamics Investigation: High-resolution laser ranging of the Earth's forests and topography. *Sci. Remote Sens.* **2020**, *1*, 100002. [CrossRef]
35. Neuenschwander, A.; Pitts, K. The ATL08 land and vegetation product for the ICESat-2 Mission. *Remote Sens. Environ.* **2019**, *221*, 247–259. [CrossRef]
36. Liang, X.; Kankare, V.; Hyyppä, J.; Wang, Y.; Kukko, A.; Haggrén, H.; Yu, X.; Kaartinen, H.; Jaakkola, A.; Guan, F.; et al. Terrestrial laser scanning in forest inventories. *ISPRS-J. Photogramm. Remote Sens.* **2016**, *115*, 63–77. [CrossRef]
37. Zhu, X.; Skidmore, A.K.; Wang, T.; Liu, J.; Darvishzadeh, R.; Shi, Y.; Premier, J.; Heurich, M. Improving leaf area index (LAI) estimation by correcting for clumping and woody effects using terrestrial laser scanning. *Agric. For. Meteorol.* **2018**, *263*, 276–286. [CrossRef]
38. White, J.C.; Coops, N.C.; Wulder, M.A.; Vastaranta, M.; Hilker, T.; Tompalski, P. Remote Sensing Technologies for Enhancing Forest Inventories: A Review. *Can. J. Remote Sens.* **2016**, *42*, 619–641. [CrossRef]
39. Yao, T.; Yang, X.Y.; Zhao, F.; Wang, Z.S.; Zhang, Q.L.; Jupp, D.; Lovell, J.; Culvenor, D.; Newnham, G.; Ni-Meister, W.; et al. Measuring forest structure and biomass in New England forest stands using Echidna ground-based lidar. *Remote Sens. Environ.* **2011**, *115*, 2965–2974. [CrossRef]
40. Jupp, D.L.; Culvenor, D.; Lovell, J.L.; Newnham, G.; Coops, N.; Saunder, G.; Webster, M. Evaluation and validation of canopy laser radar (LIDAR) systems for native and plantation forest inventory. *Final Rep. Prep. For. Wood Prod. Res. Dev. Corp.* **2005**, *20*, 150.
41. Strahler, A.H.; Schaaf, C.; Woodcock, C.; Jupp, D.; Culvenor, D.; Newnham, G.; Dubayah, R.O.; Yao, T.; Zhao, F.; Yang, X. *Echidna Lidar Campaigns: Forest Canopy Imagery and Field Data, U.S.A., 2007–2009*; ORNL Distributed Active Archive Center: Oak Ridge, TN, USA, 2011. [CrossRef]
42. Vincent, G.; Antin, C.; Laurans, M.; Heurtebize, J.; Durrieu, S.; Lavalley, C.; Dauzat, J. Mapping plant area index of tropical evergreen forest by airborne laser scanning. A cross-validation study using LAI2200 optical sensor. *Remote Sens. Environ.* **2017**, *198*, 254–266. [CrossRef]

43. Zhao, K.G.; Garcia, M.; Liu, S.; Guo, Q.H.; Chen, G.; Zhang, X.S.; Zhou, Y.Y.; Meng, X.L. Terrestrial lidar remote sensing of forests: Maximum likelihood estimates of canopy profile, leaf area index, and leaf angle distribution. *Agric. For. Meteorol.* **2015**, *209*, 100–113. [CrossRef]
44. Hyyppa, J.; Hyyppa, H.; Leckie, D.; Gougeon, F.; Yu, X.; Maltamo, M. Review of methods of small-footprint airborne laser scanning for extracting forest inventory data in boreal forests. *Int. J. Remote Sens.* **2008**, *29*, 1339–1366. [CrossRef]
45. Blair, J.B.; Rabine, D.L.; Hofton, M.A. The Laser Vegetation Imaging Sensor: A medium-altitude, digitisation-only, airborne laser altimeter for mapping vegetation and topography. *ISPRS J. Photogramm. Remote Sens.* **1999**, *54*, 115–122. [CrossRef]
46. Blair, J.B.; Rabine, D.L.; Hofton, M.A. Processing of NASA LVIS Elevation and Canopy (LGE, LCE and LGW) Data Products. 2018. Available online: http://lvis.gsfc.nasa.gov (accessed on 30 August 2019).
47. Laurin, G.V.; Chen, Q.; Lindsell, J.A.; Coomes, D.A.; Del Frate, F.; Guerriero, L.; Pirotti, F.; Valentini, R. Above ground biomass estimation in an African tropical forest with lidar and hyperspectral data. *ISPRS J. Photogramm. Remote Sens.* **2014**, *89*, 49–58. [CrossRef]
48. Zwally, H.J.; Schutz, B.; Abdalati, W.; Abshire, J.; Bentley, C.; Brenner, A.; Bufton, J.; Dezio, J.; Hancock, D.; Harding, D.; et al. ICESat's laser measurements of polar ice, atmosphere, ocean, and land. *J. Geodyn.* **2002**, *34*, 405–445. [CrossRef]
49. Abshire, J.B.; Sun, X.L.; Riris, H.; Sirota, J.M.; McGarry, J.F.; Palm, S.; Yi, D.H.; Liiva, P. Geoscience Laser Altimeter System (GLAS) on the ICESat mission: On-orbit measurement performance. *Geophys. Res. Lett.* **2005**, *32*, 4. [CrossRef]
50. Narine, L.L.; Popescu, S.; Neuenschwander, A.; Zhou, T.; Srinivasan, S.; Harbeck, K. Estimating aboveground biomass and forest canopy cover with simulated ICESat-2 data. *Remote Sens. Environ.* **2019**, *224*, 1–11. [CrossRef]
51. Xu, Z.; Zheng, G.; Moskal, L.M. Stratifying Forest Overstory for Improving Effective LAI Estimation Based on Aerial Imagery and Discrete Laser Scanning Data. *Remote Sens.* **2020**, *12*, 2126. [CrossRef]
52. Cao, L.; Coops, N.C.; Sun, Y.; Ruan, H.H.; Wang, G.B.; Dai, J.S.; She, G.H. Estimating canopy structure and biomass in bamboo forests using airborne LiDAR data. *ISPRS-J. Photogramm. Remote Sens.* **2019**, *148*, 114–129. [CrossRef]
53. Wilkes, P.; Lau, A.; Disney, M.; Calders, K.; Burt, A.; Gonzalez de Tanago, J.; Bartholomeus, H.; Brede, B.; Herold, M. Data acquisition considerations for Terrestrial Laser Scanning of forest plots. *Remote Sens. Environ.* **2017**, *196*, 140–153. [CrossRef]
54. Wang, Y.; Lehtomäki, M.; Liang, X.; Pyörälä, J.; Kukko, A.; Jaakkola, A.; Liu, J.; Feng, Z.; Chen, R.; Hyyppä, J. Is field-measured tree height as reliable as believed–A comparison study of tree height estimates from field measurement, airborne laser scanning and terrestrial laser scanning in a boreal forest. *ISPRS J. Photogramm. Remote Sens.* **2019**, *147*, 132–145. [CrossRef]
55. Zhang, W.M.; Chen, Y.M.; Wang, H.T.; Chen, M.; Wang, X.Y.; Yan, G.J. Efficient registration of terrestrial LiDAR scans using a coarse-to-fine strategy for forestry applications. *Agric. For. Meteorol.* **2016**, *225*, 8–23. [CrossRef]
56. Cheng, L.; Chen, S.; Liu, X.Q.; Xu, H.; Wu, Y.; Li, M.C.; Chen, Y.M. Registration of Laser Scanning Point Clouds: A Review. *Sensors* **2018**, *18*, 1641. [CrossRef]
57. Dong, Z.; Liang, F.; Yang, B.; Xu, Y.; Zang, Y.; Li, J.; Wang, Y.; Dai, W.; Fan, H.; Hyyppä, J.; et al. Registration of large-scale terrestrial laser scanner point clouds: A review and benchmark. *ISPRS J. Photogramm. Remote Sens.* **2020**, *163*, 327–342. [CrossRef]
58. Henning, J.G.; Radtke, P.J. Detailed stem measurements of standing trees from ground-based scanning lidar. *For. Sci.* **2006**, *52*, 67–80.
59. Ni, W.J.; Sun, G.Q.; Guo, Z.F.; Huang, H.B. A method for the registration of multiview range images acquired in forest areas using a terrestrial laser scanner. *Int. J. Remote Sens.* **2011**, *32*, 9769–9787. [CrossRef]
60. Meng, X.; Wang, L.; Silván-Cárdenas, J.L.; Currit, N. A multi-directional ground filtering algorithm for airborne LIDAR. *ISPRS J. Photogramm. Remote Sens.* **2009**, *64*, 117–124. [CrossRef]
61. Meng, X.L.; Currit, N.; Zhao, K.G. Ground Filtering Algorithms for Airborne LiDAR Data: A Review of Critical Issues. *Remote Sens.* **2010**, *2*, 833–860. [CrossRef]
62. Sithole, G.; Vosselman, G. Experimental comparison of filter algorithms for bare-Earth extraction from airborne laser scanning point clouds. *ISPRS J. Photogramm. Remote Sens.* **2004**, *59*, 85–101. [CrossRef]

63. Zhang, W.M.; Qi, J.B.; Wan, P.; Wang, H.T.; Xie, D.H.; Wang, X.Y.; Yan, G.J. An Easy-to-Use Airborne LiDAR Data Filtering Method Based on Cloth Simulation. *Remote Sens.* **2016**, *8*, 501. [CrossRef]
64. Vosselman, G. Slope based filtering of laser altimetry data. *IAPRS* **2000**, *33*, 935–942.
65. Shan, J.; Sampath, A. Urban DEM generation from raw lidar data: A labeling algorithm and its performance. *Photogramm. Eng. Remote Sens.* **2005**, *71*, 217–226. [CrossRef]
66. Axelsson, P. DEM generation from laser scanner data using adaptive TIN models. *Int. Arch. Photogramm. Remote Sens.* **2000**, *33*, 110–117.
67. Chen, Y.; Zhang, W.; Hu, R.; Qi, J.; Shao, J.; Li, D.; Wan, P.; Qiao, C.; Shen, A.; Yan, G. Estimation of forest leaf area index using terrestrial laser scanning data and path length distribution model in open-canopy forests. *Agric. For. Meteorol.* **2018**, *263*, 323–333. [CrossRef]
68. Duong, V.H.; Lindenbergh, R.; Pfeifer, N.; Vosselman, G. Single and two epoch analysis of ICESat full waveform data over forested areas. *Int. J. Remote Sens.* **2008**, *29*, 1453–1473. [CrossRef]
69. Chen, Q. Retrieving vegetation height of forests and woodlands over mountainous areas in the Pacific Coast region using satellite laser altimetry. *Remote Sens. Environ.* **2010**, *114*, 1610–1627. [CrossRef]
70. Sun, G.; Ranson, K.J.; Kimes, D.S.; Blair, J.B.; Kovacs, K. Forest vertical structure from GLAS: An evaluation using LVIS and SRTM data. *Remote Sens. Environ.* **2008**, *112*, 107–117. [CrossRef]
71. Wang, Y.; Ni, W.; Sun, G.; Chi, H.; Zhang, Z.; Guo, Z. Slope-adaptive waveform metrics of large footprint lidar for estimation of forest aboveground biomass. *Remote Sens. Environ.* **2019**, *224*, 386–400. [CrossRef]
72. Chi, H.; Sun, G.Q.; Huang, J.L.; Guo, Z.F.; Ni, W.J.; Fu, A.M. National Forest Aboveground Biomass Mapping from ICESat/GLAS Data and MODIS Imagery in China. *Remote Sens.* **2015**, *7*, 5534–5564. [CrossRef]
73. Yu, Y.; Yang, X.; Fan, W. Estimates of forest structure parameters from GLAS data and multi-angle imaging spectrometer data. *Int. J. Appl. Earth Obs. Geoinf.* **2015**, *38*, 65–71. [CrossRef]
74. Lefsky, M.A.; Harding, D.J.; Keller, M.; Cohen, W.B.; Carabajal, C.C.; Espirito-Santo, F.D.; Hunter, M.O.; de Oliveira, R. Estimates of forest canopy height and aboveground biomass using ICESat. *Geophys. Res. Lett.* **2005**, *32*, 4. [CrossRef]
75. Nie, S.; Wang, C.; Zeng, H.; Xi, X.; Xia, S. A revised terrain correction method for forest canopy height estimation using ICESat/GLAS data. *ISPRS J. Photogramm. Remote Sens.* **2015**, *108*, 183–190. [CrossRef]
76. Gwenzi, D.; Lefsky, M.A. Modeling canopy height in a savanna ecosystem using spaceborne lidar waveforms. *Remote Sens. Environ.* **2014**, *154*, 338–344. [CrossRef]
77. Lee, S.; Ni-Meister, W.; Yang, W.Z.; Chen, Q. Physically based vertical vegetation structure retrieval from ICESat data: Validation using LVIS in White Mountain National Forest, New Hampshire, USA. *Remote Sens. Environ.* **2011**, *115*, 2776–2785. [CrossRef]
78. Yang, X.; Wang, C.; Pan, F.; Nie, S.; Xi, X.; Luo, S. Retrieving leaf area index in discontinuous forest using ICESat/GLAS full-waveform data based on gap fraction model. *ISPRS J. Photogramm. Remote Sens.* **2019**, *148*, 54–62. [CrossRef]
79. Danson, F.M.; Hetherington, D.; Morsdorf, F.; Koetz, B.; Allgower, B. Forest canopy gap fraction from terrestrial laser scanning. *IEEE Geosci. Remote Sens. Lett.* **2007**, *4*, 157–160. [CrossRef]
80. Ramirez, F.A.; Armitage, R.P.; Danson, F.M. Testing the Application of Terrestrial Laser Scanning to Measure Forest Canopy Gap Fraction. *Remote Sens.* **2013**, *5*, 3037–3056. [CrossRef]
81. Nilson, T. A theoretical analysis of the frequency of gaps in plant stands. *Agric. Meteorol.* **1971**, *8*, 25–38. [CrossRef]
82. Solberg, S.; Naesset, E.; Hanssen, K.H.; Christiansen, E. Mapping defoliation during a severe insect attack on Scots pine using airborne laser scanning. *Remote Sens. Environ.* **2006**, *102*, 364–376. [CrossRef]
83. Luo, S.Z.; Chen, J.M.; Wang, C.; Gonsamo, A.; Xi, X.H.; Lin, Y.; Qian, M.J.; Peng, D.L.; Nie, S.; Qin, H.M. Comparative Performances of Airborne LiDAR Height and Intensity Data for Leaf Area Index Estimation. *IEEE J. Sel. Top. Appl. Earth Observ. Remote Sens.* **2018**, *11*, 300–310. [CrossRef]
84. Luo, S.Z.; Wang, C.; Li, G.C.; Xi, X.H. Retrieving leaf area index using ICESat/GLAS full-waveform data. *Remote Sens. Lett.* **2013**, *4*, 745–753. [CrossRef]
85. Lefsky, M.A.; Cohen, W.B.; Acker, S.A.; Parker, G.G.; Spies, T.A.; Harding, D. Lidar remote sensing of the canopy structure and biophysical properties of Douglas-fir western hemlock forests. *Remote Sens. Environ.* **1999**, *70*, 339–361. [CrossRef]
86. Takeda, T.; Oguma, H.; Sano, T.; Yone, Y.; Fujinuma, Y. Estimating the plant area density of a Japanese larch (Larix kaempferi Sarg.) plantation using a ground-based laser scanner. *Agric. For. Meteorol.* **2008**, *148*, 428–438. [CrossRef]

87. Lovell, J.L.; Jupp, D.L.B.; Culvenor, D.S.; Coops, N.C. Using airborne and ground-based ranging lidar to measure canopy structure in Australian forests. *Can. J. Remote Sens.* **2003**, *29*, 607–622. [CrossRef]
88. Alonzo, M.; Bookhagen, B.; McFadden, J.P.; Sun, A.; Roberts, D.A. Mapping urban forest leaf area index with airborne lidar using penetration metrics and allometry. *Remote Sens. Environ.* **2015**, *162*, 141–153. [CrossRef]
89. Garcia, M.; Popescu, S.; Riano, D.; Zhao, K.G.; Neuenschwander, A.; Agca, M.; Chuvieco, E. Characterization of canopy fuels using ICESat/GLAS data. *Remote Sens. Environ.* **2012**, *123*, 81–89. [CrossRef]
90. Hofton, M.A.; Minster, J.B.; Blair, J.B. Decomposition of laser altimeter waveforms. *IEEE Trans. Geosci. Remote Sens.* **2000**, *38*, 1989–1996. [CrossRef]
91. Tang, H.; Brolly, M.; Zhao, F.; Strahler, A.H.; Schaaf, C.L.; Ganguly, S.; Zhang, G.; Dubayah, R. Deriving and validating Leaf Area Index (LAI) at multiple spatial scales through lidar remote sensing: A case study in Sierra National Forest, CA. *Remote Sens. Environ.* **2014**, *143*, 131–141. [CrossRef]
92. Van der Zande, D.; Hoet, W.; Jonckheere, L.; van Aardt, J.; Coppin, P. Influence of measurement set-up of ground-based LiDAR for derivation of tree structure. *Agric. For. Meteorol.* **2006**, *141*, 147–160. [CrossRef]
93. Beland, M.; Widlowski, J.L.; Fournier, R.A.; Cote, J.F.; Verstraete, M.M. Estimating leaf area distribution in savanna trees from terrestrial LiDAR measurements. *Agric. For. Meteorol.* **2011**, *151*, 1252–1266. [CrossRef]
94. Jensen, J.L.R.; Humes, K.S.; Vierling, L.A.; Hudak, A.T. Discrete return lidar-based prediction of leaf area index in two conifer forests. *Remote Sens. Environ.* **2008**, *112*, 3947–3957. [CrossRef]
95. Korhonen, L.; Korpela, I.; Heiskanen, J.; Maltamo, M. Airborne discrete-return LIDAR data in the estimation of vertical canopy cover, angular canopy closure and leaf area index. *Remote Sens. Environ.* **2011**, *115*, 1065–1080. [CrossRef]
96. Qu, Y.H.; Shaker, A.; Silva, C.A.; Klauberg, C.; Pinage, E.R. Remote Sensing of Leaf Area Index from LiDAR Height Percentile Metrics and Comparison with MODIS Product in a Selectively Logged Tropical Forest Area in Eastern Amazonia. *Remote Sens.* **2018**, *10*, 970. [CrossRef]
97. Gastellu-Etchegorry, J.P.; Yin, T.G.; Lauret, N.; Grau, E.; Rubio, J.; Cook, B.D.; Morton, D.C.; Sun, G.Q. Simulation of satellite, airborne and terrestrial LiDAR with DART (I): Waveform simulation with quasi-Monte Carlo ray tracing. *Remote Sens. Environ.* **2016**, *184*, 418–435. [CrossRef]
98. North, P.R.J.; Rosette, J.A.B.; Suarez, J.C.; Los, S.O. A Monte Carlo radiative transfer model of satellite waveform LiDAR. *Int. J. Remote Sens.* **2010**, *31*, 1343–1358. [CrossRef]
99. Ni-Meister, W.; Yang, W.Z.; Lee, S.; Strahler, A.H.; Zhao, F. Validating modeled lidar waveforms in forest canopies with airborne laser scanning data. *Remote Sens. Environ.* **2018**, *204*, 229–243. [CrossRef]
100. Sun, G.Q.; Ranson, K.J. Modeling lidar returns from forest canopies. *IEEE Trans. Geosci. Remote Sens.* **2000**, *38*, 2617–2626.
101. Bye, I.J.; North, P.R.J.; Los, S.O.; Kljun, N.; Rosette, J.A.B.; Hopkinson, C.; Chasmer, L.; Mahoney, C. Estimating forest canopy parameters from satellite waveform LiDAR by inversion of the FLIGHT three-dimensional radiative transfer model. *Remote Sens. Environ.* **2017**, *188*, 177–189. [CrossRef]
102. Koetz, B.; Morsdorf, F.; Sun, G.; Ranson, K.J.; Itten, K.; Allgower, B. Inversion of a lidar waveform model for forest biophysical parameter estimation. *IEEE Geosci. Remote Sens. Lett.* **2006**, *3*, 49–53. [CrossRef]
103. Ma, H.; Song, J.L.; Wang, J.D.; Xiao, Z.Q.; Fu, Z. Improvement of spatially continuous forest LAI retrieval by integration of discrete airborne LiDAR and remote sensing multi-angle optical data. *Agric. For. Meteorol.* **2014**, *189*, 60–70. [CrossRef]
104. Ni-Meister, W.; Jupp, D.L.B.; Dubayah, R. Modeling lidar waveforms in heterogeneous and discrete canopies. *IEEE Trans. Geosci. Remote Sens.* **2001**, *39*, 1943–1958. [CrossRef]
105. Ma, H.; Song, J.L.; Wang, J.D. Forest Canopy LAI and Vertical FAVD Profile Inversion from Airborne Full-Waveform LiDAR Data Based on a Radiative Transfer Model. *Remote Sens.* **2015**, *7*, 1897–1914. [CrossRef]
106. Coren, F.; Sterzai, P. Radiometric correction in laser scanning. *Int. J. Remote Sens.* **2006**, *27*, 3097–3104. [CrossRef]
107. Tian, J.Y.; Wang, L.; Li, X.J.; Shi, C.; Gong, H.L. Differentiating Tree and Shrub LAI in a Mixed Forest With ICESat/GLAS Spaceborne LiDAR. *IEEE J. Sel. Top. Appl. Earth Observ. Remote Sens.* **2017**, *10*, 87–94. [CrossRef]
108. Qu, Y.H.; Shaker, A.; Korhonen, L.; Silva, C.A.; Jia, K.; Tian, L.; Song, J.L. Direct Estimation of Forest Leaf Area Index based on Spectrally Corrected Airborne LiDAR Pulse Penetration Ratio. *Remote Sens.* **2020**, *12*, 217. [CrossRef]

109. Woodgate, W.; Jones, S.D.; Suarez, L.; Hill, M.J.; Armston, J.D.; Wilkes, P.; Soto-Berelov, M.; Haywood, A.; Mellor, A. Understanding the variability in ground-based methods for retrieving canopy openness, gap fraction, and leaf area index in diverse forest systems. *Agric. For. Meteorol.* **2015**, *205*, 83–95. [CrossRef]
110. Calders, K.; Origo, N.; Disney, M.; Nightingale, J.; Woodgate, W.; Armston, J.; Lewis, P. Variability and bias in active and passive ground-based measurements of effective plant, wood and leaf area index. *Agric. For. Meteorol.* **2018**, *252*, 231–240. [CrossRef]
111. Zhu, X.; Liu, J.; Skidmore, A.K.; Premier, J.; Heurich, M. A voxel matching method for effective leaf area index estimation in temperate deciduous forests from leaf-on and leaf-off airborne LiDAR data. *Remote Sens. Environ.* **2020**, *240*, 111696. [CrossRef]
112. Fang, H.L.; Zhang, Y.H.; Wei, S.S.; Li, W.J.; Ye, Y.C.; Sun, T.; Liu, W.W. Validation of global moderate resolution leaf area index (LAI) products over croplands in northeastern China. *Remote Sens. Environ.* **2019**, *233*, 19. [CrossRef]
113. Jensen, J.L.R.; Humes, K.S.; Hudak, A.T.; Vierling, L.A.; Delmelle, E. Evaluation of the MODIS LAI product using independent lidar-derived LAI: A case study in mixed conifer forest. *Remote Sens. Environ.* **2011**, *115*, 3625–3639. [CrossRef]
114. Tang, H.; Dubayah, R. Light-driven growth in Amazon evergreen forests explained by seasonal variations of vertical canopy structure. *Proc. Natl. Acad. Sci. USA* **2017**, *114*, 2640–2644. [CrossRef] [PubMed]
115. Li, Y.M.; Guo, Q.H.; Su, Y.J.; Tao, S.L.; Zhao, K.G.; Xu, G.C. Retrieving the gap fraction, element clumping index, and leaf area index of individual trees using single-scan data from a terrestrial laser scanner. *ISPRS J. Photogramm. Remote Sens.* **2017**, *130*, 308–316. [CrossRef]
116. Wang, K.; Kumar, P. Characterizing relative degrees of clumping structure in vegetation canopy using waveform LiDAR. *Remote Sens. Environ.* **2019**, *232*. [CrossRef]
117. Garcia, M.; Gajardo, J.; Riano, D.; Zhao, K.G.; Martin, P.; Ustin, S. Canopy clumping appraisal using terrestrial and airborne laser scanning. *Remote Sens. Environ.* **2015**, *161*, 78–88. [CrossRef]
118. Luo, S.Z.; Chen, J.M.; Wang, C.; Xi, X.H.; Zeng, H.C.; Peng, D.L.; Li, D. Effects of LiDAR point density, sampling size and height threshold on estimation accuracy of crop biophysical parameters. *Opt. Express* **2016**, *24*, 1578–1593. [CrossRef]
119. Morsdorf, F.; Kotz, B.; Meier, E.; Itten, K.I.; Allgower, B. Estimation of LAI and fractional cover from small footprint airborne laser scanning data based on gap fraction. *Remote Sens. Environ.* **2006**, *104*, 50–61. [CrossRef]
120. Peduzzi, A.; Wynne, R.H.; Fox, T.R.; Nelson, R.F.; Thomas, V.A. Estimating leaf area index in intensively managed pine plantations using airborne laser scanner data. *For. Ecol. Manag.* **2012**, *270*, 54–65. [CrossRef]
121. Pearse, G.D.; Morgenroth, J.; Watt, M.S.; Dash, J.P. Optimising prediction of forest leaf area index from discrete airborne lidar. *Remote Sens. Environ.* **2017**, *200*, 220–239. [CrossRef]
122. Pang, Y.; Lefsky, M.; Sun, G.; Ranson, J. Impact of footprint diameter and off-nadir pointing on the precision of canopy height estimates from spaceborne lidar. *Remote Sens. Environ.* **2011**, *115*, 2798–2809. [CrossRef]
123. Yang, W.Z.; Ni-Meister, W.; Lee, S. Assessment of the impacts of surface topography, off-nadir pointing and vegetation structure on vegetation lidar waveforms using an extended geometric optical and radiative transfer model. *Remote Sens. Environ.* **2011**, *115*, 2810–2822. [CrossRef]
124. Heiskanen, J.; Korhonen, L.; Hietanen, J.; Pellikka, P.K.E. Use of airborne lidar for estimating canopy gap fraction and leaf area index of tropical montane forests. *Int. J. Remote Sens.* **2015**, *36*, 2569–2583. [CrossRef]
125. Hopkinson, C.; Chasmer, L. Testing LiDAR models of fractional cover across multiple forest ecozones. *Remote Sens. Environ.* **2009**, *113*, 275–288. [CrossRef]
126. Cifuentes, R.; Van der Zande, D.; Farifteh, J.; Salas, C.; Coppin, P. Effects of voxel size and sampling setup on the estimation of forest canopy gap fraction from terrestrial laser scanning data. *Agric. For. Meteorol.* **2014**, *194*, 230–240. [CrossRef]
127. Seidel, D.; Fleck, S.; Leuschner, C. Analyzing forest canopies with ground-based laser scanning: A comparison with hemispherical photography. *Agric. For. Meteorol.* **2012**, *154*, 1–8. [CrossRef]
128. Zheng, G.; Ma, L.X.; He, W.; Eitel, J.U.H.; Moskal, L.M.; Zhang, Z.Y. Assessing the Contribution of Woody Materials to Forest Angular Gap Fraction and Effective Leaf Area Index Using Terrestrial Laser Scanning Data. *IEEE Trans. Geosci. Remote Sens.* **2016**, *54*, 1475–1487. [CrossRef]
129. Yun, T.; An, F.; Li, W.; Sun, Y.; Cao, L.; Xue, L. A Novel Approach for Retrieving Tree Leaf Area from Ground-Based LiDAR. *Remote Sens.* **2016**, *8*, 942. [CrossRef]

130. Zhu, X.; Skidmore, A.K.; Darvishzadeh, R.; Niemann, K.O.; Liu, J.; Shi, Y.F.; Wang, T.J. Foliar and woody materials discriminated using terrestrial LiDAR in a mixed natural forest. *Int. J. Appl. Earth Obs. Geoinf.* **2018**, *64*, 43–50. [CrossRef]
131. Milenković, M.; Schnell, S.; Holmgren, J.; Ressl, C.; Lindberg, E.; Hollaus, M.; Pfeifer, N.; Olsson, H. Influence of footprint size and geolocation error on the precision of forest biomass estimates from space-borne waveform LiDAR. *Remote Sens. Environ.* **2017**, *200*, 74–88. [CrossRef]
132. Lei, L.; Qiu, C.; Li, Z.; Han, D.; Han, L.; Zhu, Y.; Wu, J.; Xu, B.; Feng, H.; Yang, H.; et al. Effect of Leaf Occlusion on Leaf Area Index Inversion of Maize Using UAV–LiDAR Data. *Remote Sens.* **2019**, *11*, 1067. [CrossRef]
133. Zheng, G.; Moskal, L.M. Spatial variability of terrestrial laser scanning based leaf area index. *Int. J. Appl. Earth Obs. Geoinf.* **2012**, *19*, 226–237. [CrossRef]
134. Calders, K.; Armston, J.; Newnham, G.; Herold, M.; Goodwin, N. Implications of sensor configuration and topography on vertical plant profiles derived from terrestrial LiDAR. *Agric. For. Meteorol.* **2014**, *194*, 104–117. [CrossRef]
135. Hosoi, F.; Nakai, Y.; Omasa, K. Estimation and Error Analysis of Woody Canopy Leaf Area Density Profiles Using 3-D Airborne and Ground-Based Scanning Lidar Remote-Sensing Techniques. *IEEE Trans. Geosci. Remote Sens.* **2010**, *48*, 2215–2223. [CrossRef]
136. Ballhorn, U.; Jubanski, J.; Siegert, F. ICESat/GLAS Data as a Measurement Tool for Peatland Topography and Peat Swamp Forest Biomass in Kalimantan, Indonesia. *Remote Sens.* **2011**, *3*, 1957–1982. [CrossRef]
137. Simard, M.; Pinto, N.; Fisher, J.B.; Baccini, A. Mapping forest canopy height globally with spaceborne lidar. *J. Geophys. Res. Biogeosci.* **2011**, *116*, 12. [CrossRef]
138. Grau, E.; Durrieu, S.; Fournier, R.; Gastellu-Etchegorry, J.P.; Yin, T.G. Estimation of 3D vegetation density with Terrestrial Laser Scanning data using voxels. A sensitivity analysis of influencing parameters. *Remote Sens. Environ.* **2017**, *191*, 373–388. [CrossRef]
139. Zou, J.; Zhuang, Y.; Chianucci, F.; Mai, C.; Lin, W.; Leng, P.; Luo, S.; Yan, B. Comparison of Seven Inversion Models for Estimating Plant and Woody Area Indices of Leaf-on and Leaf-off Forest Canopy Using Explicit 3D Forest Scenes. *Remote Sens.* **2018**, *10*, 1297. [CrossRef]
140. Ma, L.X.; Zheng, G.; Eitel, J.U.H.; Magney, T.S.; Moskal, L.M. Determining woody-to-total area ratio using terrestrial laser scanning (TLS). *Agric. For. Meteorol.* **2016**, *228*, 217–228. [CrossRef]
141. Guo, Q.; Su, Y.; Hu, T.; Zhao, X.; Wu, F.; Li, Y.; Liu, J.; Chen, L.; Xu, G.; Lin, G.; et al. An integrated UAV-borne lidar system for 3D habitat mapping in three forest ecosystems across China. *Int. J. Remote Sens.* **2017**, *38*, 2954–2972. [CrossRef]
142. Su, Y.; Guo, Q.; Jin, S.; Guan, H.; Sun, X.; Ma, Q.; Hu, T.; Wang, R.; Li, Y. The Development and Evaluation of a Backpack LiDAR System for Accurate and Efficient Forest Inventory. *IEEE Geosci. Remote Sens. Lett.* **2020**. [CrossRef]

Publisher's Note: MDPI stays neutral with regard to jurisdictional claims in published maps and institutional affiliations.

© 2020 by the authors. Licensee MDPI, Basel, Switzerland. This article is an open access article distributed under the terms and conditions of the Creative Commons Attribution (CC BY) license (http://creativecommons.org/licenses/by/4.0/).

MDPI
St. Alban-Anlage 66
4052 Basel
Switzerland
Tel. +41 61 683 77 34
Fax +41 61 302 89 18
www.mdpi.com

Remote Sensing Editorial Office
E-mail: remotesensing@mdpi.com
www.mdpi.com/journal/remotesensing

www.ingramcontent.com/pod-product-compliance
Lightning Source LLC
LaVergne TN
LVHW070507100526
838202LV00014B/1808